Frank D. Petruzella

PROGRAMMABLE LOGIC CONTROLLERS

Second Edition

 **Glencoe McGraw-Hill**

New York, New York
Columbus, Ohio
Mission Hills, California
Peoria, Illinois

A Division of The McGraw·Hill Companies

Cover Photo and Section Opening Photos

Cover (l) Art Montes de Oca/FPG International; Cover (r) Andy Sacks/Tony Stone Images; 3 Michael Rosenfeld/Tony Stone Images; 19 Telegraph Colour Library/FPG International; 49 Telegraph Colour Library/FPG International; 71 Michael Rosenfeld/Tony Stone Images; 93 Superstock; 121 Tom Tracy/FPG International; 165 Superstock; 197 Telegraph Colour Library/FPG International; 220 Superstock; 257 Art Montes de Oca/FPG International; 289 Superstock; 329 Superstock; 351 Superstock; 375 Michael Rosenfeld/Tony Stone Images; 401 Superstock.

Library of Congress Cataloging-in-Publication Data

Petruzella, Frank D.
    Programmable logic controllers/Frank D. Petruzella. -- 2nd ed.
       p. cm.
    Includes index.
    ISBN 0-02-802661-6 (case)
    1. Programmable controllers. I. Title.
  TJ223.P76P48 1996
  629.8'9--DC20
                                          96-42934
                                             CIP

Programmable Logic Controllers, Second Edition

Send all inquiries to:
Glencoe/McGraw-Hill
8787 Orion Place
Columbus, OH 43240-4027

ISBN 0-02-802661-6

Printed in the United States of America

7 8 9 0 004 02 01

Contents

Contents

Preface

Programmable logic controllers (PLCs) are used in almost every segment of industry where automation is required. They represent one of the fastest growing sectors of the industrial electronics field. Since their inception, PLCs have proven to be the salvation of many manufacturing plants that had relied previously on electromechanical control systems.

The second edition of Programmable Logic Controllers offers beginning students a first course in PLCs. It focuses on the underlying principles of how PLCs work and provides practical information about installing, programming, and maintaining a PLC system. No previous knowledge of PLC systems or programming is assumed. Employment opportunities in this field are good, and this book is designed to help students acquire the necessary qualifications for these jobs.

In many instances, the only source of information concerning programmable controllers is the user's manual published by the manufacturer. This textbook is not intended to replace that user's manual, but rather to complement and expand on information found in it. The second edition of Programmable Logic Controllers discusses PLCs in a generic sense. Although the content is broad enough to allow the information to be used with a wide range of PLC models, this book uses the popular Allen-Bradley PLC-2, PLC-5, and SLC-500 controller instruction set for programming examples.

The text is written in an easy-to-read and understandable language, with many clear illustrations to assist the student in comprehending the fundamentals of a PLC system. Objectives are listed at the beginning of each chapter to inform students what they should learn. This list is followed by the subject material. The relay equivalent of the programmed instruction is explained first, followed by the appropriate PLC instruction. Each chapter concludes with a set of review questions and problems. The review questions are closely related to the chapter objectives and will help students evaluate their understanding of the chapter. The problems range from easy to difficult, thus challenging students at various levels of competence.

All topics are covered in small segments to develop a firm foundation for each concept and operation before the student advances to the next. An entire chapter is devoted to logic circuits as they apply to PLCs. PLC safety procedures and considerations are stressed throughout the text. Technical terms are defined when they are first used, and an extensive glossary provides easy referral to PLC terms.

General troubleshooting procedures and techniques are stressed, and the student is instructed in how to analyze PLC problems systematically.

The second edition has been revised to include:

◆ The Allen-Bradley PLC-5 and SLC-500 instruction set

◆ Newly developed functions

◆ Additional practical programming examples

◆ Two new chapters:
  -Process Control and Data Acquisition Systems
  -Computer-Controlled Machines and Processes

◆ An expanded glossary

Both an activities manual and an instructor's manual are available for use with Programmable Logic Controllers, Second Edition. The activities manual contains true/false, completion, matching, and multiple-choice test questions for each chapter of the text. The best way to fully understand any given PLC is to work with that PLC. Therefore, in the activities manual, each chapter contains a wide range of generic programming assignments and exercises for student practice with the PLC. The instructor's guide contains answers to all textbook review questions and problems, as well as answers to the activities manual test questions.

I hope that you will find the material presentation in Programmable Logic Controllers, Second Edition; the activities manual; and the instructor's manual simple and easy to read and understand, as well as informative.

Finally, I would like to thank Nazar M. Karzay, Steve Krywy, Harry E. Myles, and Dan Siddall for lending their time and expertise to the Programmable Logic Controllers, Second Edition program. Their insightful technical reviews of the manuscript were very helpful. I would also like to thank Brian Mackin, Linda Jefferson, and Rob Ciccotelli of Glencoe for their hard work and cooperation in the preparation of this text.

Frank D. Petruzella

1

Programmable Logic Controllers (PLCs): An Overview

After completing this chapter, you will be able to:

◆ Define what a programmable logic controller (PLC) is and list its advantages over relay systems

◆ Identify the main parts of a PLC and describe their functions

◆ Outline the basic sequence of operation for a PLC

◆ Identify the general classifications of PLCs

This chapter gives a brief history of the evolution of a programmable logic controller, or PLC. The reasons for changing from relay control systems to PLCs are discussed. You will learn about the basic parts of a PLC, how a PLC is used to control a process, and about the different kinds of PLCs and their applications. The ladder logic language, which was developed to simplify the task of programming PLCs, is introduced.

.1 PROGRAMMABLE LOGIC CONTROLLERS

A *programmable logic controller (PLC)* is a solid-state device designed to perform logic functions previously accomplished by electro-mechanical relays (Fig. 1-1).

The design of most PLCs is similar to that of a computer. Basically, the PLC is an assembly of solid-state digital logic elements designed to make logical decisions and provide outputs. Programmable logic controllers are used for the control and operation of manufacturing process equipment and machinery.

The programmable logic controller is then basically a computer designed for use in machine control. Unlike a computer, it has been designed to operate in the industrial environment and is equipped with special input/output interfaces and a control programming language. The common abbreviation used in industry for these devices, PC, can be confusing because it is also the abbreviation for "personal computer." Therefore, some manufacturers refer to their programmable controller as a PLC, which stands for "programmable logic controller."

Initially the PLC was used to replace relay logic, but its ever-increasing range of functions means that it is found in many and more complex applications. Because the structure of a PLC is based on the same principles as those employed in computer architecture, it is capable not only of performing relay switching tasks, but also of performing other applications such as counting, calculating, comparing, and the processing of analog signals.

Programmable controllers offer several advantages over a conventional relay type of control. Relays have to be hardwired to perform a specific function (Fig. 1-2). When the system requirements change, the relay wiring has to be changed or modified. In extreme cases, such as in the auto industry, complete control panels had to be replaced since it was not economically feasible to

Fig. 1-1

Programmable logic controller. (Courtesy of Allen-Bradley Company, Inc.)

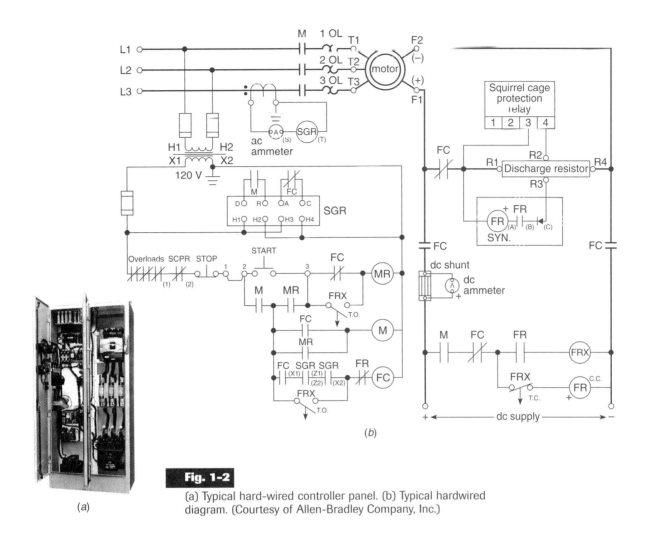

Fig. 1-2

(a) Typical hard-wired controller panel. (b) Typical hardwired diagram. (Courtesy of Allen-Bradley Company, Inc.)

rewire the old panels with each model changeover. The programmable controller has eliminated much of the hand wiring associated with conventional relay control circuits. It is small and inexpensive compared to equivalent relay-based process control systems. Programmable controllers also offer solid-state reliability, lower power consumption, and ease of expandability.

If an application has more than a half-dozen relays, it probably will be less expensive to install a PLC. Simulating a hundred relays, timers, and counters is not a problem even on small PLCs. Programmable logic controls are easy to program and install. Access to PLCs can be restricted by hardware features such as keylocks, and by software features such as passwords. Problem solving with PLCs is a major

advantage over relay-type control systems. PLCs can be designed with communications capabilities that allow them to converse with other computer systems or to provide human interfaces. The programmable controller is an event-driven device, which means that an event taking place in the field will result in an operation or output taking place.

•2 PARTS OF A PLC

A typical PLC can be divided into parts, as illustrated in Fig. 1-3. These components are the *central processing unit (CPU),* the *input/output (I/O)* section, the *power supply,* and the *programming device.*

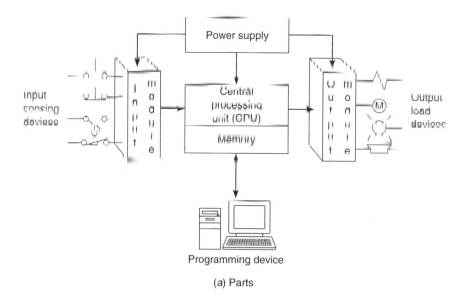

Power supply

Input sensing devices

Central processing unit (CPU)

Memory

Output load devices

Programming device

(a) Parts

PROGRAM

REAL WORLD

MEMORY

INPUT/OUTPUT

ADDRESS

DATA

CONTROL

CENTRAL PROCESSOR UNIT

The structure of a PLC is based on the same principles as those employed in computer architecture.

(b) Architecture

Fig. 1-3

There are two ways in which I/O is incorporated into the PLC: fixed and modular. *Fixed I/O* (Fig. 1-4a) is typical of small PLCs that come in one package with no separate, removable units. The processor and I/O are packaged together, and the I/O terminals are available but cannot be changed. The main advantage of this type of packaging is lower cost. The number of available I/O points varies and usually can be expanded by buying additional units of fixed I/O. One disadvantage of fixed I/O is its lack of flexibility; you are limited in what you can get in the quantities and types dictated by the packaging. Also, for some models, if any part in the unit fails, the whole unit has to be replaced.

Modular I/O (Fig. 1-4b) is divided by compartments into which separate modules can be plugged. This feature greatly increases your options and the unit's flexibility. You can choose from the modules available from the manufacturer and mix them any way you desire. The basic modular controller consists of a rack, power supply, processor module (CPU), input/output (I/O modules), and an operator interface for programming and monitoring. The modules plug into a rack. When a module is slid into the rack, it makes an electrical connection with a series of contacts called the *backplane,* located at the rear of the rack. The PLC processor is also connected to the backplane and can communicate with all the modules in the rack.

The *power supply* supplies dc power to other modules that plug into the rack. For large PLC systems, this power supply does not normally supply power to the field devices. With larger systems, power to field devices is provided by external alternating current (ac) or direct current (dc) supplies. For small and micro PLC systems, the power supply is used to power field devices.

Common Power Bus

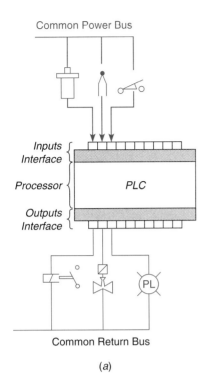

Inputs Interface

Processor

PLC

Outputs Interface

Common Return Bus

(a)

Fig. 1-4

I/O configurations: (a) fixed I/O;
(b) modular I/O.

Processor module

Combination I/O module

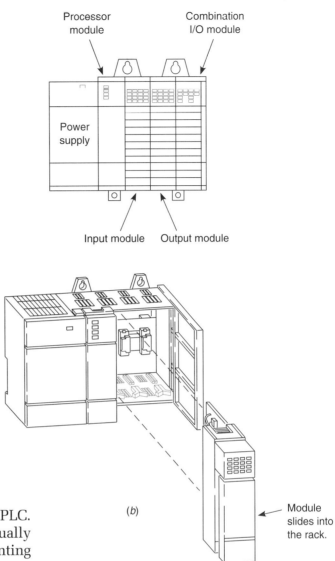

Power supply

Input module   Output module

(b)

Module slides into the rack.

The *processor* (CPU) is the "brain" of the PLC. A typical processor (Fig. 1-5 on p. 8) usually consists of a microprocessor for implementing the logic and controlling the communications among the modules. The processor requires memory for storing the results of the logical operations performed by the microprocessor. Memory is also required for the program EPROM or EEPROM plus RAM.

The CPU is designed so that the user can enter the desired circuit in relay ladder logic. The processor accepts (reads) input data from various sensing devices, executes the stored user program from memory, and sends appropriate output commands to control devices. A direct current (dc) power source is required to produce the low-level voltage used by the processor. This power supply can be housed in the CPU unit or may be a separately mounted module, depending on the PLC system manufacturer.

The *I/O section* consists of input modules and output modules (Fig. 1-6). The I/O system forms the interface by which field devices are connected to the controller. The purpose of this interface is to condition the various signals received from or sent to external field devices. Input devices such as pushbuttons, limit switches, sensors, selector switches, and thumbwheel switches are hard-wired to terminals on the input modules. Output devices such as small motors, motor starters, solenoid valves, and indicator lights are hard-wired to the terminals on the output modules. These devices are also referred to as "field" or "real world" inputs and outputs. The terms *field* or *real world* are used to distinguish actual external devices

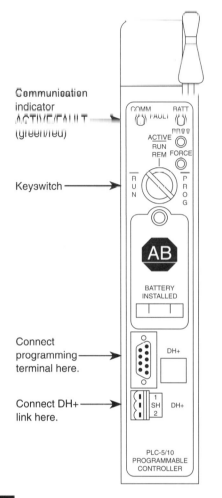

Fig. 1-5

Typical processor module. (Courtesy of Allen-Bradley Company, Inc.)

may be a hand-held unit with a light-emitting diode (LED) display (Fig. 1-7a) or an industrial terminal unit with a video display unit (Fig. 1-7b).

With some small hand-held programming devices, the program is entered using BOOLEAN operators (AND, OR, and NOT functions) individually or in combination to form logical statements. Larger PLC systems normally use either a PC (personal computer) or programming terminal along with relay ladder logic. The trend is to use a PC for programming.

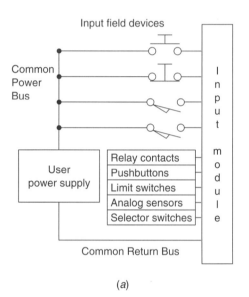

(a)

that exist and must be physically wired from the internal user program that duplicates the function of relays, timers, and counters.

The *programming device* or *terminal,* is used to enter the desired program into the memory of the processor. This program is entered using *relay ladder logic.* The program determines the sequence of operation and ultimate control of the equipment or machinery. The programming device must be connected to the controller only when entering or monitoring the program. By designing the controller to be "electrician friendly," the PLC can be programmed by people without extensive computer programming experience. Actual programming is usually achieved by pushing keys on a keyboard. The programming device

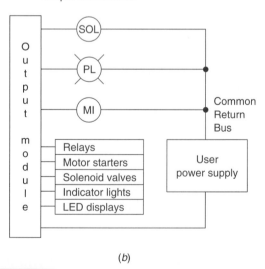

(b)

Fig. 1-6

(a) Typical input module. (b) Typical output module.

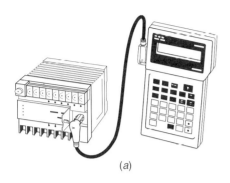

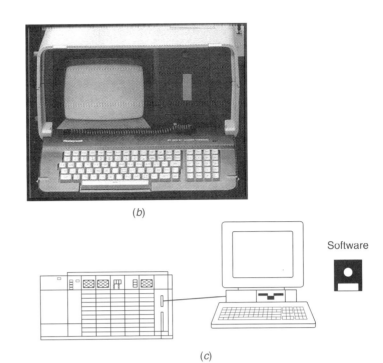

(a)

(b)

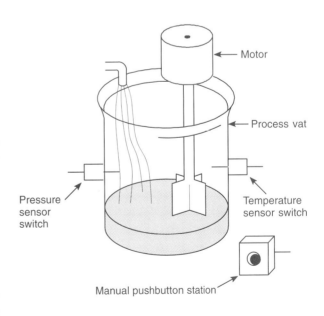

Software

(c)

Programming devices: (a) hand-held unit with light-emitting diode (LED) display; (b) industrial terminal video unit (Courtesy of Honeywell, Inc.); (c) personal computer with appropriate software.

The video display like the unit shown in Fig. 1-7b offers the advantage of displaying large amounts of logic on the screen and thus simplifying the interpretation of the program. The programming unit communicates with the processor via a serial or parallel data communications link. If the programming unit is not in use, it may be unplugged and removed. Removing the programming unit will not affect the operation of the user program. A personal computer with appropriate software can also act as a program terminal (Fig. 1-7c), making it possible to carry out the programming away from the physical location of the programmable logic controller. When the program is complete, it is saved to some form of mass storage and downloaded to the programmable logic controller when required.

a vat when the temperature and pressure reach preset values. In addition, direct manual operation of the motor is provided by means of a separate pushbutton station. The process is monitored with temperature and pressure sensor switches that close their respective contacts when conditions reach their preset values.

•3 PRINCIPLES OF OPERATION

To get an idea of how a PLC operates, consider the simple process control problem illustrated in Fig. 1-8. Here a mixer motor is to be used to automatically stir the liquid in

Motor

Process vat

Pressure sensor switch

Temperature sensor switch

Manual pushbutton station

Mixer process control problem.

This control problem can be solved using the relay method for motor control shown in the relay ladder diagram of Fig. 1-9. The motor starter coil (M) is energized when both the pressure and temperature switches are closed or when the manual pushbutton is pressed.

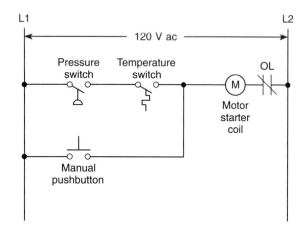

Fig. 1-9

Process control relay ladder diagram.

Now let's look at how a PLC might be used for this application. The same input field devices (pressure switch, temperature switch, and pushbutton) are used. These devices would be hard-wired to an appropriate input module according to the manufacturer's labeling scheme. Typical wiring connections for a 120-V ac input module are shown in Fig. 1-10.

The same output field device (motor starter coil) would also be used. This device would be hard-wired to an appropriate output module according to the manufacturer's labeling scheme. Typical wiring connections for a 120-V ac output module are shown in Fig. 1-11.

Next, the PLC ladder logic diagram would be constructed and programmed into the memory of the CPU. A typical ladder logic diagram for this process is shown in Fig. 1-12 (on p. 12). The format used is similar to the layout of the hard-wired relay ladder circuit. The individual symbols represent *instructions*

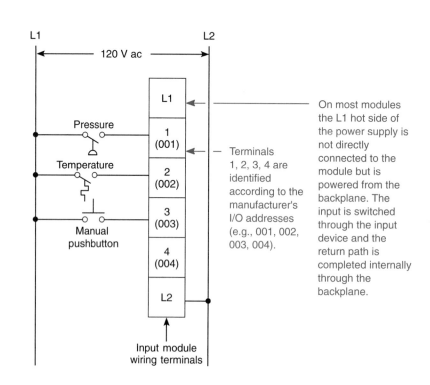

On most modules the L1 hot side of the power supply is not directly connected to the module but is powered from the backplane. The input is switched through the input device and the return path is completed internally through the backplane.

Terminals 1, 2, 3, 4 are identified according to the manufacturer's I/O addresses (e.g., 001, 002, 003, 004).

Input module wiring terminals

Fig. 1-10

Typical input module wiring connections.

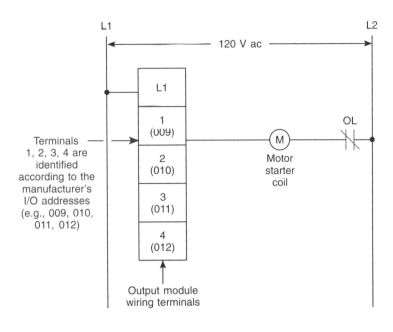

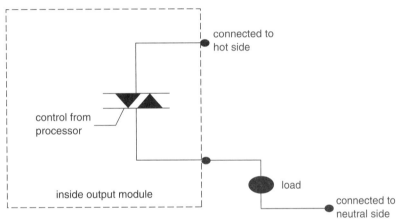

On the output module the neutral does not need to be connected. The hot is switched by the output module to the load. Then the other side of the load is connected to neutral to complete the path.

Fig. 1-11

Typical output module wiring connections.

while the numbers represent the instruction *addresses.* When programming the controller, these instructions are entered one by one into the processor memory from the operator terminal keyboard. Instructions are stored in the user program portion of the processor memory.

To operate the program, the controller is placed in the RUN mode, or operating cycle. During each operating cycle, the controller examines the status of input devices, executes the user program, and changes outputs accordingly. Each - | | - can be thought of as a set of normally open (NO) contacts. The -( )- can be considered to represent a coil that, when energized, will close a set of contacts. In the ladder logic diagram of Fig. 1-12, the coil 009 is energized when contacts 001 and 002 are closed or when contact 003 is closed. Either of these conditions provides a continuous path from left to right across the rung that includes the coil.

The RUN operation of the controller can be described by the following sequence of

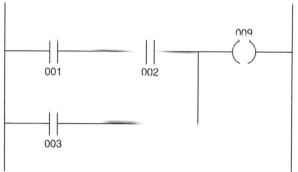

009
001
002
003
Note: Numbers 001, 002, 003, and 009 are identified with the pressure switch, temperature switch, manual pushbutton, and motor starter coil, respectively.

Fig. 1-12
Process control PLC ladder logic diagram.

120 V ac
Pressure switch
Temperature switch
OL
M
Motor starter coil
Manual pushbutton

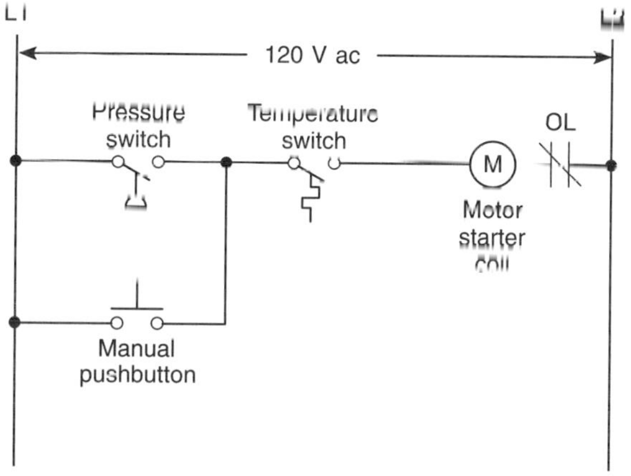

Fig. 1-13
Relay ladder diagram for modified process.

events. First, the inputs are examined and their status is recorded in the controller's memory (a closed contact is recorded as a signal that is called a logic 1 and an open contact by a signal that is called a logic 0). Then the ladder diagram is evaluated, with each internal contact given OPEN or CLOSED status according to the record. If these contacts provide a current path from left to right in the diagram, the output coil memory location is given a logic 1 value and the output module interface contacts will close. If there is no conducting path on the program rung, the output coil memory location is set to logic 0 and the output module interface contacts will be open. The completion of one cycle of this sequence by the controller is called a *scan.* The *scan time,* the time required for one full cycle, provides a measure of the speed of response of the PLC. Generally, the output memory location is updated during the scan but the actual output is not updated until the end of the program scan during the I/O scan.

1•4 MODIFYING THE OPERATION

As mentioned, one of the important features of a PLC is the ease with which the program can be changed. For example, assume that our original process control circuit for the mixing operation must be modified as shown in the *relay* ladder diagram of Fig. 1-13. The change requires that the manual pushbutton control should be permitted to operate at any pressure *but not unless* the specified temperature setting has been reached.

If a relay system were used, it would require some rewiring of the system (as shown in Fig. 1-13) to achieve the desired change. However, if a PLC system were used, *no* rewiring would be necessary. The inputs and outputs are still the same. All that is required is to change the PLC ladder logic diagram as shown in Fig. 1-14.

009
001
002
003

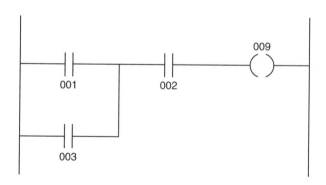

Fig. 1-14
PLC ladder logic diagram for modified process.

.5 PLCs VERSUS COMPUTERS

The architecture of a PLC is basically the same as that of a general-purpose computer. A personal computer can be made into a programmable logic controller if you provide some way for the computer to receive information from devices such as pushbuttons or switches. You also need a program to process the inputs and decide the means of turning load devices OFF and ON.

However, some important characteristics distinguish PLCs from general-purpose computers. First, unlike computers, the PLC is designed to operate in the industrial environment (Fig. 1-15) with wide ranges of ambient temperature and humidity. A well-designed PLC is not usually affected by the electrical noise inherent in most industrial locations.

A second distinction between PLCs and computers is that the hardware and software of PLCs are designed for easy use by plant electricians and technicians. Unlike the computer, the PLC is programmed in *relay ladder logic* or other easily learned languages. The PLC comes with its program language built into its permanent memory, whereas a personal computer requires a disk

Fig. 1-15

PLC in an industrial environment. (Courtesy of Westinghouse Electric Corporation.)

operating system (DOS). Which PLC you buy can limit you to the language it comes with, unless it is a modular type that enables you to plug in a language module. PLCs have large user memory. Several megabytes, which enable you to load and use software stored on a disk, is normal. Any language capability can be made available via this software, which can be loaded into the memory via these disks. PLCs are not designed with this kind of flexibility but are meant to be specialized computers for control, and they feature interfacing and control with external devices.

Computers are complex computing machines capable of executing several programs or tasks simultaneously and in any order. Most PLCs, on the other hand, execute a single program in an orderly and sequential fashion from first to last instruction.

Perhaps the most significant difference between a PLC and a computer is the fact that PLCs have been designed for installation and maintenance by plant electricians who are not required to be highly skilled computer technicians. Troubleshooting is simplified by the design of most PLCs because they include fault indicators and written fault information displayed on the programmer screen. The modular interfaces for connecting the field devices are actually a part of the PLC and are easily connected and replaced.

Just as it has transformed the way the rest of the world does business, the personal computer has infiltrated the PLC control industry. Software written and run on the PC has changed how people work with PLCs. Basically, PLC software run on a PC falls into the following two categories:

◇ PLC software that allows the user to program and document gives the user the tools to write a PLC program—using ladder logic or another programming language—and document or explain the program in as much detail as is necessary.

◇ PLC software that allows the user to monitor and control the process is also called man-machine, or operator, interface. It enables the user to view a process—or a graphical representation of a process—on a CRT, determine how the system is running, trend values, and receive alarm conditions.

1•6 PLC SIZE AND APPLICATION

There is much variation in the size identification of PLCs. Typically PLCs are divided into three major size categories: small (also known as micro), medium, and large, each with distinct operating features. The small size category covers units with up to 128 I/Os and memories up to 2 K bytes. These PLCs are capable of providing simple to advanced levels of machine control.

Medium-size PLCs have up to 2048 I/Os and memories up to 32 K bytes. Special I/O modules make medium PLCs adaptable to temperature, pressure, flow, weight, position, or any type of analog function commonly encountered in process control applications.

Large PLCs, of course, are the most sophisticated units of the PLC family. They have up to 16,000 I/Os and memories up to 2 M bytes. PLCs of this size have almost unlimited applications. Large PLCs can control individual production processes or entire plants.

The key factor in selecting a PLC is establishing exactly what the unit is supposed to do. In general it is not advisable to buy a PLC system that is larger than current needs dictate. However, future conditions should be anticipated to ensure that the system is the proper size to fill the current and possibly future requirements of an application.

There are three major types of PLC application: single-ended, multitask, and control management. A *single-ended* PLC application involves one PLC controlling one process. This would be a stand-alone unit and would not be used for communicating with other computers or PLCs. The size and sophistication of the process being controlled is obviously a factor in determining which PLC to select. The applications could dictate a large processor, but usually this category requires a small PLC.

A *multitask* PLC application usually calls for a medium-size PLC and involves one PLC controlling several processes. Adequate I/O capacity is a significant factor in this type of installation. In addition, if the PLC will be a subsystem of a larger process and have to communicate with a central PLC or computer, provisions for a data communications network are also required.

A *control management* PLC application involves one PLC controlling several others. This kind of application requires a large PLC processor designed to communicate with other PLCs and possibly with a computer. The control management PLC supervises several PLCs by downloading programs that tell the other PLCs what has to be done. It must be capable of connection to all the PLCs so it can communicate with any one it wishes, by proper addressing.

The *memory size* of a PLC ranges from 1 K to 2 M. The amount of memory required depends on the application. Factors affecting the memory size needed for a particular PLC installation include:

◆ Number of I/O points used
◆ Size of control program
◆ Data-collecting requirements
◆ Supervisory functions required
◆ Future expansion

The *instruction set* for a particular PLC type lists the different types of instructions supported. Typically, this ranges from 15 instructions on smaller units up to 100 instructions on larger, more powerful units (see Table 1-1).

Since its invention, the PLC has been successfully applied in virtually every segment of industry. This list includes steel mills, paper and pulp plants, chemical and automotive and power plants. Programmable logic controllers perform a great variety of control tasks, from repetitive ON/OFF control of a simple machine to sophisticated manufacturing and process control.

Table 1-1

TYPICAL PLC INSTRUCTIONS

Instruction	Operation
XIC (Examine ON)	Examine a bit for an ON condition
XIO (Examine OFF)	Examine a bit for an OFF condition
OTE (Output Energize)	Turn ON a bit (nonretentive)
OTL (Output Latch)	Latch a bit (retentive)
OTU (Output Unlatch)	Unlatch a bit (retentive)
TOF (Timer Off-Delay)	Turn an output ON or OFF after its rung has been OFF for a preset time interval
TON (Timer On-Delay)	Turn an output ON or OFF after its rung has been ON for a preset time interval
CTD (Count Down)	Use a software counter to count down from a specified value
CTU (Count Up)	Use a software counter to count up to a specified value

Questions

1. Define *programmable logic controller*.

2. List four advantages that PLCs offer over the conventional relay-type of control system.

3. List four tasks in addition to relay switching that PLCs are capable of performing.

4. State two ways in which I/O is incorporated into the PLC.

5. Describe how the input/output (I/O) modules connect to the processor in a modular-type PLC configuration.

6. Describe the main function of each of the following main component parts of a PLC.

    a. Processor module (CPU)
    b. I/O modules
    c. Programming device
    d. Power supply module

7. List three common types of PLC programming devices.

8. Answer the following with reference to the unit process control relay ladder diagram of Fig. 1-9 (on p. 10):

    a. When do the pressure switch contacts close?

    b. When do the temperature switch contacts close?

    c. How are the pressure and temperature switches connected with respect to each other?

    d. Describe the two conditions under which the motor starter coil will become energized.

    e. What is the approximate value of the voltage drop across each of the following when their contacts are open?

       1. Pressure switch
       2. Temperature switch
       3. Manual pushbutton

9. Answer the following with reference to the unit process control PLC ladder logic diagram of Fig. 1-12 (on p. 12):

    a. What do the individual symbols represent?
    b. What do the numbers represent?
    c. What field device is the number 002 identified with?
    d. What field device is the number 009 identified with?
    e. What two conditions will provide a continuous path from left to right across the rung?
    f. Describe the sequence of operation of the controller for one scan of the program.

10. Compare the method by which the process control operation is changed in a relay system to the method for a PLC system.

11. Compare the PLC and general-purpose computer with regard to:

    a. Operating environment
    b. Method of programming
    c. Execution of program
    d. Maintenance

12. Describe two categories of software written and run on a personal computer and used in conjunction with a PLC.

13. **A.** Explain the three size classifications for PLCs and state one general application for each size.

    B. What are the two key factors in selecting the size of a PLC?

14. Compare the single-ended, multitask, and control management type of PLC applications.

15. List five factors affecting the memory size needed for a particular PLC installation.

16. What does the instruction set for a particular PLC refer to?

Problems

1. Given two single-pole switches, write a program that will turn on an output when both switch A and switch B are closed.

2. Given two single-pole switches, write a program that will turn on an output when either switch A or switch B is closed.

3. Given four NO (Normally Open) pushbuttons (A-B-C-D), write a program that will turn a lamp on if pushbuttons A and B or C and D are closed.

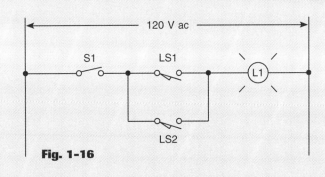

Fig. 1-16

4. Write a program for the relay ladder diagram shown in Fig. 1-16.

5. Write a program for the relay ladder diagram shown in Fig. 1-17.

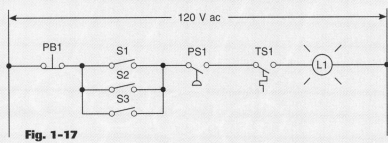

Fig. 1-17

PLC Hardware Components

After completing this chapter, you will be able to:

◆ List and describe the function of the hardware components used in PLC systems

◆ Describe the basic circuitry and applications for discrete and analog I/O modules, and interpret typical I/O and CPU specifications

◆ Explain I/O addressing

◆ Describe the general classes and types of PLC memory devices

◆ List and describe the different types of PLC peripheral support devices available

This chapter will expose you to the details of PLC hardware and modules that make up a PLC control system. Illustrations of the various subparts of a PLC, as well as general connection paths, are shown. The CPU and memory hardware components, including the various types of memory that are available, are discussed. The hardware of the input/output section, including the difference between the discrete and analog types of modules, is described.

2.1 THE I/O SECTION

The input and output interface modules provide the equivalents of eyes, ears, and tongue to the brain of a PLC: the CPU. The I/O section consists of an I/O rack and individual I/O modules similar to that shown in Fig. 2-1. Input interface modules accept signals from the machine or process devices and convert them into signals that can be used by the controller. Output interface modules convert controller signals into external signals used to control the machine or process. A typical PLC has room for several I/O modules, allowing it to be customized for a particular application by selecting the appropriate modules. A slot in the PLC can hold any type of I/O module.

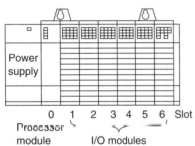

A typical PLC has room for several I/O modules that plug into slots in a rack.

Fig. 2-1

I/O section.

The I/O system provides an interface between the hard-wired components in the field and the CPU. The input interface allows *status information* regarding processes to be communicated to the CPU, and thus allows the CPU to communicate *operating signals* through the output interface to the process devices under its control.

A *chassis* is a physical hardware assembly that houses devices such as I/O modules, processor modules, and power supplies. Chassis come in different sizes according to the number of slots they contain. In general, they can have 4, 8, 12, or 16 slots.

A logical *rack* is an addressable unit consisting of 128 input points and 128 output points. A rack uses 8 words in the input image table file and 8 words in the output image table file. A word in the output image table file and its corresponding word in the input image table file are called an *I/O group*. A rack can contain a maximum of 8 I/O groups (numbered from 0 through 7) for up to 128 discrete I/O (Fig. 2-2). There can be more than one rack in a chassis and more than one chassis in a rack.

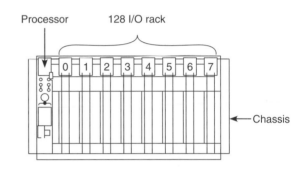

	Output Image Table		
O:00	0 1 2 3 4 5 6 7	10 11 12 13 14 15 16 17
O:01	0 1 2 3 4 5 6 7	10 11 12 13 14 15 16 17
O:02	0 1 2 3 4 5 6 7	10 11 12 13 14 15 16 17
Words O:03	0 1 2 3 4 5 6 7	10 11 12 13 14 15 16 17
O:04	0 1 2 3 4 5 6 7	10 11 12 13 14 15 16 17
O:05	0 1 2 3 4 5 6 7	10 11 12 13 14 15 16 17
O:06	0 1 2 3 4 5 6 7	10 11 12 13 14 15 16 17
O:07	0 1 2 3 4 5 6 7	10 11 12 13 14 15 16 17

	Input Image Table		
I:00	0 1 2 3 4 5 6 7	10 11 12 13 14 15 16 17
I:01	0 1 2 3 4 5 6 7	10 11 12 13 14 15 16 17
I:02	0 1 2 3 4 5 6 7	10 11 12 13 14 15 16 17
Words I:03	0 1 2 3 4 5 6 7	10 11 12 13 14 15 16 17
I:04	0 1 2 3 4 5 6 7	10 11 12 13 14 15 16 17
I:05	0 1 2 3 4 5 6 7	10 11 12 13 14 15 16 17
I:06	0 1 2 3 4 5 6 7	10 11 12 13 14 15 16 17
I:07	0 1 2 3 4 5 6 7	10 11 12 13 14 15 16 17

Note: Input and output addresses for Allen-Bradley are in the octal numbering system.

Fig. 2-2

Logical rack.

One benefit of a PLC system is the ability to locate the I/O modules near the field devices to minimize the amount of wiring required. This rack (Fig. 2-3) is referred to as a *remote*

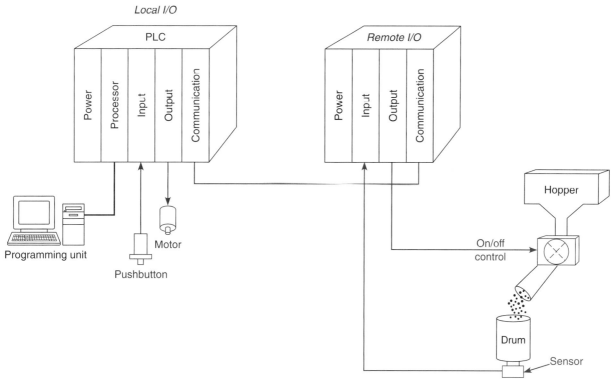

Local I/O

PLC

Power | Processor | Input | Output | Communication

Remote I/O

Power | Input | Output | Communication

Programming unit

Pushbutton

Motor

Hopper

On/off control

Drum

Sensor

The processor receives signals from the remote input modules and sends signals back to their output modules via the communications module.

Fig. 2-3

Remote I/O rack.

rack when it is located away from the processor module. To communicate with the processor, the remote rack uses a special communications network. Each remote rack requires a unique station number to distinguish one from another. The remote racks are linked to the local rack through a *communications module.* Cables connect the modules with each other. If fiber optic cable is used between the CPU and I/O rack, it is possible to operate I/O points from distances greater than 20 miles with no voltage drop. Coaxial cable will allow remote I/O to be installed at distances greater than two miles. Fiber optic cable will not pick up noise due to adjacent high power lines or equipment normally found in an industrial environment. Coaxial cable is more susceptible to this type of noise.

The location of a module within a rack and the terminal number of a module to which

an input or output device is connected will determine the device's address (Fig. 2-4 on p. 22). Each input and output device must have a specific address. This address is used by the processor to identify where the device is located to monitor or control it. In addition, there is some means of connecting field wiring on the I/O module housing. Connecting the field wiring to the I/O housing allows easier disconnection and reconnection of the wiring to change modules. Lights are also added to each module to indicate the ON or OFF status of each I/O circuit. Most output modules also have blown fuse indicators.

In general, basic addressing elements include:

◆**Type**
The type determines if an input or output is being addressed.

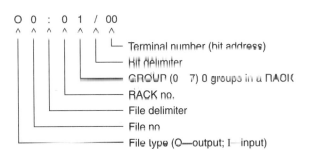

Examples:
I1: 27/17 – Input, File 1, Rack 2, Group 7, Bit 17
O0:34/07 – Output, File 0, Rack 3, Group 4, Bit 7
I1:0/0 – Input, File 1, Rack 0, Group 0, Bit 0 (Short Form Blank = 0)
O0:1/1 – Output, File 0, Rack 0, Group 1, Bit 1

(a) Allen-Bradley PLC 5 addressing format.

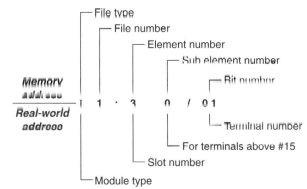

Examples:
O0:4.0/15–Output module in slot 4, terminal 15
I1:3.0/8–Input module in slot 3, terminal 8
O0:6.0–Output module, slot 6
I1:5.0–Input module, slot 5

(b) Allen-Bradley SLC 500 addressing format.

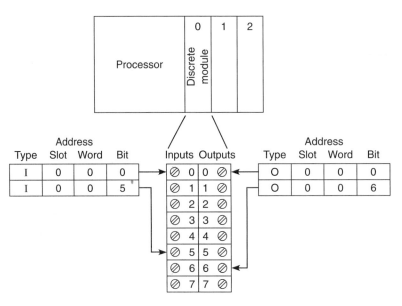

(c) Discrete I/O module addressing.

Fig. 2-4

I/O module addressing.

◆**Slot**

The slot number is the physical location of the I/O module. This may be a combination of the rack number and the slot number when using expansion racks.

◆**Word and Bit**

The word and bit are used to identify the actual terminal connection in a particular I/O module. A discrete module usually uses only one word, and each connection corresponds to a different bit that makes up the word.

The design of a PLC determines whether the system is capable of being addressed flexibly or it is rigid in its addressing method. Flexible addressing schemes allow PLC system designers to create control logic software without having to follow a sequential

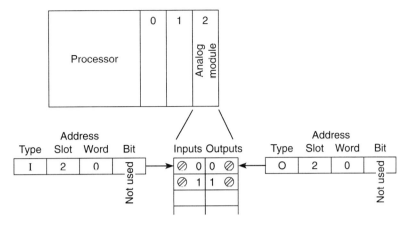

(*d*) Analog I/O module addressing. The bit part of the address is usually not used; however, bits of the digital representation of the analog value can be addressed by the programmer if necessary.

(*e*) *Symbolic addresses* are real names or codes the programmer can substitute for a logical address because they relate physically to the application. They are a physical name convention for a location in the data table. In this example, the symbolic addresses are LS-3 and pump-14, while the actual addresses are I:3/3 and O:4/14, respectively.

Fig. 2-4 (continued)

I/O module addressing.

I/O assignment, resulting in a randomly addressed and installed I/O system.

Within flexible systems, individual slot and point addresses are normally determined by the sequence in which the I/O racks are connected together. In the case of some small PLCs, the system contains one rack and therefore has I/O addressing fixed by the manufacturer. Actual address labeling varies greatly from manufacturer to manufacturer.

A standard I/O module consists of a printed circuit board and a terminal assembly similar to that shown in Fig. 2-5 (on p. 24). The printed circuit board contains the electronic circuitry used to interface the circuit of the processor with that of the input or output device. It is designed to plug into a slot or connector in the I/O rack or directly into the processor. The terminal assembly, which is attached to the front edge of the printed circuit board, is used for making field-wiring connections. The module contains terminals for each input and output connection, status lights for each of the inputs and outputs, and connections to the power supply used to power the inputs and outputs.

Most modules have plug-on wiring terminal strips. The terminal strip is plugged onto the actual module. If there is a problem with a module, the entire strip is removed, a new module is inserted, and the terminal strip is plugged into the new module. Unless otherwise specified, never install or remove I/O modules or terminal blocks while the PLC is powered. A module inserted into the wrong

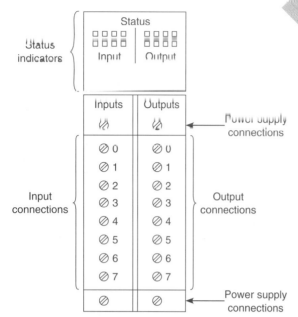

Status indicators

Input connections

Input

Output

Power supply connections

Output connections

Power supply connections

The arrangement of the terminals, status indicators, and power connections may vary with different manufacturers. Note that I/O modules can have both input and output connections in the same physical module.

Fig. 2-5

Typical combination I/O module.

slot could be damaged by improper voltages connected through the wiring arm. Most faceplates and I/O modules are keyed to prevent putting the wrong faceplate on the wrong module. In other words, an output module cannot be placed in the slot where an input module was originally located.

Input and output module cards can be placed anywhere in the rack, but they are normally grouped together for ease of wiring. I/O modules can be 8, 16, or 32 point cards. The number refers to the number of inputs or outputs available. The standard I/O module has eight inputs or outputs. A *high-density* module may have up to 32 inputs or outputs. The advantage with the high-density module is that it is possible to install 32 inputs or outputs in one slot for greater space savings. The only disadvantage is that the high-density output modules cannot handle as much current per output. The 32 point cards usually have at least four common points.

•2 DISCRETE I/O MODULES

The most common type of I/O interface module is the *discrete* type. This type of interface connects field input devices of the ON/OFF nature such as selector switches, pushbuttons, and limit switches. Likewise, output control is limited to devices such as lights, small motors, solenoids, and motor starters that require simple ON/OFF switching. The classification of *discrete I/O* covers *bit-oriented* inputs and outputs. In this type of input or output, each bit represents a complete information element in itself and provides the status of some external contact or advises of the presence or absence of power in a process circuit.

Each discrete I/O module is powered by some *field-supplied* voltage source. Since these voltages can be of different magnitude or type, I/O modules are available at various ac and dc voltage ratings, as listed in Table 2-1. They receive their module voltage and current for proper operation from the backplane of the rack enclosure into which they are inserted. Power from this supply is used to power the electronics, both active and passive, that reside on the I/O module circuit board. The relatively higher currents required by the loads of an output module are supplied by user-supplied power. Module power supplies may be rated for 3 A, 4 A, 12 A, or 16 A depending on the type and number of modules used.

Table 2-1

COMMON RATINGS FOR DISCRETE I/O INTERFACE MODULES

Input Interfaces	Output Interfaces
24 V ac/dc	12–48 V ac
48 V ac/dc	120 V ac
120 V ac/dc	230 V ac
230 V ac/dc	120 V dc
5 V dc (TTL level)	230 V dc
	5 V dc (TTL level)
	24 V dc

Figure 2-6 shows block diagrams for one input of a typical alternating current (ac) *discrete input module.* The input circuit is composed of two basic sections: the *power* section and the *logic* section. The power and logic sections are normally coupled together with a circuit, which electrically separates the two.

A simplified schematic and wiring diagram for one input of a typical ac input module is shown in Fig. 2-7a and b. When the pushbutton is closed, 120 V ac is applied to the bridge rectifier through resistors R1 and R2. This produces a low-level direct current (dc) voltage, which is applied across the LED of the optical isolator. The zener diode (Z_D) voltage rating sets the minimum level of voltage that can be detected. When light from the LED strikes the phototransistor, it switches into conduction and the status of the pushbutton is communicated in logic or low-level dc voltage to the processor. The

optical isolator not only separates the higher ac input voltage from the logic circuits, but also prevents damage to the processor due to line voltage transients. Optical isolation also helps reduce the effects of electrical noise, common in the industrial environment, which can cause erratic operation of the processor. Coupling and isolation can also be accomplished by use of a pulse transformer.

Input modules typically have light-emitting diodes (LEDs) for monitoring the inputs. There is one LED for every input. If the input is ON, the LED is ON. The LEDs on the modules are very useful for troubleshooting.

Input modules perform four tasks in the PLC control system. They

◆ sense when a signal is received from a sensor on the machine

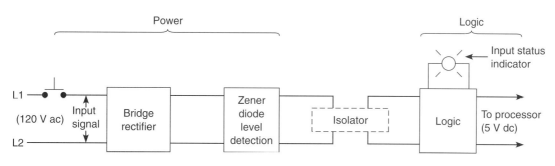

Block diagrams of a discrete input module.

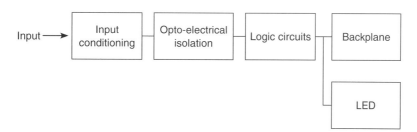

The input circuit responds to an input signal in the following manner:
• An input filter removes false signals due to contact bounce or electrical interference
• Opto-electrical isolation protects the input circuit and backplane circuits by isolating logic circuits from input signals
• Logic circuits process the signal
• An input LED turns ON or OFF, indicating the status of the corresponding input device

Fig. 2-6

AC discrete input module.

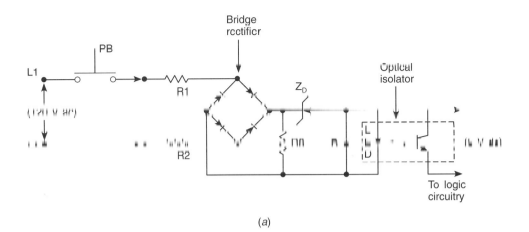

(a)

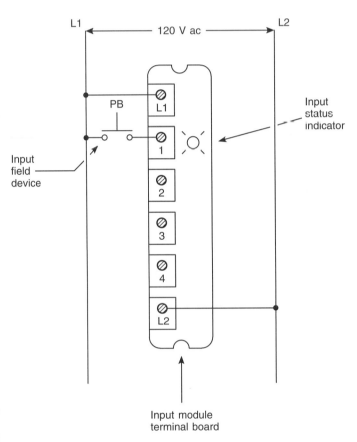

Fig. 2-7

(a) Simplified schematic
for an ac input module.
(b) Typical input module
wiring connection.

(b)

◆ convert the input signal to the correct
voltage level for the particular PLC

◆ isolate the PLC from fluctuations in the
input signal's voltage or current

◆ send a signal to the processor indicating which sensor originated the signal

Figure 2-8 shows block diagrams for one output of a typical *discrete output module.* Like
the input module, it is composed of two
basic sections: the *power* section and the
logic section, coupled by an *isolation* circuit.
The output interface can be thought of as a
simple electronic switch to which power is
applied to control the output device.

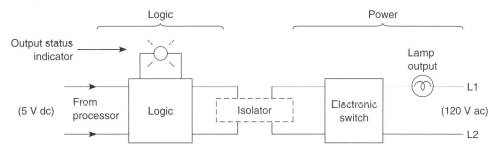

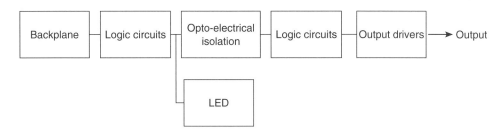

(a) Block diagram of a discrete output module.

(b) The output circuit controls the output signal in the following manner:
  • Logic circuits determine the output status
  • An output LED indicates the status of the output signal
  • Opto-electrical isolation separates output circuit logic and backplane circuits from field signals
  • The output driver turns the corresponding output ON or OFF.

Fig. 2-8

AC discrete output module.

A simplified schematic and wiring diagram for one output of a typical ac output module is shown in Fig. 2-9a. As part of its normal operation, the processor sets the output status according to the logic program. When the processor calls for an output, a voltage is applied across the LED of the isolator. The LED then emits light, which switches the phototransistor into conduction. This in turn switches the *triode ac semiconductor switch (triac)* into conduction, which, in turn, turns on the lamp. Since the triac conducts in either direction, the output to the lamp is alternating current. The triac, rather than having ON and OFF status, actually has LOW and HIGH resistance levels, respectively. In its OFF state (HIGH resistance), a small leakage current of a few milliamperes still flows through the triac.

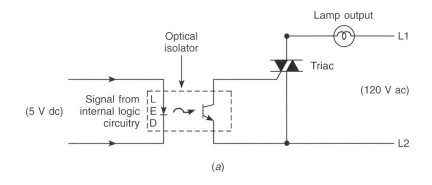

(a)

Fig. 2-9

(a) Simplified schematic for an ac output module.

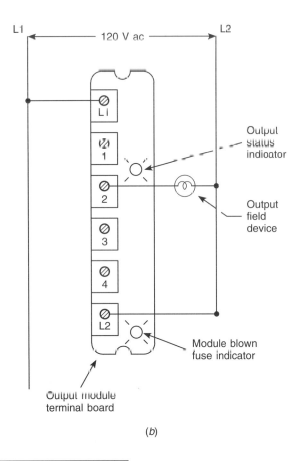

(b)

Fig. 2-9 (continued)

(b) Typical output module wiring connection.

As with input circuits, the output interface is usually provided with LEDs that indicate the status of each output. Fuses are generally required for the output module, and they are provided on a per circuit basis, thus allowing for each circuit to be protected and operated separately. Some modules also provide visual indicators for fuse status (see Fig. 2-9*b*).

Individual ac outputs are usually limited by the size of the triac to 1 or 2 amperes (A). The maximum current load for any one module is also specified. To protect the output module circuits, specified current ratings should not be exceeded. For controlling larger loads, such as large motors, a standard control relay is connected to the output module. The contacts of the relay can then be used to control a larger load or motor starter, as shown in Fig. 2-10. When a control relay is used in this manner, it is called an *interposing* relay.

Discrete output modules are used to turn real-world output devices either ON or OFF. These modules can be used to control any two-state device, and they are available in ac and dc versions and in various voltage ranges and current ratings. Output modules can be purchased with *transistor, triac,* or *relay* output. Triac outputs can be used only for control of ac devices, while transistor outputs can be used only for control of dc devices. Relay outputs can be used with ac or dc devices, but they have a much slower switching time compared to solid-state outputs. Allen-Bradley modules are color-coded for each identification as follows:

Color	Type of I/O
Red . . . . ac inputs/outputs
Blue . . . . dc inputs/outputs
Orange. . . . . . . Relay outputs
Green. . . . . Specialty modules

The design of *dc field devices* typically requires that they be used in a specific sinking or sourcing circuit, depending on the internal circuitry of the device. *Sinking* and *sourcing* references are terms used to describe a current signal flow relationship between field input and output devices in a control system and their power supply (Fig. 2-11). DC input and output field circuits are commonly used with field devices that have some form of internal solid-state circuitry that need a dc signal voltage to function. Field devices connected to the positive side (+V) of the field power supply are sourcing field devices. Field devices connected to the negative side (dc common) of the field power supply are sinking field devices. To maintain electrical compatibility between field devices and the programmable controller system, this definition is extended to the input/output circuits on the discrete dc I/O modules.

❖ Sourcing I/O circuits supply (source) current to sinking field devices

❖ Sinking I/O circuits receive (sink) current from sourcing field devices

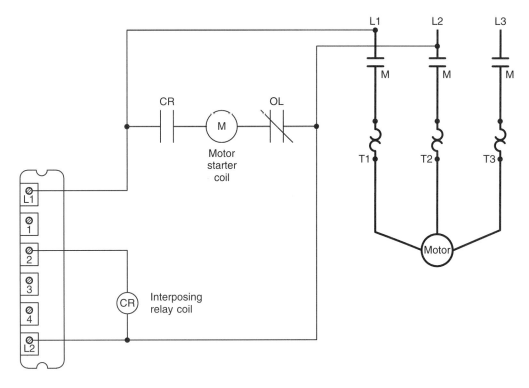

Fig. 2-10

Interposing relay connection.

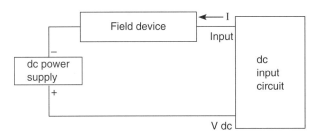

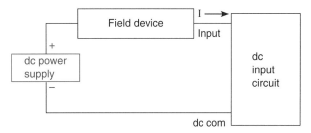

(a) Sinking device with sourcing input module circuit. The field device is on the negative side of the power supply, between the supply and the input terminal. When the field device is activated, it sinks current from the input circuit.

(b) Sourcing device with sinking input module circuit. The field device is on the positive side of the power supply, between the supply and the input terminal. When the field device is activated, it souces current to the input circuit.

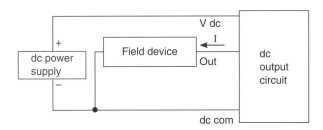

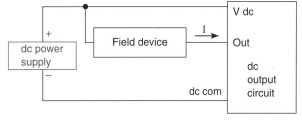

(c) Sinking device with sourcing output module circuit. The field device is on the negative side of the power supply, between the supply and the output terminal. When the output is activated, it sources current to the field device. A current source output uses a PNP transistor.

(d) Sourcing device with sinking output module circuit. The field device is on the positive side of the power supply, between the supply and the output terminal. When the output is activated, it sinks current from the field device. A current sink output uses an NPN transistor.

Fig. 2-11

Sinking and sourcing references.

PLC Hardware Components

2·3 ANALOG I/O MODULES

The earlier PLCs were limited to discrete I/O interfaces, which allowed only ON/OFF-type devices to be connected. This limitation meant that the PLC could have only partial control of many process applications. Today, however, a complete range of both discrete and analog interfaces are available that will allow controllers to be applied to practically any type of control process.

Analog input interface modules contain the circuitry necessary to accept analog voltage or current signals from analog field devices. These inputs are converted from an analog to a digital value by an *analog-to-digital (A/D)* converter circuit. The conversion value, which is proportional to the analog signal, is expressed as a 12-bit binary or as a 3-digit binary-coded decimal (BCD) for use by the processor. Analog input sensing devices include temperature, light, speed, pressure, and position transducers. Figure 2-12 shows a typical analog input interface module connection to a thermocouple. A varying dc voltage in the millivolt range, proportional to the temperature being monitored, is produced by the thermocouple. This voltage is amplified and digitized by the analog input module and then sent to the processor on command from a program instruction. Because of the low voltage level of the input signal, a shielded cable is used in wiring the circuit to reduce unwanted electrical noise signals that can be induced in the conductors from other wiring. This noise can cause temporary operating errors that can lead to hazardous or unexpected machine operation.

There are two basic types of analog input modules available: *current* sensing and *voltage* sensing. Voltage input modules are available in two types: unipolar and bipolar. Unipolar modules can accept only one polarity for input. For example, if the application requires the card to measure 0 to +10 V, a unidirectional card would be used. The bipolar card will accept input of positive and negative

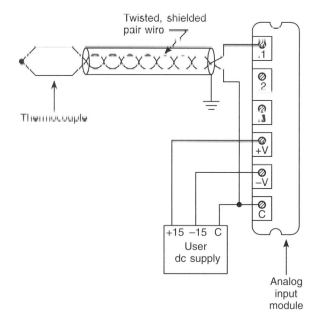

Fig. 2-12

Typical thermocouple connection to an analog input module.

polarity. For example, if the application produces a voltage between −10 V and +10 V, a bidirectional input card would be used because the measured voltage could be negative or positive. Current input modules are normally designed to measure current in the 4-mA to 20-mA range.

The *analog output interface module* receives from the processor digital data, which are converted into a proportional voltage or current to control an analog field device. The digital data is passed through a *digital-to-analog (D/A)* converter circuit to produce the necessary analog form. Analog output devices include small motors, valves, analog meters, and seven-segment displays.

Figure 2-13 illustrates the use of analog I/O modules in a typical PLC control system. In this application the PLC controls the amount of fluid placed in a holding tank by adjusting the percentage of the valve opening. The valve is initially open 100%. As the fluid level in the tank approaches the preset point, the processor modifies the output, which adjusts the valve to maintain a set point.

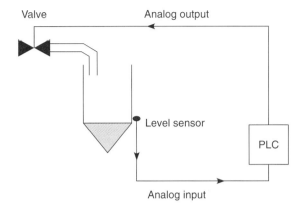

Fig. 2-13

Typical analog I/O control system.

2•4 SPECIAL I/O MODULES

Special I/O modules have been developed to meet several needs. These include:

◆**High-Speed Counter Module**

The high-speed counter module is used to provide an interface for applications requiring counter speeds that surpass the capability of the PLC ladder program. High-speed counter modules are used to count pulses from sensors, encoders, and switches at very high speeds. Modules that can count pulses up to 75 kHz are common.

◆**Thumb-Wheel Module**

The thumb-wheel module allows the use of thumb-wheel switches for feeding information in parallel to the PLC to be used in the control program. The thumb-wheel information is usually in binary coded decimal (BCD) form, and enables a person to change set points or preset points externally without modifying the control program.

◆**TTL Module**

The TTL module allows the transmitting and receiving of TTL signals for communication with the PLC's processor. TTL-level signals are in a form the processor can accept, and only buffering is required.

◆**Encoder-Counter Module**

The encoder-counter module allows continual monitoring of an incremental or absolute encoder. Encoders keep track of the position of shafts or axes. Gray code is common for absolute encoders, with position determined by decoding the gray code.

◆**BASIC or ASCII Module**

The ASCII module allows the transmitting and receiving of ASCII files. These files are usually programs or manufacturing data. The modules are normally programmed with BASIC commands. The user writes a program in the *BASIC language.* BASIC modules can be used to output text to a printer or terminal to update an operator.

◆**Stepper-Motor Module**

The stepper-motor module provides pulse trains to a stepper-motor translator, which enables control of a stepper motor. The commands for the module are determined by the control program in the PLC.

◆**BCD-Output Module**

The BCD-output module enables a PLC to operate devices that require BCD-coded (binary coded decimal) signals such as seven-segment displays.

Some special modules are referred to as *intelligent I/O* because they have their own microprocessors on board that can function in parallel with the PLC. These include:

◆**PID Module**

The proportional-integral-derivative (PID) module is used in process control applications that incorporate PID algorithms. An algorithm is a complex program based on mathematical calculations. A PID module allows process control to take place outside the CPU. This arrangement prevents the CPU from being burdened with complex calculations. The microprocessor in the PID module processes data, compares the data to setpoints provided by the CPU, and determines the appropriate output signal.

◆ Servo Module

The servo module is used in closed-loop process control applications. Closed-loop control is accomplished via feedback from the device. The programming of this module is done through the PLC; but once programmed, it can control a device independently without interfering with the PLC's normal operation.

◆ Communication Module

As different systems are integrated, data must be shared throughout the system. PLCs must be able to communicate with computers, computer numerical control (CNC) machines, robots, and other PLCs. This module allows the user to connect the PLC to high-speed local networks that may be different from the network communication provided with the PLC.

◆ Language Module

The language module enables the user to write programs in a high-level language. Via a high-level-language interpreter, it converts the high-level commands into machine language understandable to a PLC's processor. BASIC is the most popular language module. Other language modules available include C, Forth, and PASCAL.

◆ Speech Module

Speech modules typically are used to digitize a human voice pronouncing the desired word, phrase, or sentence. The digitized sound is stored in the module's memory. Each word, phrase, or sentence is given a number. Ladder logic is used to output the appropriate message at the appropriate time.

·5 I/O SPECIFICATIONS

Manufacturer's specifications provide much information about how an interface device is correctly and safely used. The specifications place certain limitations, not only on the module, but also on the field equipment that it can operate. The following is a list of some typical manufacturers' I/O specifications, along with a short description of what is specified.

◆ Nominal Input Voltage

This ac or dc value specifies the magnitude and type of voltage signal that will be accepted.

◆ On-State Input Voltage Range

This value specifies the voltage at which the input signal is recognized as being absolutely on.

◆ Nominal Current per Input

This value specifies the minimum input current that the input devices must be capable of driving to operate the input circuit.

◆ Ambient Temperature Rating

This value specifies what the maximum temperature of the air surrounding the I/O modules should be for best operating conditions.

◆ Input Delay

Also known as *response time,* this value specifies the time duration for which the input signal must be on before being recognized as a valid input. This delay is a result of filtering circuitry provided to protect against contact bounce and voltage transients. This input delay is typically in the 9-ms to 25-ms range.

◆ Nominal Output Voltage

This ac or dc value specifies the magnitude and type of voltage source that can be controlled by the output.

◆ Output Voltage Range

This value specifies the minimum and maximum output operating voltages. An output circuit rated at 120 V ac, for example, may have an absolute working range of 92 V ac (min.) to 138 V ac (max.).

◆ Maximum Output Current Rating per Output and Module

These values specify the maximum current that a single output and the module as a whole can safely carry under load (at rated voltage). For example, the specification may give each output a current limit of 1 A. The overall rating of the module current will normally be less than the total of the individuals. The overall rating might be 6 A because each of the eight devices would not normally draw their 1 A at the same time.

◆ Maximum Surge Current per Output

This value specifies the maximum inrush current and duration (e.g., 20 A for 0.1 s) for which an output circuit can exceed its maximum continuous current rating.

◆ Off-State Leakage Current per Output

This value specifies the maximum value of leakage current that flows through the output in its OFF state.

◆ Electrical Isolation

This maximum value (volts) defines the isolation between the I/O circuit and the controller's logic circuitry. Although this isolation protects the logic side of the module from excessive input or output voltages or current, the power circuitry of the module may be damaged.

◆ Number of Inputs and Outputs per Card

This value indicates the number of field inputs or outputs that can be connected to the module. Some modules provide more than one common terminal, which allows the user to use different voltage ranges on the same card as well as to distribute the current more effectively.

◆ Backplane Current Draw

This value indicates the amount of current the module requires from the backplane. The sum of the backplane current drawn for all modules in a chassis is used to select the appropriate chassis power supply rating.

◆ Resolution

The resolution of an analog I/O module specifies how accurately an analog value can be represented digitally. The higher the resolution (typically specified in bits), the more accurately an analog value can be represented.

◆ Input Impedance and Capacitance

For analog I/Os, these values must be matched to the external device connected to the module. Typical ratings are in megohms (MΩ) and picofarads (pF).

◆ Common Mode Rejection Ratio

This specification refers to an analog module's ability to prevent noise from interfering with data integrity on a single channel and from channel to channel on the module.

•6 THE CPU

The CPU houses the processor-memory module(s), communications circuitry, and power supply. Figure 2-14a on p. 34 is a simplified illustration of the CPU. Central processing unit architectures may differ from one manufacturer to another, but in general most of them follow this organization. The power supply may be located inside the CPU enclosure or may be a separate unit mounted next to the enclosure, as shown in Fig. 2-14b.

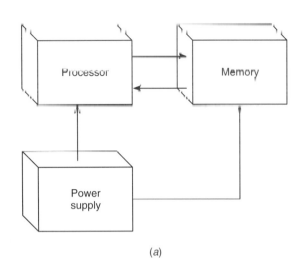

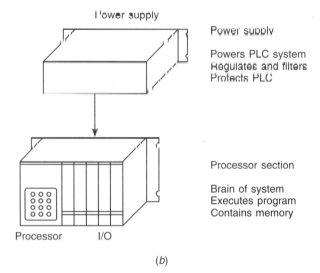

Power supply

Power supply
Powers PLC system
Regulates and filters
Protects PLC

Processor section

Brain of system
Executes program
Contains memory

Processor   I/O

(a)   (b)

Fig. 2-14

Major components of the CPU: (a) simplified illustration
of the CPU; (b) power supply mounted outside CPU
enclosure.

Depending upon the type of memory, volatile
or nonvolatile, the power supply could also
include a backup battery system. For example, on the Allen-Bradley SLC 500 system, a
battery is installed on the processor cards.

A PLC's power supply provides all the voltage levels required for operation. The power
supply converts 120 or 220 V ac into the dc
voltage required by the CPU, memory, and
I/O electronic circuitry. The PLC operates on
+5 and −5 V dc. Therefore, the power supply must be capable of rectifying the step-
ping-down of the ac input voltage to a usable
level of dc voltage.

The term *CPU* is often used interchangeably
with the term *processor*. However, by strict
definition, the *CPU* term encompasses all
the necessary elements that form the intelligence of the system. There are definite rela-
tionships between the sections that form the
CPU and the constant interaction among
them. The processor is continually interact-
ing with the system memory to interpret and
execute the user program that controls the
machine or process. The system power supply provides all the necessary voltage levels
to ensure proper operation of all processor
and memory components.

The CPU contains the same type of microprocessor found in a personal computer. The
difference is that the program used with the
microprocessor is written to accommodate
just ladder logic instead of other programming
languages. The CPU executes the operating
system, manages memory, monitors inputs,
evaluates the user logic (ladder diagram), and
turns on the appropriate outputs. The CPU of
a PLC system may contain more than one
microprocessor. The advantage of using multiprocessing is that control and communication
tasks can be divided up, and the overall operating speed is improved. For example, some
PLC manufacturers use a control microprocessor and a logic microprocessor. The control
microprocessor carries out the more complex
computations and data manipulations. The
logic microprocessor executes the timing,
logic, and counting operations, as well as
looks after the applications program.

Associated with the processor unit will be a
number of status LED indicators to provide
system diagnostic information to the operator
(Fig. 2-15). Also, a keyswitch may be provided that allows you to select one of the following three modes of operation (otherwise
these modes must be accessed from the programming device): RUN, PROG, and REM.

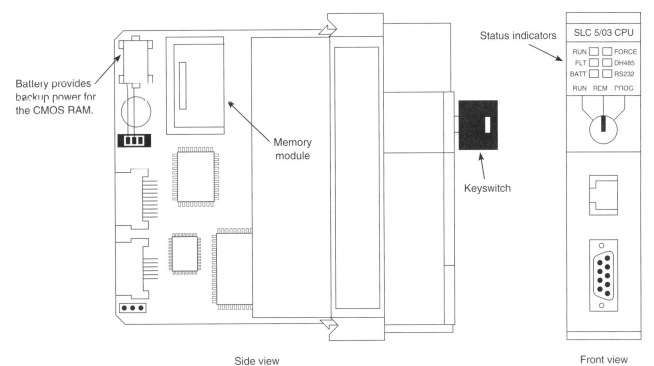

Battery provides backup power for the CMOS RAM.

Memory module

Status indicators

SLC 5/03 CPU

RUN ☐ ☐ FORCE
FLT ☐ ☐ DH485
BATT ☐ ☐ RS232

RUN   REM   PROG

Keyswitch

Side view

Front view

Fig. 2-15

Typical processor unit.

RUN Position

◆ Places the processor in the Run mode

◆ Executes the ladder program and energizes output devices

◆ Prevents you from performing online program editing in this position

◆ Prevents you from using a programmer/operator interface device to change the processor mode

PROG Position

◆ Places the processor in the Program mode

◆ Prevents the processor from scanning or executing the ladder program, and the controller outputs are de-energized

◆ Allows you to perform program entry and editing

◆ Prevents you from using a programmer/operator interface device to change the processor mode

REM Position

◆ Places the processor in the Remote mode: either the REMote Run, REMote Program, or REMote Test mode

◆ Allows you to change the processor mode from a programmer/operator interface device

◆ Allows you to perform online program editing

The processor module also contains circuitry to communicate with the programming device. Somewhere on the module you will find a connector that allows the PLC to be connected to an external programming device.

In recent years the decision-making capabilities of PLC processors have gone far beyond simple logic processing. The processor may perform other functions such as timing, counting, latching, comparing, and complicated math beyond the basic four functions of addition, subtraction, multiplication, and division.

PLC Hardware Components

Many electronic components found in processors and other types of modules are sensitive to *electrostatic* voltages. These voltages can be as low as 30 V and the current as low as 0.001 A, far less than you can feel, hear, or see (Fig. 2-16). You need to build up only 3,500 V to feel the effects of electrostatic discharge, only 4,500 V to hear them, and only 5,000 V to see a spark. The normal movements of someone around a work bench can generate up to 6,000 V. The charge that builds up on someone who walks across a nylon carpet in dry air can reach 35,000 V. The following static control procedures should be followed when handling and working with static-sensitive devices and modules:

◆ Ground yourself by touching a conductive surface before handling static-sensitive components

◆ Wear a wrist strap that grounds you during work

◆ Be careful not to touch the backplane connector or connector pins of the PLC system (always handle the circuit cards by the edge if possible)

◆ Be careful not to touch other circuit components in a module when you configure or replace its internal components

◆ Create a static-safe work area by covering your work bench and floor area with a conductive surface that is grounded

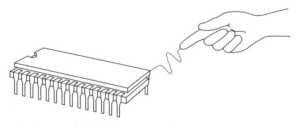

Electrostatic discharge can produce currents that could destroy a sensitive device or degrade performance.

Fig. 2-16

Electrostatic damage.

2•7 MEMORY DESIGN

Memory is where the control plan or program is held or stored in the controller. The information stored in the memory relates to how the input and output data should be processed.

The complexity of the program determines the amount of memory required. Memory elements store individual pieces of information called *bits* (for *binary digits*). The amount of memory capacity is specified in increments of 1000 or in "K" increments, where 1 K is 1024 *bytes* of memory storage (a byte is 8 bits).

The program is stored in the memory as 1's and 0's, which are assembled in the form of 16-bit or 8-bit words in an 8-bit microprocessor system. The memory size varies from as small as 256 words for small systems to 128,000 words for very large systems. *Memory sizes* are expressed in thousands of words that can be stored in the system; thus 2 K is a memory of 2000 words, 64 K is a memory of 64,000 words.

Memory location refers to an address in the CPU's memory where a binary word can be stored. A word consists usually of 8 or 16 bits. A single contact may use one location in the machine's memory. The total number of binary digits, or bits, that can be stored in the RAM memory of a PLC is called the memory capacity. Memory capacity is most often expressed in terms of 8-bit groups, or *bytes*. Words may be 1, 2, or 4 *bytes* wide.

Memory utilization refers to the number of memory locations required to store each type of instruction. A rule of thumb for memory locations is one location per coil or contact. One K of memory would then allow a program containing 1000 coils and contacts to be stored in RAM. If a timer takes two bytes of memory, we could have 512 timers in 1 K of memory.

The memory of a PLC is broken into sections that have specific functions. Sections of memory used to store the status of inputs and outputs typically are called *input tables* and *output tables* (Fig. 2-17). These terms simply refer to a location where the status of an input or output device is stored. Each bit is either a 1 or 0, depending on whether the input is open or closed. A closed contact would have a binary 1 stored in its respective location in the input table, while an open contact would have a 0 stored. A lamp that is ON would have a 1 stored in its respective location in the output table, while a lamp that is OFF would have a 0 stored. Input and output image tables are constantly being revised by the CPU. Each time a memory location is examined, the table changes if the contact or coil has changed state.

Status table files store system information such as scan time, fault status, fault codes, and watch dog timer; and some have precision timing bits for use in the control program. *Timer files* are usually three words long. One word contains timer status information; another contains the preset value or setpoint; and the last contains the accumulated value. *Counter files,* also three words long, have the same configuration as the timer. *Bit, control,* and *integer files* are also used to allow more programming flexibility and allow for more complex instructions.

Although there are many types, memory can be placed into two general categories: *volatile* and *nonvolatile.* Volatile memory will lose its stored information if all operating power is lost or removed. Volatile memory is easily altered and quite suitable for most applications when supported by battery backup.

Nonvolatile memory has the ability to retain stored information when power is removed accidentally or intentionally. Although nonvolatile memory generally is unalterable, there are special types used in which the stored information can be changed.

PLCs execute memory-checking routines to be sure that the PLC memory has not been corrupted. This memory checking is undertaken for safety reasons. It helps ensure that the PLC will not execute if memory is corrupted. The program normally runs from RAM for fastest speed and is transferred from EEPROM or EPROM to RAM at power up.

2·8 MEMORY TYPES

Today's PLCs make use of many different types of volatile and nonvolatile memory devices. Some of the PLC memory is used to hold system memory and some is used to hold user memory. Data are stored in memory locations by a process called *writing.* Data

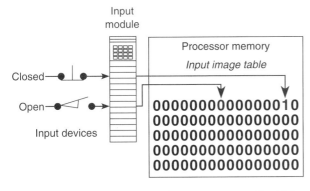

(*a*) Input image table. Each input has one corresponding bit in memory. If the input is closed, the bit is set to 1; if the input is open, the bit is reset to zero.

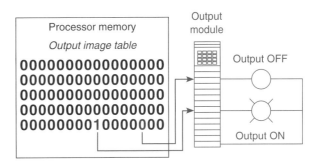

(*b*) Output image table. Each output has one corresponding bit in memory. If the bit is a 1, the output will be ON, if the bit is a 0, the output will be OFF.

Fig. 2-17

Input and output image tables.

are retrieved from memory by what is referred to as *reading*. Following is a generalized description of a few of the more common types of memory devices. Details for specific memory types can be obtained from the specification sheets provided as part of the software package for a controller.

Read-Only Memory (ROM)

Read-only memory (ROM) is designed so that information stored in memory can only be read, and under ordinary circumstances cannot be changed. Information found in the ROM is placed there by the manufacturer, for the internal use and operation of the PLC. Read-only memories are nonvolatile; they retain their information when power is lost and do not require battery backup.

ROM memory is used by the PLC for the *operating system*. The operating system is burned into ROM memory by the PLC manufacturer and controls the system software that the user uses to program the PLC. The ladder logic that the programmer creates is a high-level language. The operating system software must convert the ladder diagram to instructions that the microprocessor can understand.

Random Access Memory (RAM or R/W)

Random access memory (RAM), often referred to as *read-write (R/W) memory*, is designed so that information can be written into or read from the memory. RAM is a type of solid-state memory contained in an integrated circuit and is commonly used for *user memory*. The user's program, timer/counter values, input/output status, etc., are stored in RAM. When the program is executed, the microprocessor, under the control of the program, allows input to the RAM, which can change the stored information.

RAM is memory that could be lost if the power is turned off. Volatile RAM is usually protected by a battery (Fig. 2-18). A PLC must *not* lose its programming if power is

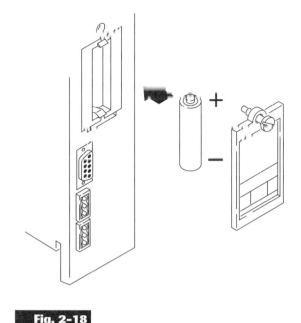

Fig. 2-18

Battery used to protect processor RAM memory.

interrupted briefly, and this is accomplished with battery backup for those times when ac power is lost. The battery takes over when the PLC is shut off. Most PLCs use CMOS-RAM technology for user memory. CMOS-RAM chips have very low current draw and can maintain memory with a lithium battery for an extended time, two to five years in many cases. Some processors have a capacitor that provides at least 30 minutes of battery backup when the battery is disconnected and power is OFF. Data in RAM are not lost if the battery is replaced within 30 minutes.

Today's controllers, for the most part, use the CMOS-RAM with battery support for user program memory. Random access memory provides an excellent means for easily creating and altering a program. The CMOS-RAM is very popular because it has a very low current drain (15-μA range) when not being accessed, and the information stored in its memory can be retained by as little as 2 V dc.

Programmable Read-Only Memory (PROM)

The *programmable read-only memory (PROM)* is a special type of ROM. Programmable read-only memory allows initial and/or additional information to be written into the chip. Programmable read-only memory may be written into only once after being received from the manufacturer. Programming is accomplished by pulses of current that melt fusible links in the chip, preventing it from being reprogrammed. Very few controllers use PROM for program memory because any program change would require a new set of PROM chips.

Erasable Programmable Read-Only Memory (EPROM)

The *erasable programmable read-only memory (EPROM)* is a specially designed PROM that can be reprogrammed after being entirely erased with the use of an ultraviolet light source. Also called an *ultraviolet PROM (UVPROM),* this chip has a quartz window over a silicon material that contains the integrated circuits (Fig. 2-19). This window is normally covered by an opaque material. When the opaque material is removed and the circuitry is exposed to ultraviolet light for approximately 20 minutes, the memory content can be erased. Once erased, the EPROM chip can be reprogrammed using the programming device.

EPROM or UVPROM memory is used to back up, store, or transfer PLC programs. The PLC processor can only read from this type of memory device. An external PROM programmer is used to program (write to) the device. The UVPROM is a nonvolatile memory device and does not require battery backup.

Electrically Erasable Programmable Read-Only Memory (EEPROM)

Electrically erasable programmable read-only memory (EEPROM) is a nonvolatile memory that offers the same programming flexibility as does RAM. The EEPROM can be electrically overwritten with new data instead of being erased with ultraviolet light. The EEPROM is also nonvolatile memory, so it does not require battery backup. It provides permanent storage of the program and can be changed easily using standard programming devices.

Ultraviolet or EPROM chip. (Courtesy of Allen-Bradley Company, Inc.)

Typically, an EEPROM memory module is used to store, back up, or transfer PLC programs. The PLC processor can read and write to an EEPROM. Do *not* expose the processor to surfaces or other areas that may typically hold an electrostatic charge. Electrostatic charges can alter or destroy memory.

2•9 PROGRAMMING DEVICES

Easy-to-use programming equipment is one of the important features of programmable controllers. The programming device provides the primary means by which the user can communicate with the circuits of the controller (Fig. 2-20 on p. 40).

The programming unit can be a liquid crystal display (LCD) hand-held terminal, a single-line LED display unit, or a keyboard and video display unit. The video display offers the advantage of displaying large amounts of logic on the screen, simplifying the interpretation of the program. The programming unit communicates with the processor via a serial or parallel data communications link. If the programming unit is not in use, it may be unplugged and removed. Removing the programming unit will not affect the operation of the user program.

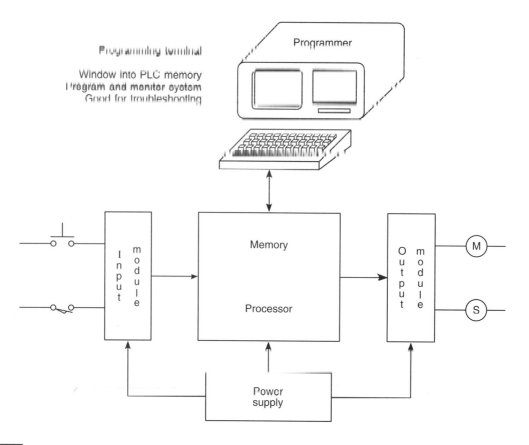

Programmer

Programming terminal

Window into PLC memory
Program and monitor system
Good for troubleshooting

Fig. 2-20

User communications with PLC circuits.

The programming device allows the user to *enter, change,* or *monitor* a PLC controller program. Easy-to-use programming equipment is an important feature of the PLC. *Industrial CRT terminals* can be used for programming the controller. These terminals are self-contained video display units with a keyboard and the electronics necessary to communicate with the CPU and to display data. The CRT, like the video display, offers the advantage of displaying large amounts of logic on the screen, which simplifies the interpretation of the program (Fig. 2-21).

Industrial terminals are dedicated to one brand of PLC. In most cases they must be attached to the PLC to be able to program. Dedicated terminals can be used to troubleshoot ladder logic while the PLC is running. They can force inputs and outputs on or off for troubleshooting. The main disadvantage of these terminals is that one is required

for each different brand of PLC an industry might have. They are also quite expensive and are used to program only one brand of PLC.

Miniprogrammers, also known as *hand-held programmers,* are an inexpensive and portable means for programming small PLCs. The display is usually an LED or liquid-crystal display (LCD) type, and the keyboard consists of numeric keys, programming instruction keys, and special function keys (Fig. 2-22).

Hand-held programmers are inexpensive and easy to use. They typically have membrane keypads that are immune to the contaminants in the factory environment. Hand-held programmers are classified as dumb devices. This means there is no intelligence in the device; so all of the brains for operation of the PLC and for the programming system are located in the PLC. *They must be attached to*

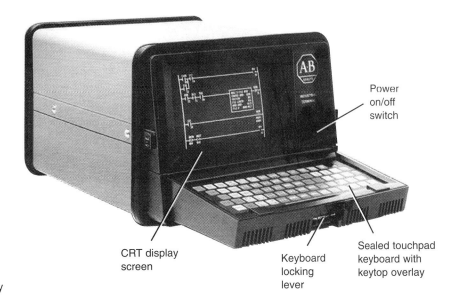

Fig. 2-21

Power
on/off
switch

CRT display
screen

Keyboard
locking
lever

Sealed touchpad
keyboard with
keytop overlay

Fig. 2-21

A CRT industrial terminal.
(Courtesy of Allen-Bradley
Company, Inc.)

a PLC to be used. They are handy for trouble-shooting and can easily be carried out to the manufacturing system and plugged into the PLC. Once plugged in, they can be used to monitor the status of inputs, outputs, variables, counters, timers, etc. This eliminates the need to carry a large programming device out onto the factory floor.

A *personal computer* is currently the most commonly used programming device (Fig. 2-23). It is converted into a programmer for programmable controllers with the aid of programming software supplied as a diskette set. The same computer can program any brand of PLC that has software available for

it. This makes it possible to carry out the programming away from the physical location of the programmable controller. When the program is complete, it is saved to some form of mass storage and downloaded to the programmable controller when required.

The computer can also be used to document the PLC program. Notes for technicians can be added and the ladder diagram can be output to a printer to obtain a hard copy. This documentation is invaluable for understanding and troubleshooting ladder diagrams. The programmer can add notes, names of input or output devices, and comments that may be useful for troubleshooting and maintenance.

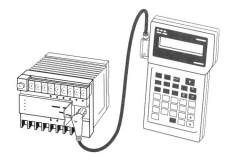

Fig. 2-22

Hand-held programming device.

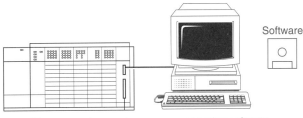

Software

The personal computer uses programming software
to convert it into a programmer for PLCs.

Fig. 2-23

Programming device.

2·10 RECORDING AND RETRIEVING DATA

Printers are used to provide hard-copy printouts of the processor's memory in ladder diagram format. Lengthy ladder programs cannot be shown completely on a screen. Typically, a screen shows a maximum of five rungs at a time. A printout can show programs of any length and analyze the complete program.

Program loaders are used to record and store the user program or to load preprogrammed instructions into the processor. Recording the user program provides a backup program in the event the processor program is lost as a result of memory failure or accidental erasure.

Some older PLC systems use a magnetic cassette recorder to record and store the user program. These tape systems have been superseded by computer *disk drives.* The advantages of using a floppy disk to record and store programs include faster speed, rapid program accessibility, and greater quantity of data that can be stored (Fig. 2-24).

The PLC can have only *one* program in memory at a time. To change the program in the PLC, it is necessary either to enter a new program directly from the keyboard or to download one from the computer hard disk. Some PLCs use internal EEPROMS or EPROM *memory modules* (Fig. 2-25) that can store a backup to the program entered in the PLC. If the PLC were to lose its program, the program in the EEPROM or EPROM would be downloaded quickly to the PLC's memory.

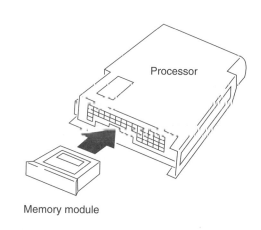

Processor

Memory module

Fig. 2-25

An EEPROM or UVPROM memory module installed in the processor is used as backup to the program entered in the PLC.

With a memory module, you can:

◆ Save the contents of the processor RAM for storage purposes

◆ Load the contents of the EEPROM and EPROM memory into the processor RAM

◆ Use the EPROM memory module when program security is required because the program in the EPROM cannot be altered when it is installed in the controller

◆ Configure the PLC to automatically download the program on power up or if there is a memory error

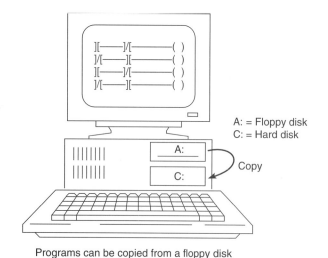

A: = Floppy disk
C: = Hard disk

Copy

Programs can be copied from a floppy disk to a hard disk or from a hard disk to EEPROM.

Fig. 2-24

Copying programs.

2•11 PLC WORKSTATIONS

A PLC *workstation* or *operator interface* can be connected to communicate with a PLC and to replace pushbuttons, pilot lights, thumb-wheels, and other operator control panel devices (Fig. 2-26). Luminescent touch-screen keypads provide an operator interface that operates like traditional hard-wired control panels.

Through personal computer–based setup software, you can configure display screens to:

◆ Replace hard-wired pushbuttons and pilot lights with realistic-looking icons.

The machine operator needs only to touch the display panel to activate the pushbuttons.

◆ Show operations in graphic format for easier viewing

◆ Allow the operator to change timer and counter presets by touching the numeric keypad graphic on the touch-screen

◆ Show alarms, complete with time of occurrence and location

◆ Display variables as they change over time

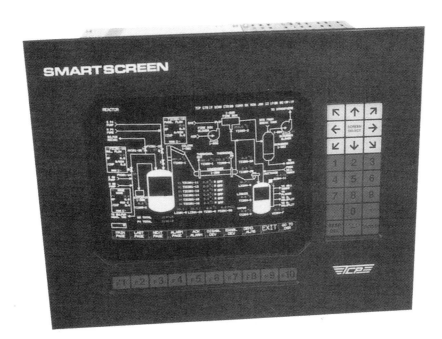

Fig. 2-26

PLC workstation. (Courtesy Total Control Products, Inc., Melrose Park, Illinois).

Questions

1. What is the function of a PLC input module?

2. What is the function of a PLC output module?

3. Define the term *logical rack*.

4. **A.** What is a remote rack?

   B. Why are remote racks used?

5. How does the processor identify the location of a specific input or output device?

6. Describe three basic elements of an I/O address.

7. What connections must be made to the terminals of an I/O module?

8. Compare a standard I/O module with a high-density type.

9. What types of field input devices are suitable for use with discrete input modules?

10. What types of field output devices are suitable for use with discrete output modules?

11. List three functions of the optical isolator circuit used in I/O module circuits.

12. Name the two basic sections of an I/O module.

13. List four tasks performed by an input module.

14. What electronic component is often used as the switching device for 120-V ac output interface module?

15. **A.** What is the maximum current rating for a typical 120-V ac output interface module?

    B. Explain how outputs with larger current requirements are handled.

16. What electronic component is used as the switching device for dc output modules?

17. What type of output devices can be controlled by an output module that uses relays for the switching device?

18. Compare the connection of dc sourcing and sinking field devices.

19. **A.** Compare discrete and analog I/O modules with respect to the type of input or output devices with which they can be used.

B. Explain the function of the A/D converter circuit used in analog input modules.

C. List three common types of analog input sensing devices.

D. Why is shielded cable often used when wiring low-voltage analog sensing devices?

20. State one application for each of the following special I/O modules:

   a. High-speed counter module
   b. Thumb-wheel module
   c. TTL module
   d. Encoder-counter module
   e. BASIC or ASCII module
   f. Stepper-motor module
   g. BCD-output module

21. List four types of intelligent I/O modules that have their own microprocessors on board.

22. Write a short description for each of the following I/O specifications:

   a. Nominal input voltage
   b. On-state input voltage range
   c. Nominal current per input
   d. Nominal output voltage
   e. Output voltage range
   f. Maximum output current rating
   g. OFF-state leakage current per output
   h. Electrical isolation
   i. Number of points
   j. Backplane current draw
   k. Resolution
   l. Input impedance and capacitance
   m. Common mode rejection ratio

23. Explain the basic function of each of the three major parts of the CPU.

24. List three typical modes of operation that can be selected by the keyswitch of a processor unit.

25. State three other functions, in addition to simple logic processing, that PLC processors are capable of performing.

26. What steps can be taken to prevent damage to static-sensitive PLC components?

27. A. What information is stored in input and output tables?

   B. How is this information stored in memory?

28. Compare the memory storage characteristics of volatile and nonvolatile memory elements.

29. Why do PLCs execute memory checking routines?

30. Compare ROM and RAM memory design with regard to:

   a. How information is placed into the memory
   b. How information in the memory is changed
   c. Classification as volatile or nonvolatile

31. A. How is initial and/or additional information written into a PROM chip?

   B. What is the main limitation of PROM memory chips?

32. How is the program erased in the following chips?

   a. EPROM
   b. EEPROM

33. List three possible functions of a PLC programming device.

34. List three types of programming equipment available.

35. How can a personal computer be converted into a PLC programmer?

36. What information can be included as part of computer documentation of a program?

37. What are the benefits of using a printer to provide a hard-copy printout of the program?

38. List three advantages of using a floppy disk, over magnetic tape storage, to record and store PLC programs.

39. Explain the function of an EEPROM or UVPROM memory module installed in a processor.

40. Outline several functions that a PLC workstation screen can be configured to do.

Problems

1. A discrete 120-V ac output module is to be used to control a 230-V dc solenoid valve. Draw a diagram showing how this could be accomplished using an interposing relay.

2. Assume a thermocouple generates a linear voltage of from 20 mV to 50 mV when the temperature changes from 750°F to 1250°F. How much voltage will be generated when the temperature of the thermocouple is at 1000°F?

3. **A.** The input delay time of a given module is specified as 12 ms. How much is this expressed in seconds?

   B. The output leakage current of a given module is specified as 950 μA. How much is this expressed in amperes?

   C. The maximum ambient temperature for a given I/O module is specified as 60°C. How much is this expressed in degrees Fahrenheit?

4. Create a typical five-digit address (according to Fig. 2-4) for each of the following.

   a. A pushbutton connected to terminal 5 of module group 2 located on rack 1.
   b. A lamp connected to terminal 3 of module group 0 located on rack 2.

5. Assume the triac of an ac output module fails in the shorted state. How would this affect the device connected to this output?

6. A personal computer is to be used to program several different PLC models. What is required?

Number Systems and Codes

After completing this chapter, you will be able to:

◆ Define the decimal, binary, octal, and hexadecimal numbering systems and be able to convert from one numbering or coding system to another

◆ Explain the BCD, Gray, and ASCII code systems

◆ Define the terms *bit, byte, word, least significant bit (LSB),* and *most significant bit (MSB)* as they apply to binary memory locations

◆ Describe the purpose of the encoder and the decoder integrated circuits (IC)

◆ Add, subtract, multiply, and divide binary numbers

Using PLC's requires us to become familiar with other number systems besides decimal. Some PLC models and individual PLC functions use other numbering systems. This chapter deals with some of these numbering systems, including binary, octal, hexadecimal, BCD, Gray, and ASCII. The basics of each system, as well as conversion from one system to another, are explained.

3.1 DECIMAL SYSTEM

Knowledge of different number systems and digital codes is quite useful when working with PLCs or with most any type of digital computer. This is true because a basic requirement of these devices is to represent, store, and operate on numbers. In general, PLCs work on binary numbers in one form or another; these are used to represent various codes or quantities.

The *decimal system,* which is most common to us, has a base of 10. The *radix* or *base* of a number system determines the total number of different symbols or digits used by that system. For instance, in the decimal system, 10 unique numbers or digits—i.e., the digits 0 through 9—are used: the total number of symbols is the same as the base, and the symbol with the largest value is 1 less than the base.

The value of a decimal number depends on the digits that make up the number and the place value of each digit. A place (weight) value is assigned to each position that a digit would hold from right to left. In the decimal system the first position, starting from the rightmost position, is 0; the second is 1; the third is 2; and so on up to the last position. The weighted value of each position can be expressed as the base (10 in this case) raised to the power of the position. For the decimal system then, the position weights are 1, 10, 100, 1000, etc. Figure 3-1 illustrates how the value of a decimal number can be calculated by multiplying each digit by the weight of its position and summing the results.

3.2 BINARY SYSTEM

The *binary system* uses the number 2 as the base. The only allowable digits are 0 and 1. With digital circuits it is easy to distinguish between two voltage levels (i.e., +5 V and 0 V), which can be related to the binary digits 1

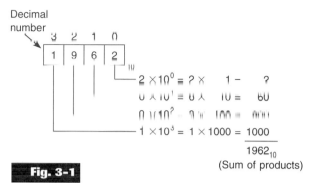

Fig. 3-1

Weighted value in the decimal system.

and 0 (Fig . 3-2). Therefore the binary system can be applied quite easily to PLCs and computer systems. Most PLC timers and counters operate in binary.

Since the binary system uses only two digits, each position of a binary number can go through only two changes, and then a 1 is carried to the immediate left position. Note that Table 3-1 shows a comparison among four common number systems: decimal (base 10), octal (base 8), hexadecimal (base 16), and binary (base 2). Note that all numbering systems start at zero.

The decimal equivalent of a binary number is calculated in a manner similar to that used for a decimal number. This time the weighted values of the positions are 1, 2, 4, 8, 16, 32, 64, etc. Instead of being 10 raised to the power of the position, the weighted value is 2 raised to the power of the position. Figure 3-3 illustrates how the binary number 10101101 is converted to its decimal equivalent: 173.

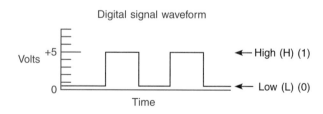

Digital signal waveform

Fig. 3-2

Digital signal waveform.

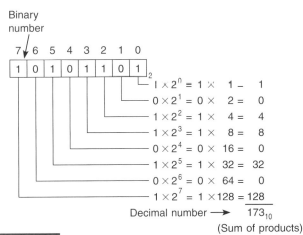

Fig. 3-3

Converting a binary number to a decimal number.

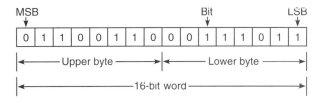

Fig. 3-4

A 16-bit word.

Each digit of a binary number is known as a *bit.* In a PLC the processor-memory element consists of hundreds or thousands of

Table 3-1

NUMBER SYSTEM COMPARISONS

Decimal	Octal	Hexadecimal	Binary
0	0	0	0
1	1	1	1
2	2	2	10
3	3	3	11
4	4	4	100
5	5	5	101
6	6	6	110
7	7	7	111
8	10	8	000
9	11	9	1001
10	12	A	1010
11	13	B	1011
12	14	C	1100
13	15	D	1101
14	16	E	1110
15	17	F	1111
16	20	10	10000
17	21	11	10001
18	22	12	10010
19	23	13	10011
20	24	14	10100

locations. These locations, or *registers,* are referred to as *words.* Each word is capable of storing data in the form of *binary digits,* or *bits.* The number of bits that a word can store depends on the type of PLC system used. Eight-bit and sixteen-bit words are the most common. As the technology continues to develop, 32-bit words or larger are possible. Bits can also be grouped within a word into *bytes.* Usually a group of 8 bits is a byte, and a group of 1 or more bytes is a word.

Figure 3-4 illustrates a 16-bit word made up of 2 bytes. The *least significant bit (LSB)* is the digit that represents the smallest value, and the *most significant bit (MSB)* is the digit that represents the largest value. A bit within the word can exist only in two states: a logical 1 (or ON) condition, or a logical 0 (or OFF) condition.

The size of the programmable controller *memory* relates to the amount of user program that can be stored. If a memory size is 884 words, then it can actually store 14,144 (884 × 16) bits of information using 16-bit words or 7072 (884 × 8) using an 8-bit word. Therefore, when comparing different PLC systems, one must know the number of bits per word of memory to determine the relative capacity of the systems' memories. Normally, programmable controllers do not require storage space above 128 K and, in many instances, need a memory size of only 1 K to 2 K.

To convert a decimal number to its binary equivalent, we must perform a series of divisions by 2. Figure 3-5 illustrates the conversion of the decimal number 47 to binary. We start by dividing the decimal number by 2. If there is a remainder, it is placed in the LSB of the binary number. If there is no remainder, a 0 is placed in the LSB. The result of the division is brought down, and the process is repeated until the result of successive divisions has been reduced to 0.

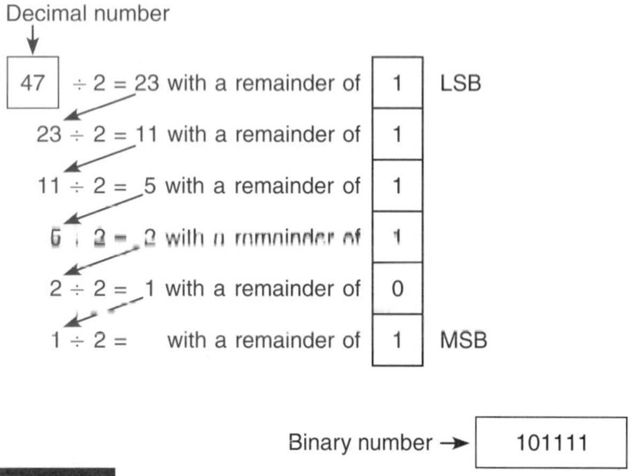

Decimal number

Fig. 3-5

Converting a decimal number to a binary number.

Table 3-2

SIGNED BINARY NUMBERS

Magnitude			Decimal Value
Sign			
		0111	+7
		0110	+6
		0101	+5
Same as		0100	+4
binary		0011	+3
numbers		0010	+2
		0001	+1
		0000	0
		1001	−1
		1010	−2
		1011	−3
		1100	−4
		1101	−5
		1110	−6
		1111	−7

3•3 NEGATIVE NUMBERS

If a decimal number is positive, it has a plus sign; if a number is negative, it has a minus sign. In binary number systems, such as used in a PLC, it is not possible to use positive and negative symbols to represent the polarity of a number. One method of representing a binary number as either a positive or negative value is to use an extra digit, or *sign bit*, at the MSB side of the number. In the sign bit position, a 0 indicates that the number is positive, and a 1 indicates a negative number (Table 3-2).

Another method of expressing a negative number in a digital system is by using the complement of a binary number. To complement a binary number, change all the 1s to 0s and all the 0s to 1s. This is known as the 1's complement form of a binary number. For example, the 1's complement of 1001 is 0110.

The most common way to express a negative binary number is to show it as a *2's complement* number. The 2's complement is the binary number that results when 1 is added to the 1's complement. This is shown in Table 3-3. A 0 sign bit means a positive number, while a 1 sign bit means a negative number.

Using the 2's complement makes it easier for the PLC to perform mathematical operations. The correct sign bit is generated by forming the 2's complement. The PLC knows that a number retrieved from memory is a negative number if the MSB is 1. Whenever a negative number is entered from a keyboard, the PLC stores it as a 2's complement. What follows is the original number in true binary followed by its 1's complement, its 2's complement, and finally its decimal equivalent.

Table 3-3

1'S AND 2'S COMPLEMENT REPRESENTATION OF POSITIVE AND NEGATIVE NUMBERS

Signed Decimal	1's Complement	2's Complement
+7	0111	0111
+6	0110	0110
+5	0101	0101
+4	0100 Same as	0100
+3	0011 binary	0011
+2	0010 numbers	0010
+1	0001	0001
0	0000	0000
−1	1110	1111
−2	1101	1110
−3	1100	1101
−4	1011	1100
−5	1010	1011
−6	1001	1010
−7	1000	1001

Octal is used as a convenient means of handling large binary numbers. As shown in Table 3-4 on p. 54, one octal digit can be used to express three binary digits. The octal numbering system makes use of 8 digits: 0 through 7. As in all other numbering systems, each digit in an octal number has a weighted decimal value according to its position. Figure 3-7 on p. 54 illustrates how the octal number 462 is converted to its decimal equivalent: 306.

As mentioned, octal is used as a convenient means of handling large binary numbers. For example, the octal number 462 can be converted to its binary equivalent by assembling the 3-bit groups, as illustrated in Fig. 3-8 on p. 54. Thus, octal 462 is binary 100110010 and decimal 306. Notice the simplicity of the notation. The octal 462 is much easier to read and write than its binary equivalent.

3.4 OCTAL SYSTEM

To express the number in the binary system requires many more digits than in the decimal system. Too many binary digits can become cumbersome to read or write. To solve this problem, other related numbering systems are used.

The *octal numbering system,* a base 8 system, is often used in microprocessor, computer, and programmable controller systems because 8 data bits make up a byte of information that can be addressed by the PLC user or programmer. Figure 3-6 illustrates the addressing of I/O modules using the octal numbering system. The digits range from 0 to 7; therefore, numbers 8 and 9 are *not* allowed. The octal numbering system is used by PLC manufacturer Allen-Bradley for I/O addressing.

Input Module 0	Input Module 1	Input Module 2	Input Module 3
I00	I10	I20	I30
I01	I11	I21	I31
I02	I12	I22	I32
I03	I13	I23	I33
I04	I14	I24	I34
I05	I15	I25	I35
I06	I16	I26	I36
I07	I17	I27	I37

Fig. 3-6

Addressing of I/O modules using the octal numbering system.

Table 3-4

BINARY AND RELATED OCTAL CODE

Binary	Octal
000	0
001	1
010	2
011	3
100	4
101	5
110	6
111	7

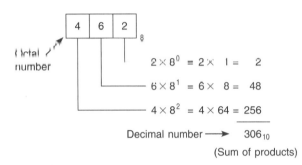

Fig. 3-7

Converting an octal number to a decimal number.

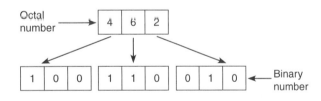

Fig. 3-8

Converting an octal number to a binary number.

3.5 HEXADECIMAL SYSTEM

The *hexadecimal (hex) numbering system* is used in programmable controllers because a word of data consists of 16 data bits, or two 8-bit bytes. The hexadecimal system is a base 16 system, with A to F used to represent decimal numbers 10 to 15 (Table 3-5).

The techniques used when converting hexadecimal to decimal and decimal to hexadecimal are the same as those used for binary and octal. To convert a hexadecimal number to its decimal equivalent, the hexadecimal digits in the columns are multiplied by the base 16 weight, depending on digit significance. Figure 3-9 illustrates how the conversion would be done for the hex number 1B7.

Like octal numbers, hexadecimal numbers can easily be converted to binary numbers. Conversion is accomplished by writing the 4-bit binary equivalent of the hex digit for each position, as illustrated in Fig. 3-10. As Figs. 3-9 and 3-10 show, the hex number 1B7 is 000110110111 in binary and 439 in decimal.

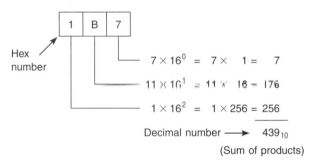

Fig. 3-9

Converting a hexadecimal number to a decimal number.

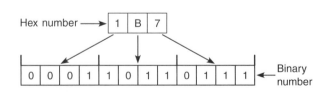

Fig. 3-10

Converting a hexadecimal number to a binary number.

Table 3-5

HEXADECIMAL NUMBERING SYSTEM

Hexadecimal	Binary	Decimal
0	0000	0
1	0001	1
2	0010	2
3	0011	3
4	0100	4
5	0101	5
6	0110	6
7	0111	7
8	1000	8
9	1001	9
A	1010	10
B	1011	11
C	1100	12
D	1101	13
E	1110	14
F	1111	15

The BCD representation of a decimal number is obtained by replacing each decimal digit by its BCD equivalent. To distinguish the BCD numbering system from a binary system, a BCD designation is placed to the right of the units digit. The BCD representation of the decimal number 7863 is shown in Fig. 3-11 on p. 56.

Scientific calculators are available to convert numbers back and forth between decimal, binary, octal, and hexadecimal. They are inexpensive and easy to use, for example, in converting a number displayed in decimal to one in binary. This simply involves one keystroke to change the display mode from decimal to binary. In addition, most PLCs contain number conversion functions, as illustrated in Fig. 3-12 on p. 56. As shown in Fig. 3-12(a), BCD-to-binary conversion is required for the input. Binary-to-BCD conversion is required for the output. Note that in Fig. 3-12(b) the convert-to-decimal

3•6 BCD SYSTEM

The BCD system provides a convenient way of handling large numbers that need to be input to or output from a PLC. As you can see from going through the number systems, there is no easy way to go from binary to decimal and back. The BCD system provides a means of converting a code readily handled by humans (decimal) to a code readily handled by the equipment (binary). PLC thumb-wheel switches and LED displays are examples of PLC devices that make use of the BCD number system.

The BCD system uses 4 bits to represent each decimal digit. The 4 bits used are the binary equivalents of the numbers from 0 to 9. In the BCD system, the largest decimal number that can be displayed by any four digits is 9 (see Table 3-6).

Table 3-6

BCD CODE WITH BINARY AND DECIMAL EQUIVALENTS

Decimal	BCD	Binary
0	0000	0
1	0001	1
2	0010	10
3	0011	11
4	0100	100
5	0101	101
6	0110	110
7	0111	111
8	1000	1000
9	1001	1001
10	0001 0000	1010
11	0001 0001	1011
12	0001 0010	1100
13	0001 0011	1101
14	0001 0100	1110
15	0001 0101	1111

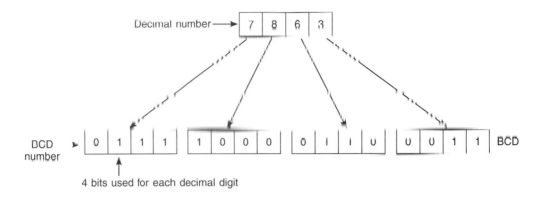

Fig. 3-11

The BCD representation of a decimal number.

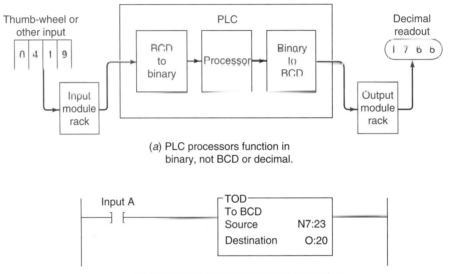

(a) PLC processors function in binary, not BCD or decimal.

(b) Example of convert-to-decimal instruction.

Fig. 3-12

PLC number conversion.

instruction will convert the binary bit pattern at the source address, N7:23, into a BCD bit pattern of the same decimal value as the destination address, O:20. The instruction executes every time it is scanned and the instruction is true.

Many PLCs allow you to change the format of the data that the data monitor displays. For example, the *change radix* function found on Allen-Bradley controllers allows you to change the display format of data to binary, octal, decimal, hexadecimal, or ASCII.

3.7 GRAY CODE

The *Gray code* is a special type of binary code that does *not* use position weighting. In other words, each position does not have a definite weight. The Gray code is set up so that as we progress from one number to the next, only one bit changes. This can be quite confusing for counting circuits, but it is ideal for encoder circuits. For example, *absolute encoders* are position transducers that use the Gray code to determine angular position.

The Gray code has the advantage that for each "count" (each transition from one number to the next) *only one* digit changes. Table 3-7 shows the Gray code and the binary equivalent for comparison. In binary, as many as four digits could change for a single "count." For example, the transition from binary 0111 to 1000 (decimal 7 to 8) involves a change in all four digits. This kind of change increases the possibility for error in certain digital circuits. For this reason, the Gray code is considered to be an error-minimizing code. Because only one bit changes at a time, the speed of transition for the Gray code is considerably faster than that for codes such as BCD.

3•8 ASCII CODE

ASCII stands for American Standard Code for Information Interchange. It is an alphanumeric code because it includes letters as well as numbers. The characters accessed by the ASCII code include 10 numeric digits; 26 lowercase and 26 uppercase letters of the alphabet; and about 25 special characters, including those found on a standard typewriter, i.e., @, #, $, %, *, etc. Table 3-8 on p. 58 shows a partial listing of the ASCII code. It is used to interface the PLC CPU with alphanumeric keyboards and printers.

3•9 ENCODING AND DECODING

The PLC, for the most part, uses digital integrated-circuit (IC) chips for conversion from one number system or code to another. An *encoder* IC is used to convert from *decimal* to *binary* numbers. Figure 3-13 on p. 59 shows an experimental encoder circuit used to convert decimal numbers 0 through 9 to binary numbers. Numeric signals can take on any of a range of values. Most PLCs use a 16-bit integer to encode numeric values, so the range of possible values is from −32,768 to 32,767.

Table 3-7

GRAY CODE AND BINARY EQUIVALENT

Gray Code	Binary
0000	0000
0001	0001
0011	0010
0010	0011
0110	0100
0111	0101
0101	0110
0100	0111
1100	1000
1101	1001
1111	1010
1110	1011
1010	1100
1011	1101
1001	1110
1000	1111

With power applied and all input switches open, the LED lights should all be off. This indicates that the binary number 0000 is appearing at the output. Closing decimal input switch 3, for example, inputs the decimal number 3 to the encoder. This in turn will cause the 1's and 2's LEDs to come on to indicate the equivalent binary number, 0011.

A *decoder* IC is used to convert back from *binary* to *decimal* numbers. Figure 3-14 on p. 59 shows an experimental decoder circuit used to convert binary numbers 0000 through 1001 back to decimal numbers. With power applied and all input switches open, the zero output LED will light, indicating binary number 0000. Closing binary input switches 2's and 4's, for example, inputs the binary number 0110 to the decoder. This in turn will cause the decimal 6 LED to come on, thus indicating the equivalent decimal number, 6.

Table 3-8

PARTIAL LISTING OF ASCII CODE

Character	7-Bit ASCII	Character	7-Bit ASCII
A	100 0001	X	101 1000
B	100 0010	Y	101 1001
C	100 0011	Z	101 1010
D	100 0100	0	011 0000
E	100 0101	1	011 0001
F	100 0110	2	011 0010
G	100 0111	3	011 0011
H	100 1000	4	011 0100
I	100 1001	5	011 0101
J	100 1010	6	011 0110
K	100 1011	7	011 0111
L	100 1100	8	011 1000
M	100 1101	9	011 1001
N	100 1110	blank	010 0000
O	100 1111	.	010 1110
P	101 0000	,	010 1100
Q	101 0001	+	010 1011
R	101 0010	−	010 1101
S	101 0011	#	010 0011
T	101 0100	(	010 1000
U	101 0101	%	010 0101
V	101 0110	=	011 1101
W	101 0111		

3.10 PARITY BIT

Some PLC communication systems use a binary digit to check the accuracy of data transmission. For example, when data are transferred between PLCs, one of the binary digits may be accidentally changed from a 1 to a 0. This can happen due to a transient or a noise, or by a failure in some portion of the transmission network. A *parity bit* is used to detect errors that may occur while a word is moved.

Parity is a system where each character transmitted contains one additional bit. That bit is known as a parity bit. The bit may be a binary 0 or binary 1, depending on the number of 1's and 0's in the character itself.

Two systems of parity are normally used: odd and even. *Odd* parity means that the total number of binary 1 bits in the character, including the parity bit, is odd. *Even* parity means that the number of binary 1 bits in the character, including the parity bit, is even. Examples of odd and even parity are shown in Table 3-9 on p. 60.

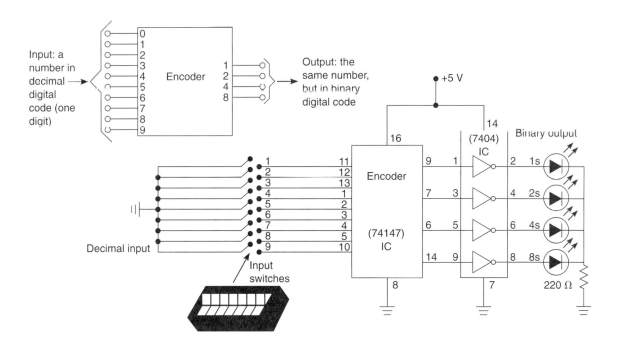

Fig. 3-13

Decimal-to-binary encoder.

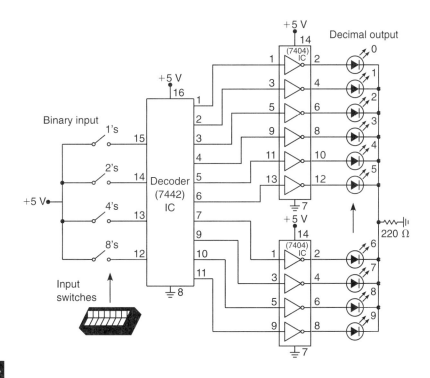

Fig. 3-14

Binary-to-decimal decoder.

Table 3-9

ODD AND EVEN PARITY

Character	Even Parity Bit	Odd Parity Bit
0000	0	1
0001	1	0
0010	1	0
0011	0	1
0100	1	0
0101	0	1
0110	0	1
0111	1	0
1000	1	0
1001	0	1

3.11 BINARY ARITHMETIC

Arithmetic circuit units form a part of the CPU. Mathematical operations include addition, subtraction, multiplication, and division. Binary addition follows rules similar to decimal addition. When adding with binary numbers, there are only four conditions that can occur:

$$
\begin{array}{cccc}
0 & 1 & 0 & 1 \\
+0 & +0 & +1 & +1 \\
\hline
0 & 1 & 1 & 0 \text{ carry } 1
\end{array}
$$

As in the case of adding decimals, the first three conditions are easy, but the last condition is slightly different. In decimal, 1 + 1 = 2. In binary, a 2 is written 10. Therefore, in binary, 1 + 1 = 0, with a carry of 1 to the next most significant place value.

When adding larger binary numbers, the resulting 1's are carried into higher-order columns, as shown in the following examples.

Decimal	Equivalent binary
5 +2 7	101 + 10 111
10 + 3 13	carry 1 10|10 + |11 11|01
26 +12 38	carry 1 |1 carry 1|1010 + |1100 1|0|0110

Manufacturers produce several arithmetic ICs. Figure 3-15 shows a 4-bit parallel adder circuit that uses a 7483 IC chip. The circuit will add two 4-bits and produce the sum at its output.

In arithmetic functions, the initial numeric quantities that are to be combined by subtraction are the *minuend* and *subtrahend*. The result of the subtraction process is called the *difference,* represented as:

$$
\begin{array}{l}
A \text{ (minuend)} \\
-B \text{ (subtrahend)} \\
\hline
C \text{ (difference)}
\end{array}
$$

To subtract from larger binary numbers, subtract column by column, borrowing from the adjacent column when necessary. Remember

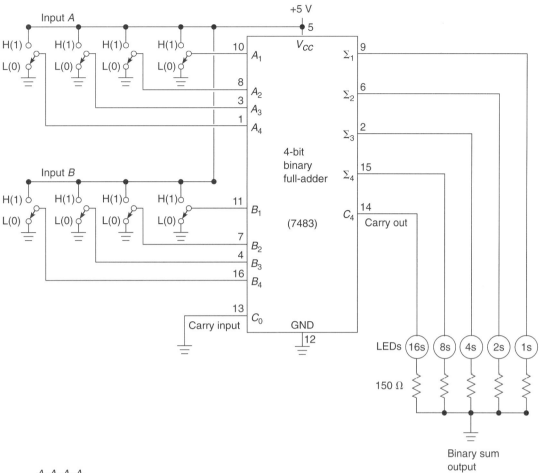

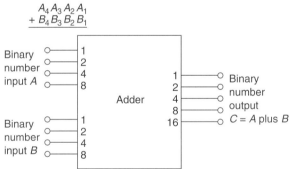

Fig. 3-15

Four-bit adder circuit.

that when borrowing from the adjacent column, there are now two digits, i.e., 0 borrow 1 gives 10.

EXAMPLE

Subtract 1001 from 1101.

$$
\begin{array}{r}
1101 \\
-1001 \\
\hline
0100
\end{array}
$$

Subtract 0111 from 1011.

$$
\begin{array}{r}
1011 \\
-0111 \\
\hline
0100
\end{array}
$$

Binary numbers can also be negative. The procedure for this calculation is identical to that of decimal numbers because the smaller value is subtracted from the larger value and a negative sign is placed in front of the result.

EXAMPLE

Subtract 111 from 100.

$$
\begin{array}{r}
111 \\
-100 \\
\hline
-011
\end{array}
$$

Subtract 11011 from 10111.

$$
\begin{array}{r}
11011 \\
-10111 \\
\hline
-00100
\end{array}
$$

There are other methods available for doing subtraction:

1's complement
2's complement

The procedure for subtracting numbers using the 1's complement is as follows.

Step 1 Change the subtrahend to 1's complement.

Step 2 Add the two numbers

Step 3 Remove the last carry and add it to the number (end-around carry).

Decimal	Binary	
10	1010	1010
− 6	−0110	1's complement → +1001
4	100	10011
		End around carry → ⌐ 1
		100

When there is a carry at the end of the result, the result is positive. When there is no carry, then the result is negative and a minus sign has to be placed in front of it.

EXAMPLE

Subtract 11011 from 01101.

$$
\begin{array}{r}
01101 \\
+ \;\, \oslash 00100 \\
\hline
10001
\end{array}
$$

The 1's complement There is no carry, so we take the 1's complement and add the minus sign:

$$-01110$$

For subtraction using the 2's complement, the 2's complement is added instead of subtracting the numbers. In the result, if the carry is a 1, then the result is positive; if the carry is a 0, then the result is negative and requires a minus sign.

Subtract 101 from 111.

$$
\begin{array}{rl}
111 & \\
+ \ \oslash\,011 & \text{The 2's complement} \\
\overline{1010} & \text{The first 1 indicates that} \\
& \text{the result is positive, so it} \\
& \text{is disregarded:} \\
\underline{010} &
\end{array}
$$

Subtract 11011 from 01101.

$$
\begin{array}{rl}
01101 & \\
+ \ \oslash\,00101 & \text{The 2's complement} \\
\overline{10010} & \text{There is no carry, so the} \\
& \text{result is negative; therefore} \\
& \text{a I has to be subtracted} \\
& \text{and the 1's complement} \\
& \text{taken to give the result:}
\end{array}
$$

subtract 1 $10010 - 1 = 10001$

1's complement -01110

Figure 3-16 on p. 64 shows the 7483 adder IC rewired for a 4-bit parallel subtracter. The operation is accomplished by using 1's complement. The 1's complement of a binary number is obtained by negating all its digits (changing all 1's to 0's and all the 0's to 1's) using inverters.

Binary numbers are multiplied in the same manner as decimal numbers. When multiplying binary numbers, there are only four conditions that can occur:

$$
\begin{array}{l}
0 \times 0 = 0 \\
0 \times 1 = 0 \\
1 \times 0 = 0 \\
1 \times 1 = 1
\end{array}
$$

To multiply numbers with more than one digit, form partial products and add them together, as shown in the following example.

Decimal	Equivalent binary
5	101
×6	×110
30	000
	101
	101
	11110

The process for dividing one binary number by another is the same for both binary and decimal numbers, as shown in the following example.

Decimal	Equivalent binary
7	111
2)14	10)1110
	10
	11
	10
	10
	10
	00

The basic function of a *comparator* is to compare the relative magnitude of two quantities. Figure 3-17 on p. 65 shows a 4-bit comparator circuit that uses a 7485 IC chip. The circuit compares the magnitude of two binary numbers, each containing 4 digits. It has three outputs:

$$
\begin{array}{l}
A = B \ (A \text{ equals } B) \\
A > B \ (A \text{ is greater than } B) \\
A < B \ (A \text{ is less than } B)
\end{array}
$$

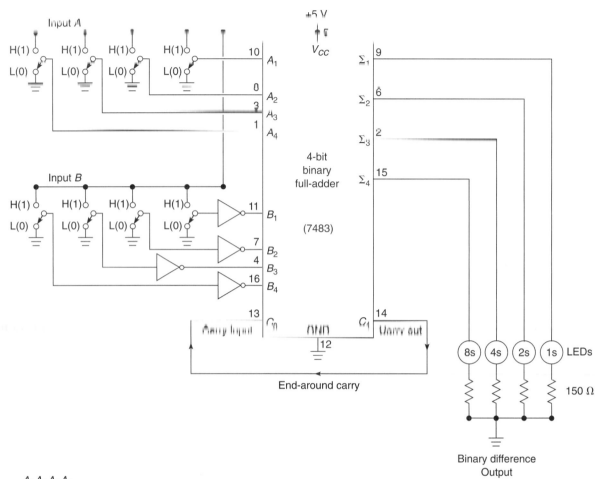

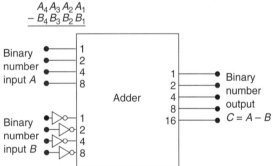

Four-bit parallel subtracter.

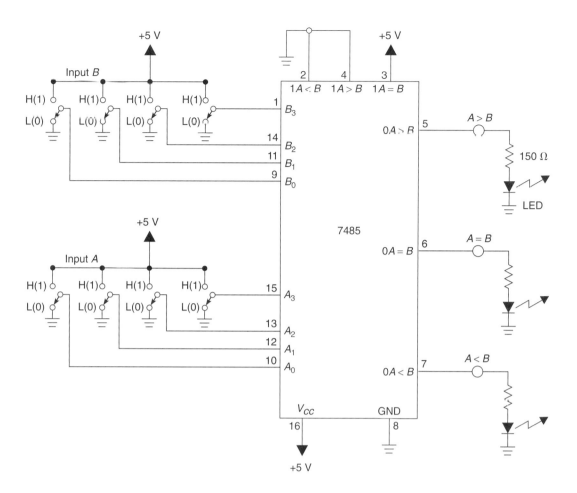

Fig. 3-17

Four-bit *A-B* comparator.

Questions

1. Convert each of the following binary numbers to decimal numbers:

   a. 10
   b. 100
   n. 111
   d. 1011
   e. 1100
   f. 10010
   g. 10101
   h. 11111
   i. 11001101
   j. 11100011

2. Convert each of the following decimal numbers to binary numbers:

   a. 7
   b. 19
   c. 28
   d. 46
   e. 57
   t. 86
   g. 94
   h. 112
   i. 148
   j. 230

3. Convert each of the following octal numbers to decimal numbers:

   a. 36
   b. 104
   c. 120
   d. 216
   e. 360
   f. 1516

4. Convert each of the following octal numbers to binary numbers:

   a. 74
   b. 130
   c. 250
   d. 1510
   e. 2551
   f. 2634

5. Convert each of the following hexadecimal numbers to decimal numbers:

   a. 5A
   b. C7
   c. 9B5
   d. 1A6

6. Convert each of the following hexadecimal numbers to binary numbers:

   a. 4C
   b. E8
   c. 6D2
   d. 31B

7. Convert each of the following decimal numbers to BCD:

   a. 146
   b. 389
   c. 1678
   d. 2502

8. What is the most important characteristic of the Gray code?

9. What is the basic function of an encoder circuit?

10. What is the basic function of a decoder circuit?

11. What makes the binary system so applicable to computer circuits?

12. Define each of the following as they apply to the binary memory locations or registers:

   a. Bit
   b. Byte
   c. Word
   d. LSB
   e. MSB

13. State the base used for each of the following number systems:

   a. Octal
   b. Decimal
   c. Binary
   d. Hexadecimal

14. Define the term *sign bit*.

15. Explain the difference between the 1's complement of a number and the 2's complement.

16. What is ASCII code?

17. Why are parity bits used?

18. Add the following binary numbers:

   a. 110 + 111
   b. 101 + 011
   c. 1100 + 1011

19. Subtract the following binary numbers:

    a. 1101 − 101
    b. 1001 − 110
    c. 10111 − 10010

20. Multiply the following binary numbers:

    a. 110 × 110
    b. 010 × 101
    c. 101 × 11

21. Divide the following unsigned binary numbers:

    a. 1010 ÷ 10
    b. 1100 ÷ 11
    c. 110110 ÷ 10

Problems

1. The following binary PLC sequencer code information is to be programmed using the hexadecimal code. Convert each piece of binary information to the appropriate hexadecimal code for entry into the PLC from the keyboard.

    a. 0001 1111
    b. 0010 0101
    c. 0100 1110
    d. 0011 1001

2. The encoder circuit shown in Fig. 3-18 is used to convert the decimal digits on the keyboard to a binary code. State the output status (HIGH/LOW) of *A-B-C-D* when decimal number

    a. 2 is pressed.
    b. 5 is pressed.
    c. 7 is pressed.
    d. 8 is pressed.

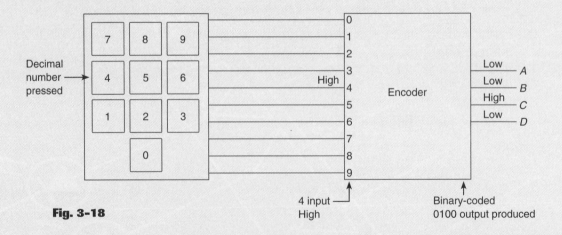

Fig. 3-18

3. If the bits of a 16-bit word or register are numbered according to the octal numbering system, beginning with 00, what consecutive numbers would be used to represent each of the bits?

4. Express the decimal number 18 in *each* of the following number codes:

   a. Binary
   b. Octal
   c. Hexadecimal
   d. BCD

Fundamentals of Logic

After completing this chapter, you will be able to:

◆ Describe the binary concept and the functions of gates

◆ Draw the logic symbol, construct a truth table, and state the Boolean equation for the AND, OR, and NOT functions

◆ Construct circuits from Boolean expressions and derive Boolean equations for given logic circuits

◆ Convert relay ladder schematics to ladder logic programs

◆ Develop elementary programs based on logic gate functions

◆ Program instructions that perform logical operations

This chapter gives an overview of digital logic gates and illustrates how to duplicate this type of control on a PLC. Boolean algebra, which is a shorthand way of writing digital gate diagrams, is discussed briefly. Some small hand-held programmers have digital logic keys, such as AND, OR, and NOT, and are programmed using Boolean expressions.

4.1 THE BINARY CONCEPT

The PLC, like all digital equipment, operates on the binary principle. The term *binary principle* refers to the idea that many things can be thought of as existing in one of *two states*. The states can be defined as "high" or "low," "on" or "off," "yes" or "no," and "1" or "0." For instance, a light can be on or off, a switch open or closed, or a motor running or stopped.

This two-state binary concept, applied to gates, can be the basis for making decisions. The *gate* is a device that has one or more inputs with which it will perform a logical decision and produce a result at its one output. Figures 4-1 and 4-2 give two examples that show how logic gate decisions are made.

Logic is the ability to make decisions when one or more different factors must be taken into account before an action is taken. This is the basis for the operation of the PLC, where it is required for a device to operate when certain conditions have been met.

4.2 AND, OR, AND NOT FUNCTIONS

The operations performed by digital equipment are based on three fundamental logic functions: AND, OR, and NOT. Each function has a rule that will determine the outcome and a *symbol* that represents the

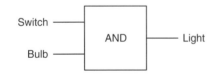

The light in a room can be turned on only when the switch is on *and* a light bulb is in the light socket.

Fig. 4-1
The logical AND.

You will be awakened from your sleep when the alarm goes off *or* the ceiling caves in *or* you fall out of bed.

Fig. 4-2
The logical OR.

operation. For the purpose of this discussion, the outcome or output is called *Y* and the signal inputs are called *A, B, C,* etc. Also, binary 1 represents the presence of a signal or the occurrence of some event, while binary 0 represents the absence of the signal or nonoccurrence of the event.

The AND Function

The symbol drawn in Fig. 4-3 is called an AND gate. An AND gate is a device with two or more inputs and one output. The AND gate output is 1 only if all inputs are 1. The *truth table* in Fig. 4-3 shows the resulting output from each of the possible input combinations.

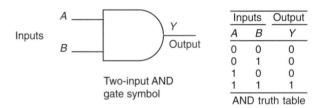

Inputs	Output
A B	Y
0 0	0
0 1	0
1 0	0
1 1	1

AND truth table

Fig. 4-3
AND gate.

Figures 4-4 and 4-5 show practical applications of the AND gate function. When switch *A* and *B* are operated, the output *Y,* or the lamp, becomes active—it turns ON. If the active state is considered to be a logical 1 and the inactive state a logical 0, a truth table can be developed for the AND function as shown. When 1 is used to depict the active

All possible input combinations

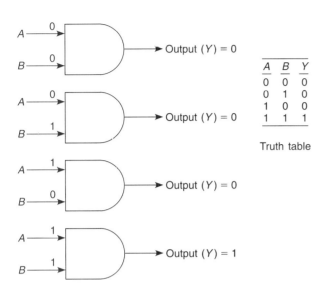

A	B	Y
0	0	0
0	1	0
1	0	0
1	1	1

Truth table

Basic rule: If all inputs are 1, the output will be 1.
If any input is 0, the output will be 0.

Fig. 4-4

AND gate function application—example 1.

state and 0 the inactive state, positive logic is being used. The AND gate operates like a *series* circuit that produces an output voltage when a voltage appears at each of its inputs.

The OR Function

The symbol drawn in Fig. 4-6 is called an OR gate. An OR gate can have any number of inputs but only one output. The OR gate output is 1 if one or more inputs are 1. The truth table in Fig. 4-6 shows the resulting output Y from each possible input combination.

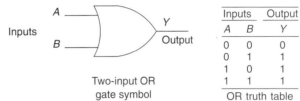

Inputs		Output
A	B	Y
0	0	0
0	1	1
1	0	1
1	1	1

OR truth table

Two-input OR gate symbol

Fig. 4-6

OR gate.

Figures 4-7 and 4-8 on p. 74 show practical applications of the OR gate function. The OR gate is essentially a *parallel* circuit that produces an output voltage when a voltage appears at any input.

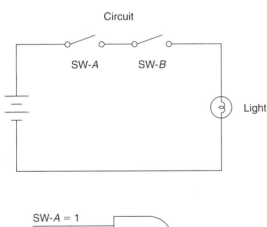

SW-A		SW-B		Light	
Open	(0)	Open	(0)	Off	(0)
Open	(0)	Closed	(1)	Off	(0)
Closed	(1)	Open	(0)	Off	(0)
Closed	(1)	Closed	(1)	On	(1)

Truth table

The light will be on only when both switch *A* and switch *B* are closed.

Fig. 4-5

AND gate function application—example 2.

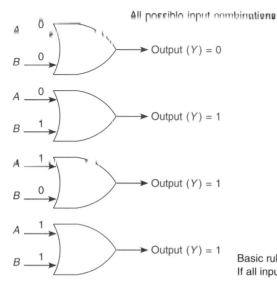

All possible input combinations

Inputs		Output
A	B	Y
0	0	0
0	1	1
1	0	1
1	1	1
	Truth table	

Basic rules: If one or more inputs are 1, the output is 1.
If all inputs are 0, the output will be 0.

Fig. 4-7

OR gate function application—example 1

The NOT Function

The symbol drawn in Fig. 4-9 is that of a NOT function. Unlike the AND and OR functions, the NOT function can have only *one* input. The NOT output is 1 if the input is 0. The output is 0 if the input is 1. The result of the NOT operation is always the inverse of the input and the NOT function is, therefore, called an *inverter*. The NOT function is often depicted by using a bar across the top of the letter, indicating an inverted output. The small circle at the output of the inverter is termed a *state indicator* and indicates that an inversion of the logical function has taken place.

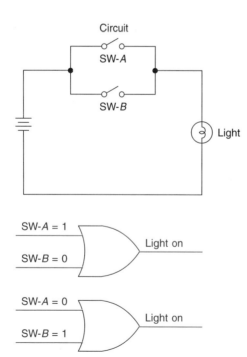

SW-A		SW-B		Light	
Open	(0)	Open	(0)	Off	(0)
Open	(0)	Closed	(1)	On	(1)
Closed	(1)	Open	(0)	On	(1)
Closed	(1)	Closed	(1)	On	(1)
		Truth table			

The light will be on if switch A or switch B is closed.

Fig. 4-8

OR gate function application—example 2.

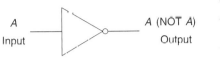

A	NOT A
0	1
1	0

NOT truth table

Fig. 4-9

NOT function symbol.

Figure 4-10 shows an example of a practical application of the NOT function, where a normally closed pushbutton is in series with the output. When the pushbutton is *not* actuated, the output is ON, and when the pushbutton is actuated, the output is OFF.

The NOT function is most often used in conjunction with the AND or the OR gate. Figure 4-11 shows the NOT function connected to one input of an AND gate.

The NOT symbol placed at the output of an AND gate would invert the normal output result. An AND gate with an inverted output is called a NAND gate. The NAND gate symbol and truth table are shown in Fig. 4-12 on p. 76. The NAND function is often used in integrated circuit logic arrays and can be used in programmable controllers to solve complex logic.

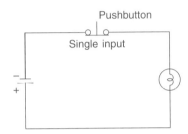

Pushbutton		Light	
Not pressed	(0)	On	(1)
Pressed	(1)	Off	(0)

Truth table

The light will be on if the pushbutton is not pressed.

Fig. 4-10

NOT gate function application—example 1.

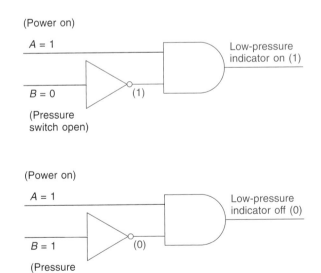

Pressure switch	Power	Pressure indicator
0	1	1
1	1	0

Truth table

In this low-pressure-warning indicator circuit, if the power is on (1) and the pressure switch is not closed (0), the warning indicator will be on.

Fig. 4-11

NOT gate function application—example 2.

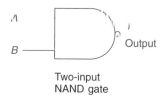

Inputs		Output
A	B	Y
0	0	1
0	1	1
1	0	1
1	1	0

NAND truth table

Two-input
NAND gate

Fig 4-12

NAND gate symbol and truth table.

The same rule applies if a NOT symbol is placed at the output of the OR gate. The normal output is inverted, and the function is referred to as a NOR gate. The NOR gate symbol and truth table are shown in Fig. 4-13.

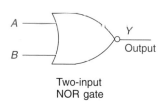

Inputs		Output
A	B	Y
0	0	1
0	1	0
1	0	0
1	1	0

NOR truth table

Two-input
NOR gate

Fig. 4-13

NOR gate symbol and truth table.

The Exclusive-OR (XOR) Function

An often used combination of gates is the exclusive-OR (XOR) function (Fig. 4-14). The output of this circuit is HIGH only when one input or the other is HIGH, but *not both*. The exclusive-OR gate is commonly used for the *comparison* of two binary numbers.

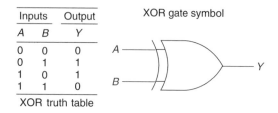

Inputs		Output
A	B	Y
0	0	0
0	1	1
1	0	1
1	1	0

XOR truth table

XOR gate symbol

Fig. 4-14

The XOR (exclusive-OR) gate symbol and truth table.

4-3 BOOLEAN ALGEBRA

The mathematical study of the binary number system and logic is called *Boolean algebra.* The purpose of this algebra is to provide a simple way of writing complicated combinations of logic statements. There are many applications where Boolean algebra could be applied to solving PLC programming problems, and in fact some programmable controllers can be programmed directly using Boolean instructions (Table 4-1). Compared to relay ladder logic (RLL), Boolean logic is more natural. Everyone knows the meanings of the words *and, or,* and *not*. Except for electricians and PLC programmers, not everyone is familiar with ladder logic.

Figure 4-15 on p. 78 summarizes the basic operators of Boolean algebra as they relate to the basic AND, OR, and NOT functions. Inputs are represented by capital letters *A, B, C,* etc., and the output by a capital *Y.* The multiplication sign ($\times$) or dot ($\cdot$) represents the AND operation, an addition sign ($+$) represents the OR operation, the circle with an addition sign ($\oplus$) represents the EXCLUSIVE OR operation, and a bar over the letter ($\overline{A}$) represents the NOT operation.

Digital systems may be designed using Boolean algebra. Circuit functions are represented by Boolean equations. See Figs. 4-16 and 4-17 on p. 78 for two examples of how the basic AND, OR, and NOT functions are used to form Boolean equations.

An understanding of the technique of writing simplified Boolean equations for complex logical statements is a useful tool when creating PLC control programs. Some laws of Boolean algebra are different from those of ordinary algebra. The three basic laws on p. 79 illustrate the close comparison between Boolean algebra and ordinary algebra, as well as one major difference between the two.

Table 4-1

Boolean Instruction and Function	Graphic Symbol
Store (STR)—Load (LD) Begins a new rung or an additional branch in a rung with a normally open contact.	—| |—
Store Not (STR NOT)—Load Not (LD NOT) Begins a new rung or an additional branch in a rung with a normally closed contact.	—|/|—
Or (OR) Logically ORs a normally open contact in parallel with another contact in a rung.	
Or Not (OR NOT) Logically ORs a normally closed contact in parallel with another contact in a rung.	
And (AND) Logically ANDs a normally open contact in series with another contact in a rung.	
And Not (AND NOT) Logically ANDs a normally closed contact in series with another contact in a rung.	
And Store (AND STR)—And Load (AND LD) Logically ANDs two branches of a rung in series.	
Or Store (OR STR)—Or Load (OR LOAD) Logically ORs two branches of a rung in parallel.	
Out (OUT) Reflects the status of the rung (on/off) and outputs the discrete (ON/OFF) state to the specified image register point or memory location.	—(OUT)— —◯—
Or Out (OR OUT) Reflects the status of the rung and outputs the discrete (ON/OFF) state to the image register. Multiple OR OUT instructions referencing the same discrete point can be used in the program.	—(OROUT)—
Output Not (OUT NOT) Reflects the status of the rung and turns the output OFF for an ON execution condition; turns the output ON for an OFF execution condition.	—⊘—

Logic symbol	Logic statement	Boolean equation
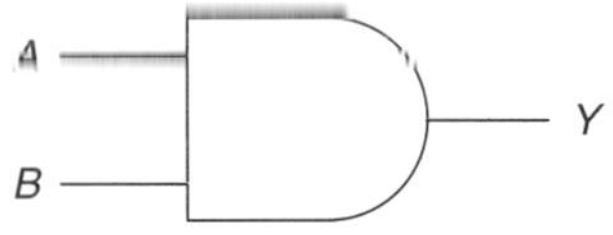 A ─── Y B ───	Y is 1 if A and B are 1	$Y = A \cdot B$ or $Y = AB$
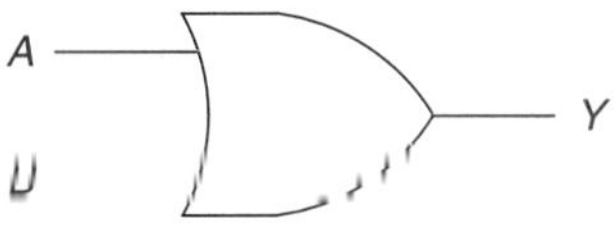 A ─── Y B ───	Y is 1 if A or B is 1	$Y = A + B$
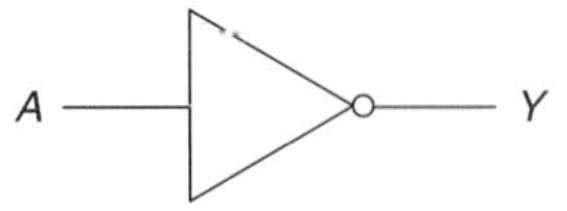 A ─── Y	Y is 1 if A is 0 Y is 0 if A is 1	$Y = \overline{A}$

Fig. 4-15

Boolean algebra as related to AND, OR, and NOT functions.

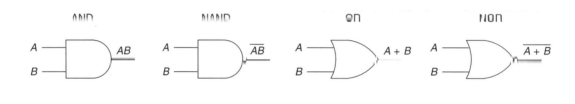

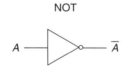

NOT

$A \longrightarrow \overline{A}$

Basic logic gates implement simple logic functions.
Each logic function can be expressed in terms of a
Boolean expression, as shown.

Fig. 4-16

Boolean equation—example 1.

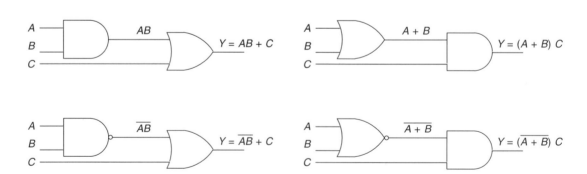

Any combination of control functions can be expressed as shown.

Fig. 4-17

Boolean equation—example 2.

COMMUTATIVE LAW

$$A + B = B + A$$
$$A \cdot B = B \cdot A$$

ASSOCIATIVE LAW

$$(A + B) + C = A + (B + C)$$
$$(A \cdot B) \cdot C = A \cdot (B \cdot C)$$

DISTRIBUTIVE LAW

$$A \cdot (B + C) = (A \cdot B) + (A \cdot C)$$
$$A + (B \cdot C) = (A + B) \cdot (A + C)$$

This law holds true only in Boolean algebra.

De Morgan's law is one of the most important results of Boolean algebra. It shows that any logical function can be implemented with AND gates and inverters or OR gates and inverters (see Fig. 4-18).

4•4 DEVELOPING CIRCUITS FROM BOOLEAN EXPRESSIONS

As logic circuits become more complex, the need to express these circuits in Boolean form becomes greater. A simple logic gate is quite straightforward in its operation. However, by grouping these gates into combinations, it becomes more difficult to determine which combinations of inputs will produce an output. Figures 4-19 and 4-20 on p. 80 illustrate the method used to develop a circuit from a Boolean expression.

4•5 PRODUCING THE BOOLEAN EQUATION FROM A GIVEN CIRCUIT

Figures 4-21 and 4-22 on p. 81 illustrate how to produce the Boolean equation from a given circuit.

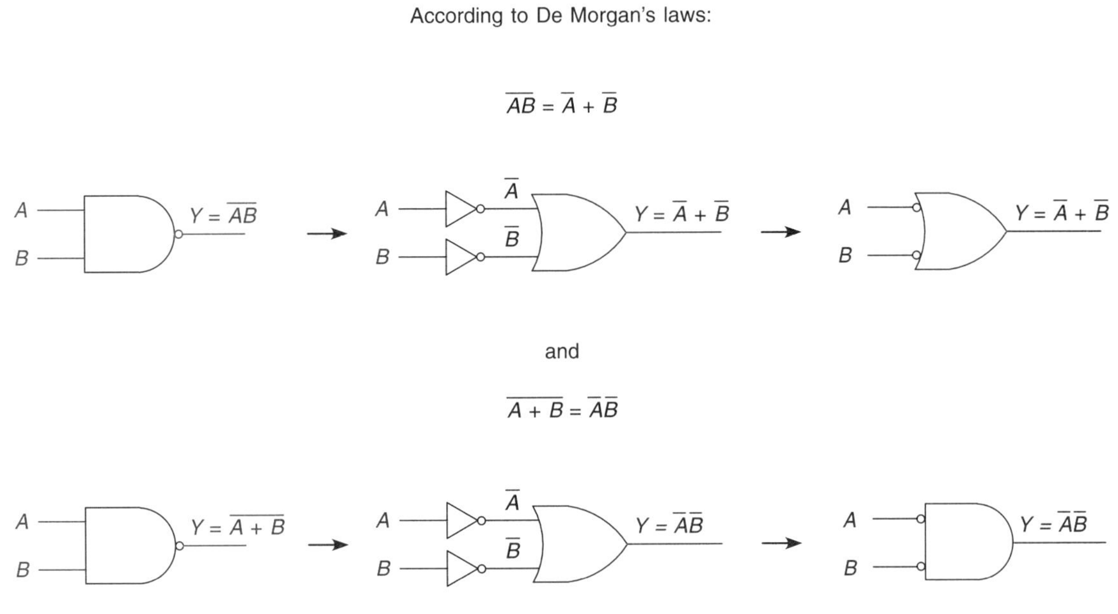

According to De Morgan's laws:
$\overline{AB} = \overline{A} + \overline{B}$
A
B
$Y = \overline{AB}$
$\overline{A}$
$\overline{B}$
$Y = \overline{A} + \overline{B}$
and
$\overline{A + B} = \overline{A}\,\overline{B}$
$Y = \overline{A + B}$
$Y = \overline{A}\,\overline{B}$

Fig. 4-18
De Morgan's laws.

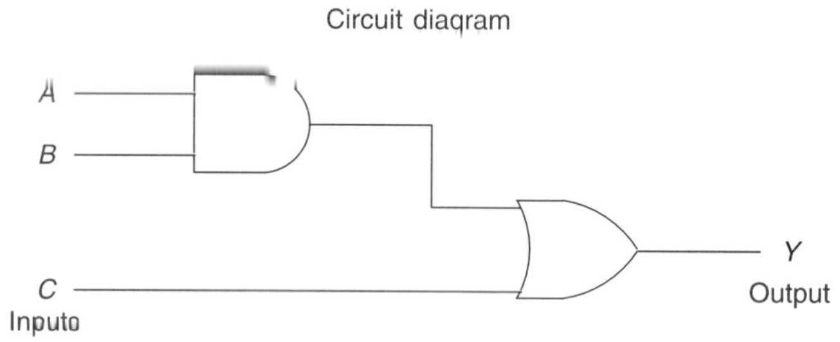

Circuit diagram

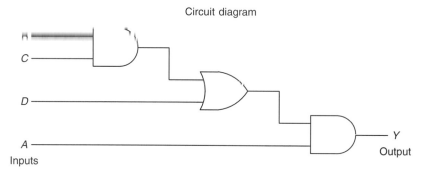

A

B

C

Inputs

Y

Output

Boolean expression: Y = AB + C
Gates required: (by inspection)
    1   AND gate with input A and B
    1   OR gate with input C and output from previous AND gate

Fig. 4-19

Circuit development using a Boolean expression—example 1.

Circuit diagram

C

D

A

Inputs

Y

Output

Boolean expression: Y = A(BC + D)
Gates required: (by inspection)
    1   AND gate with inputs B and C
    1   OR gate with inputs B • C and D
    1   AND gate with inputs A and the output
            from the OR gate

Fig. 4-20

Circuit development using a Boolean expression—example 2.

4•6 HARD-WIRED LOGIC VERSUS PROGRAMMED LOGIC

The term *hard-wired logic* refers to logic control functions that are determined by the way devices are interconnected. Hard-wired logic can be implemented using relays and relay ladder schematics. Relay ladder schematics are universally used and understood in industry. Figure 4-23 on p. 82 shows a typical relay ladder schematic of a motor STOP/START control station with pilot lights. The control scheme is drawn between two vertical supply lines. All the components are placed between these two lines, called *rails* or *legs,* connecting the two power lines with what look like *rungs* of a ladder—thus the name, *ladder logic program.*

Hard-wired logic is fixed; it is changeable only by altering the way devices are connected. In contrast, programmable control is based on the basic logic functions, which are programmable and easily changed. These functions (AND, OR, NOT) are used

Write the Boolean equation for the following circuit:

Original circuit

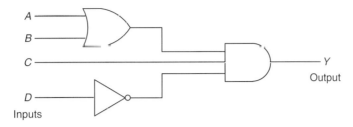

Circuit with Boolean terms

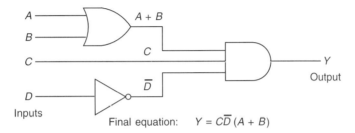

Final equation: $Y = C\overline{D}\,(A + B)$

Fig. 4-21

Producing a Boolean equation from a given circuit—example 1.

Write the Boolean equation for the following circuit:

Original circuit

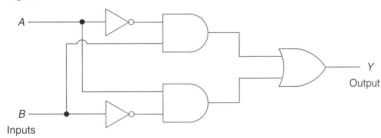

Circuit with Boolean terms

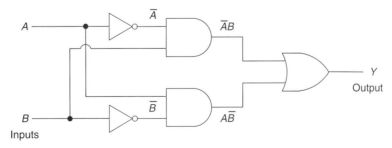

Final equation: $Y = \overline{A}B + A\overline{B}$

Fig. 4-22

Producing a Boolean equation from a given circuit—example 2.

Fundamentals of Logic

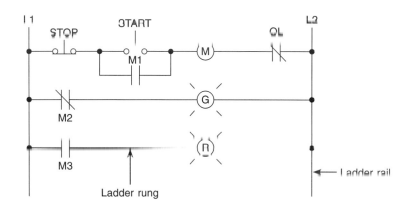

Fig. 4-23

Relay ladder schematic.

Contact symbolism is a simple way of expressing the control logic in terms of symbols that are used on relay control schematics. A rung is the contact symbolism required to control an output. Some PLCs allow a rung to have multiple outputs. A complete ladder logic program thus consists of several rungs, each of which controls an output.

Because the PLC uses ladder logic diagrams, the conversion from any existing relay logic to programmed logic is simple. Each rung is a combination of input conditions (symbols) connected from left to right, with the symbol that represents the output at the far right. The symbols that represent the inputs are connected in series, parallel, or some combination of the two to obtain the desired logic. The following group of examples illustrate the relationship between the relay ladder schematic, the ladder logic schematic program, and the equivalent logic gate circuit (see Examples 4-1 to 4-9 on pp. 83–85).

either singly or in combinations to form instructions that will determine if a device is to be switched on or off. The form in which these instructions are implemented to convey commands to the PLC is called the *language.* The most common PLC language is *ladder logic.*

In Fig. 4-24 you can see a typical ladder logic program for the relay ladder schematic of Fig. 4-23. The instructions used are the relay equivalent of normally open (NO) and normally closed (NC) contacts and coils.

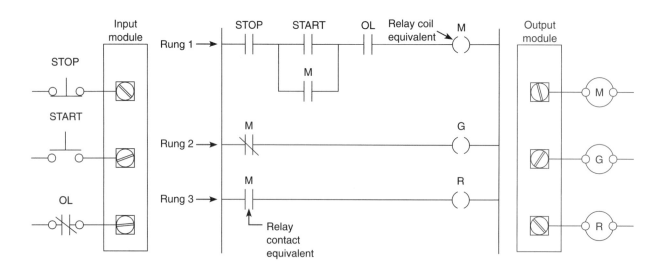

Fig. 4-24

Ladder logic program.

Example 4-1

Relay schematic
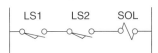
LS1 LS2 SOL

Ladder logic program
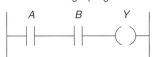
A B Y

Gate logic
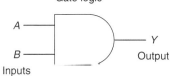
A
B
Inputs
Y
Output

Boolean equation: $AB = Y$

Example 4-1

Two limit switches connected in series and used to control a
solenoid valve.

Example 4-2

Relay schematic
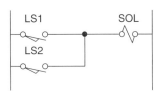
LS1 SOL
LS2

Ladder logic program
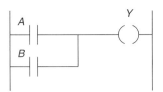
A
B
Y

Gate logic
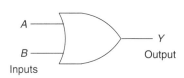
A
B
Inputs
Y
Output

Boolean equation: $A + B = Y$

Example 4-2

Two limit switches connected in parallel and used to control a
solenoid valve.

Example 4-3

Relay schematic
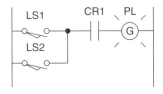
LS1 CR1 PL
LS2   (G)

Ladder logic program
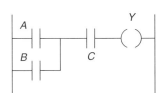
A Y
B C

Gate logic
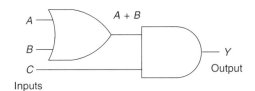
A
B
C
Inputs
A + B
Y
Output

Boolean equation: $(A + B)C = Y$

Example 4-3

Two limit switches connected in parallel with each other and in series
with a relay contact, and used to control a pilot light.

Example 4-4

Relay schematic
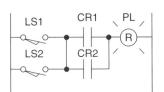
LS1 CR1 PL
LS2 CR2 (R)

Ladder logic program
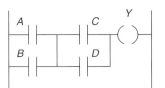
A C Y
B D

Gate logic
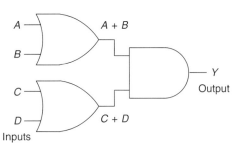
A
B
A + B
C
D
C + D
Inputs
Y
Output

Boolean equation: $(A + B)(C + D) = Y$

Example 4-4

Two limit switches connected in parallel with each other and in series
with two sets of contacts (that are connected in parallel with each
other), and used to control a pilot light.

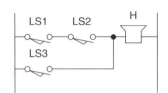

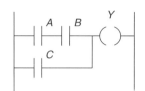

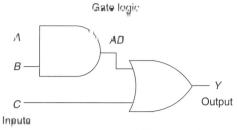

Boolean equation: $(AB) + C = Y$

Example 4-5

Two limit switches connected in series with each other and in parallel with a third limit switch, and used to control a warning horn.

Relay schematic   Ladder logic program   Gate logic

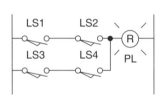

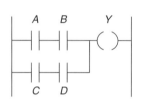

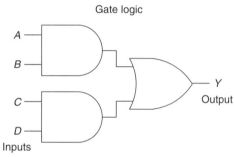

Boolean equation: $(AB) + (CD) = Y$

Example 4-6

Two limit switches connected in series with each other and in parallel with two other switches (that are connected in series with each other), and used to control a pilot light.

Relay schematic   Ladder logic program   Gate logic

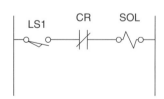

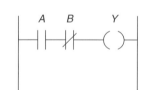

Boolean equation: $A\overline{B} = Y$

Example 4-7

One limit switch connected in series with an NC relay contact used to control a solenoid valve.

Relay schematic

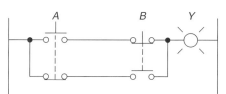

Ladder logic program

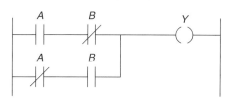

Gate logic

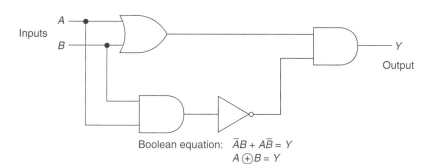

Boolean equation: $\overline{A}B + A\overline{B} = Y$
$A \oplus B = Y$

Example 4-8

Exclusive OR circuit. The output of this circuit is ON only when push-button A or B is pressed, but not both, $\oplus$ is the exclusive OR symbol.

Relay schematic

Ladder logic program

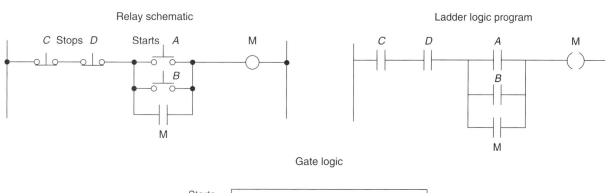

Gate logic

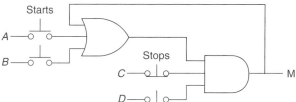

Example 4-9

A motor control circuit with two stop buttons. When the start button is depressed, the motor runs. By sealing, it continues to run when the start button is released. The stop buttons stop the motor when they are depressed.

4.7 PROGRAMMING WORD LEVEL LOGIC INSTRUCTIONS

Some PLCs provide word-level logic instructions as part of their instruction set. Table 4-2 shows how to select the correct logic instruction for different situations.

Figure 4-25 illustrates the operation of the AND instruction to perform an AND operation using the bits in the two *source* addresses. This instruction tells the processor to perform an AND operation on B3:5 and B3:7, and store the result in Destination B3:10 when input device A is true.

Figure 4-26 illustrates the operation of the OR instruction, which ORs the data in Source A, bit by bit, with the data in Source B and stores the result at the Destination address. The address of Source A is B3:1, the address of Source B is B3:2, and the Destination address is B3:20. The instruction may be programmed conditionally, with input instruction(s) preceding it, or unconditionally as shown without any input instructions preceding it.

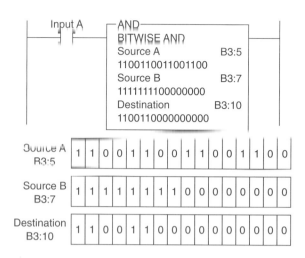

Fig. 4-25

AND instruction (destination bits are result of logical AND operation).

Figure 4-27 illustrates the operation of the XOR instruction. In this example, data from input I:1.0 is compared, bit by bit, with data from input I:3.0. Any mismatches energize the corresponding bit in word O:4. As you can see, there is a 1 in every bit location in the Destination corresponding to the bit locations where Source A and Source B are *different* and a 0 in the Destination where Source A and Source B are the same. The XOR is often used in diagnostics, where real-world inputs, such as limit switches, are compared with their desired states.

Table 4-2

SELECTING LOGIC INSTRUCTIONS

If you want to . . .	. . . use this instruction.
Know when matching bits in two different words are both ON	AND
Know when one or both matching bits in two different words are ON	OR
Know when one or the other bit of matching bits in two different words is ON	XOR
Reverse the state of bits in a word	NOT

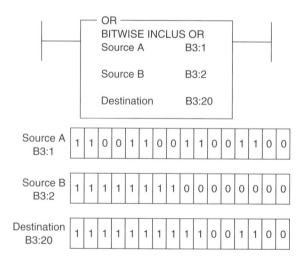

Fig. 4-26

OR instruction (destination bits are result of logical OR operation).

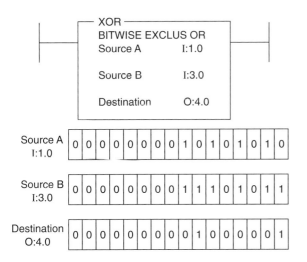

| Source A
I:1.0 | 0 | 0 | 0 | 0 | 0 | 0 | 0 | 0 | 1 | 0 | 1 | 0 | 1 | 0 | 1 | 0 |

| Source B
I:3.0 | 0 | 0 | 0 | 0 | 0 | 0 | 0 | 0 | 1 | 1 | 1 | 0 | 1 | 0 | 1 | 1 |

| Destination
O:4.0 | 0 | 0 | 0 | 0 | 0 | 0 | 0 | 0 | 0 | 1 | 0 | 0 | 0 | 0 | 0 | 1 |

Fig. 4-27

XOR instruction.

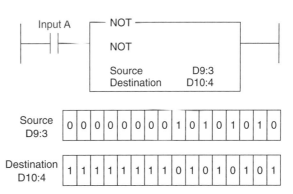

| Source
D9:3 | 0 | 0 | 0 | 0 | 0 | 0 | 0 | 0 | 1 | 0 | 1 | 0 | 1 | 0 | 1 | 0 |

| Destination
D10:4 | 1 | 1 | 1 | 1 | 1 | 1 | 1 | 1 | 0 | 1 | 0 | 1 | 0 | 1 | 0 | 1 |

Fig. 4-28

NOT instruction (destination bits are result of logical OR operation).

Figure 4-28 illustrates the operation of the NOT instruction. This instruction inverts the bits from the Source word to the Destination word. The bit pattern in D10.4 is the result of the instruction being true and is the inverse of the bit pattern in D9:3.

Questions

1. Explain the binary principle.

2. What is the purpose of an electronic gate?

3. Draw the logic symbol, construct a truth table, and state the Boolean equation for each of the following.

   a. Two-input AND gate
   b. NOT function
   c. Three-input OR gate
   d. XOR function

4. Express each of the following equations as a ladder logic program:

   a. $Y = (A + B)CD$
   b. $Y = A\overline{B}C + \overline{D} + E$
   c. $Y = [(\overline{A} + \overline{B})C] + DE$
   d. $Y = (A\overline{B}\overline{C}) + (D\overline{E}F)$

5. Write the ladder logic program, draw the logic gate circuit, and state the Boolean equation for the following two relay ladder diagrams. (see Fig. 4-29):

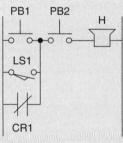

Fig. 4-29

(a)　　　　　　　　　　　　(b)

6. Develop a logic gate circuit for each of the following Boolean expressions using AND, OR, and NOT gates:

   a. $Y = ABC + D$
   b. $Y = AB + CD$
   c. $Y = (A + B)(\overline{C} + D)$
   d. $Y = \overline{A}(B + CD)$
   e. $Y = \overline{A}B + C$
   f. $Y = (ABC + D)(E\overline{F})$

7. State the logic instruction you would use when you want to:

   a. Know when one or both matching bits in two different words are ON

   b. Reverse the state of bits in a word

   c. Know when matching bits in two different words are both ON

   d. Know when one or the other bit of matching bits, but not both, in two different words is ON

Problems

1. It is required to have a pilot light come on when *all* of the following circuit requirements are met:

   o All four circuit pressure switches must be closed.

   o At least two out of three circuit limit switches must be closed.

   o The reset switch must *not* be closed.

   Using AND, OR, and NOT gates, design a logic circuit that will solve this hypothetical problem.

2. Write the Boolean equation for each of the logic gate circuits in Fig. 4-30a to f.

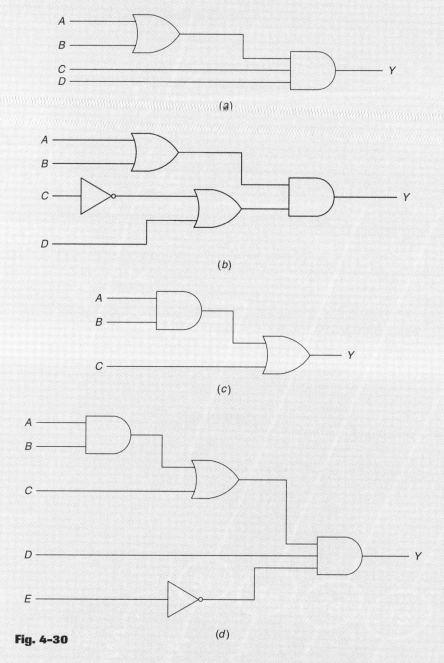

(a)

(b)

(c)

Fig. 4-30

(d)

Fundamentals of Logic

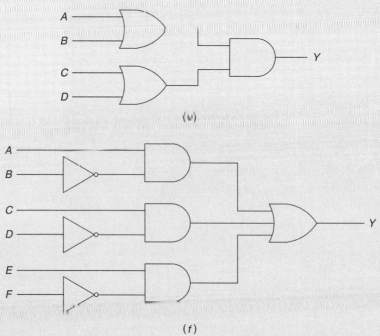

(e)

(f)

Fig. 4-30 (continued)

3. Match each of the following situations (i) to (v) with the analogous logic circuit in Fig. 4-31:

   (i) To purchase the car, you must have $10,000.00 and a trade-in, or come up with another $2,000.00.

   (ii) Two representatives from management and two from the union must be in attendance for the arbitration meeting. If a person from either group fails to show up, the meeting is called off.

   (iii) To obtain a credit in the course, you must be registered and also pass at least one of the major tests.

   (iv) A pair of kings or a pair of aces will win the hand.

   (v) To qualify as a participant, you must attend at least one morning or afternoon session of either day of the conference.

Logic circuits

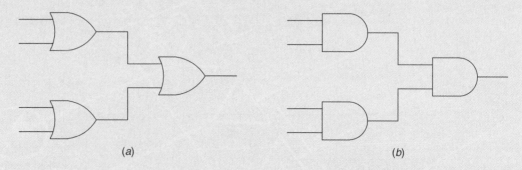

(a)   (b)

Fig. 4-31

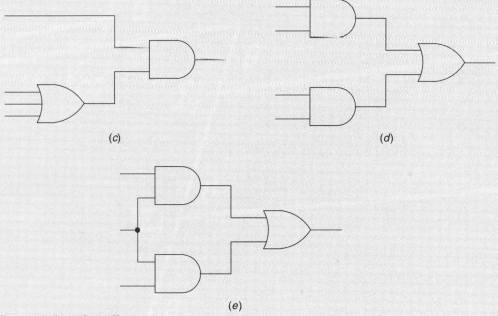

(c)

(d)

(e)

Fig. 4-31 (continued)

4. The logic circuit of Fig. 4-32 is used to activate an alarm when its output Y is logic HIGH or 1. Draw a truth table for the circuit showing the resulting output for all 16 of the possible input conditions.

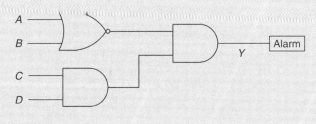

Fig. 4-32

5. What will be the data stored in the destination address of Fig. 4-33 for each of the following logical operations?
  a. AND operation
  b. OR operation
  c. XOR operation

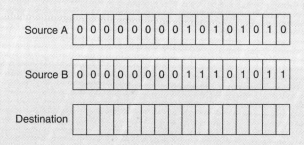

| Source A | 0 | 0 | 0 | 0 | 0 | 0 | 0 | 0 | 1 | 0 | 1 | 0 | 1 | 0 | 1 | 0 |

| Source B | 0 | 0 | 0 | 0 | 0 | 0 | 0 | 0 | 1 | 1 | 1 | 0 | 1 | 0 | 1 | 1 |

| Destination | | | | | | | | | | | | | | | | |

Fig. 4-33

6. Write the Boolean expression, and draw the gate logic diagram and typical PLC logic ladder diagram for a control system wherein a fan is to run only when all of the following conditions are met:

  ⊙ Input A is OFF

  ⊙ Input B is on or input C is on, or both B and C are ON

  ⊙ Inputs D and E are both ON

  ⊙ One or more of inputs F, G, or H is ON

Fundamentals of Logic

Basics of PLC Programming

After completing this chapter, you will be able to:

- ◆ Define and identify the functions of a PLC memory map

- ◆ Describe input and output image table files and types of data files

- ◆ Describe the PLC program scan sequence

- ◆ Understand how ladder diagram language, Boolean language, and function chart are used to communicate information to the PLC

- ◆ Define and identify the function of internal relay instructions

- ◆ Identify the common operating modes found in PLCs

- ◆ Write and enter ladder logic

Each input and output PLC module terminal is identified by a unique address. In PLCs, the internal symbol for any input is a contact. Similarly, in most cases, the internal PLC symbol for all outputs is a coil. This chapter shows how these contact/coil functions are used to program a PLC for circuit operation. This chapter covers only the basic set of instructions that perform functions similar to relay functions. You will also learn more about the program scan cycle and the scan time of a PLC.

5.1 PROCESSOR MEMORY ORGANIZATION

The term *processor memory organization* refers to how certain areas of memory in a given PLC are used. Not all PLC manufacturers organize PLC memories in the same way. Although they do not all use the same memory makeup and terminology, the principles involved are the same.

Figure 5-1 shows an illustration of the Allen-Bradley PLC-2 memory organization, known as a *memory map*. Every PLC has a memory map, but it may not be like the one illustrated. The memory space can be divided into two broad categories: the *user program* and the *data table*. The individual sections, their order, and the sections' length will vary and may be fixed or variable, depending on the manufacturer and model.

The user program is where the programmed logic ladder diagram is entered and stored. The user program will account for most of the total memory of a given PLC system. It contains the logic that controls the machine operation. This logic consists of *instructions* that are programmed in a ladder logic format. Most instructions require one *word* of memory.

The data table stores the information needed to carry out the user program. This includes information such as the status of input and output devices, timer and counter values,

data storage, and so on. Contents of the data table can be divided into two categories: *status data* and *numbers* or *codes*. Status is ON/OFF type of information represented by 1's and 0's, stored in unique bit locations. Number or code information is represented by groups of bits that are stored in unique byte or word locations.

A *processor file* is the collection of *program files* and *data files* created under a particular processor file name. It contains all the instructions, data, and configuration information pertaining to a user program. Figure 5-2 shows typical program and data file memory organization for an Allen-Bradley SLC-500 controller. The contents of each file are outlined below.

Program Files

Program files are the areas of processor memory where ladder logic programming is stored. They may include:

Fig. 5-1

Memory map for an Allen-Bradley PLC-2.

(a) General organization

Data table	Input/output locations Internal relay and timer/counter locations
User program	The user program causes the controller to operate in a particular manner
Housekeeping memory	Used to carry out functions needed to make the processor operate

(No access by user)

(b) Memory map shows how memory is organized

Section	Address
Processor work area #1	000 00 ↓ 007 17
Output image table file	010 00 ↓ 017 17
Bit/Word storage	020 00 ↓ 026 17
Reserved	027
Timer/Counter accumulated values (or bit/word storage)	030 00 ↓ 077 17
Processor work area #2	100 00 ↓ 107 17
Input image table file	110 00 ↓ 117 17
Bit/Word storage	120 00 ↓ 127 17
Timer/Counter preset values	130 00 ↓ 177 17
Expanded data table and/or user program	200 00 ↓ End of Memory

Bit address

Word address

◆ **System program** (file 0)—This file is always included and contains various system-related information and user-programmed information such as processor type, I/O configuration, processor file name, and password.

◆ **Reserved** (file 1)—This file is reserved by the processor and is not accessible to the user.

◆ **Main ladder program** (file 2)—This file is always included and contains user-programmed instructions that define how the controller is to operate.

◆ **Subroutine ladder program** (files 3–255)—These are user-created and are activated according to subroutine instructions residing in the main ladder program file.

Data Files

The data file portion of the processor's memory stores input and output status, processor status, the status of various bits, and numerical data. All this information is accessed via the ladder logic program. These files are organized by the type of data they contain and may include:

◆ **Output** (file 0)—This file stores the state of the output terminals for the controller.

◆ **Input** (file 1)—This file stores the status of the input terminals for the controller.

◆ **Status** (file 2)—This file stores controller operation information. This file is useful for troubleshooting controller and program operation.

◆ **Bit** (file 3)—This file is used for internal relay logic storage.

◆ **Timer** (file 4)—This file stores the timer accumulated and preset values and status bits.

◆ **Counter** (file 5)—This file stores the counter accumulated and preset values and status bits.

◆ **Control** (file 6)—This file stores the length, pointer position, and status bit for specific instructions such as shift registers and sequencers.

◆ **Integer** (file 7)—This file is used to store numerical values or bit information.

◆ **Reserved** (file 8)—This file is not accessible to the user.

◆ **Network Communications** (file 9)—This file is used for network communications if installed or used like files 10–255.

◆ **User-defined** (files 10–255)—These files are user-defined as bit, timer, counter, control, and/or integer data storage.

There are about 1000 program files for an Allen-Bradley PLC-5 controller. These program files may be set up in two ways: either (1) standard ladder logic programming, with

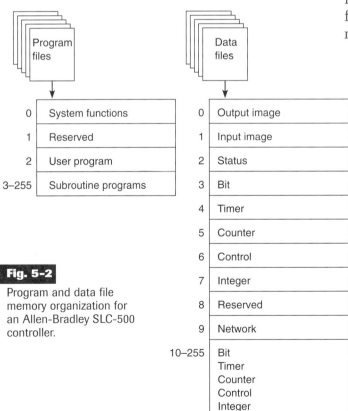

Fig. 5-2

Program and data file memory organization for an Allen-Bradley SLC-500 controller.

	Program files		Data files
0	System functions	0	Output image
1	Reserved	1	Input image
2	User program	2	Status
3–255	Subroutine programs	3	Bit
		4	Timer
		5	Counter
		6	Control
		7	Integer
		8	Reserved
		9	Network
		10–255	Bit Timer Counter Control Integer

the main program in program file 2 and program files 3 through 999 assigned, as needed, to subroutines, or (2) in sequential function charts in which files 2 through 999 are assigned steps or transitions, as required. With the processor set up for standard ladder logic, the main program will always be in program file 2, and program files 3 through 999 will be subroutines. In either case, the processor can store and execute only one program at a time.

Figure 5-3 shows typical data file memory organization for an Allen-Bradley PLC-5 controller. Each data file is made up of numerous *elements.* Each element may be one, two, or three words in length. Timer, counter, and control elements are three words in length; floating-point elements are two words in length; and all other elements are a single word in length. A *word* consists of sixteen bits, or binary digits. The processor operates on two different data types: integer and floating point. All data types,

except the floating-point files, are treated as integers or whole numbers. All element and bit addresses in the output and input data files are numbered octally. Element and bit addresses in all other data files are numbered decimally. Typical addressing formats are as follows:

◆ The addresses in the *output data file* and the *input data file* are potential locations for either input modules or output modules mounted in the I/O chassis:
  - The address O:012/15 is in the output image table file, rack 1, I/O group 2, bit 15.
  - The address I:013/17 is in the input image table file, rack 1, I/O group 3, bit 17.

◆ The *status data file* contains information about the processor status:
  - The address S:015 addresses word 15 of the status file.
  - The address S:027/09 addresses bit 9 in word 27 of the status file.

◆ The *bit data file* stores bit status. It frequently serves for storage when using internal outputs, sequencers, bit-shift instructions, and logical instructions:
  - The address B3:400 addresses word 400 of the bit file. The file number (3) must be included as part of the address. Note that the input, output, and status data files are the only files that do not require the file number designator.
  - Word 2, bit 15 is addressed as B3/47 because bit numbers are always measured from the beginning of the file. Remember that here, bits are numbered decimally (not octally, as they are in the input and output files).

◆ The *timer file* stores the timer status and timer data. A timer element consists of three words: the control word, preset word, and accumulated word. The addressing of the timer control word is the assigned timer number. Timers in file 4 are numbered starting with T4:0 and running through T4:999. The addresses for the three timer words in timer T4:0 are:

Address range		Size, in elements
O:000 – O:037	Output image file	32
I:000 – I:037	Input image file	32
S:000 – S:031	Processor status	32
B3:000 – B3:999	Bit file	1-1000
T4:000 – T4:999	Timer file	1-1000
C5:000 – C5:999	Counter file	1-1000
R6:000 – R6:999	Control file	1-1000
N7:000 – N7:999	Integer file	1-1000
F8:000 – F8:999	Floating-point file	1-1000
	Files to be assigned file nos. 9-999	1-1000 per file

Fig. 5-3

Data file memory organization for an Allen-Bradley PLC-5 controller.

Control word:	T4:0
Preset word:	T4:0.PRE
Accumulated word:	T4:0.ACC

The enable-bit address in the control word is T4:0/EN, the timer-timing-bit address is T4:0/TT, and the done-bit address is T4:0/DN.

♦ The *counter file* stores the counter status and counter data. A counter element consists of three words: the control word, preset word, and accumulated word. The addressing of the counter control is the assigned counter number. Counters in file 5 are numbered beginning with C5:0 and running through C5:999. The addresses for the three counter words in counter C5:0 are:

Control word:	C5:0
Preset word:	C5:0.PRE
Accumulated word:	C5:0.ACC

The count-up-enable-bit address in the control word is C5:0/CU, the count-down-enable-bit address is C5:0/CD, the done-bit address is C5:0/DN, the over-flow address is C5:0/OV, and the under-flow address is C5:0/UN.

♦ The *control file* stores the control element's status and data, and it is used to control various file instructions. The control element consists of three words: the control word, length word, and position word. The addressing of the control's control word is the assigned control number. Control elements in control file 6 are numbered beginning with R6:0 and running through R6:999. The addresses for the three words in control element R6:0 are:

Control word:	R6:0
Length:	R6:0.LEN
Position:	R6:0.POS

There are numerous control bits in the control word, and their function depends on the instruction in which the control element is used.

♦ The *integer file* stores integer data values, with a range from -32,768 through 32,767. Stored values are displayed in decimal form. The integer element is a single-word (16-bit) element. As many as 1000 integer elements, addressed from N7:000 through N7:999, can be stored.
  - The address N7:100 addresses word 100 of the integer file.
  - Bit addressing is decimal, from 0 through 15. For example, bit 12 in word 15 is addressed N7:015/12.

♦ The *floating-point file* element can store values in the range from $\pm 1.1754944e^{-38}$ to $\pm 3.4028237e^{+38}$. The floating-point element is a two-word (32-bit) element. As many as 1000 elements, addressed from F8:000 through F8:999, can be stored. Individual words or bits cannot be addressed in the floating-point file.

♦ Data files 9 through 999 may be assigned to different data types, as required. When assigned to a certain type, a file is then reserved for that type and cannot be used for any other type.

The bit file, integer file, or floating-point file can be used to store status or data. Which of these you use depends on the intended use of the data. If you are dealing with status rather than data, the bit file is preferable. If you are using very large or very small numbers and require a decimal point, floating point is preferable. The floating-point data type may have a restriction, however, because it may not interface well with external devices or with internal instructions such as counters and timers, which use only 16-bit words. In such a situation, it may be necessary to use the integer file type.

Figure 5-4 on p. 98 shows a typical connection of a switch to the input image table file through the input module. When the switch is closed, the processor detects a voltage at the input terminal and records that information by storing a binary 1 in the proper bit location. Each connected input has a bit in the input image table file that corresponds exactly to the terminal to which the input is connected. The input image table file is changed to reflect the current status of the switch during the I/O scan phase of operation. If the input is on (switch closed), its corresponding bit in the table is set to 1. If the input is off (switch open), the corresponding bit is "cleared," or reset to 0.

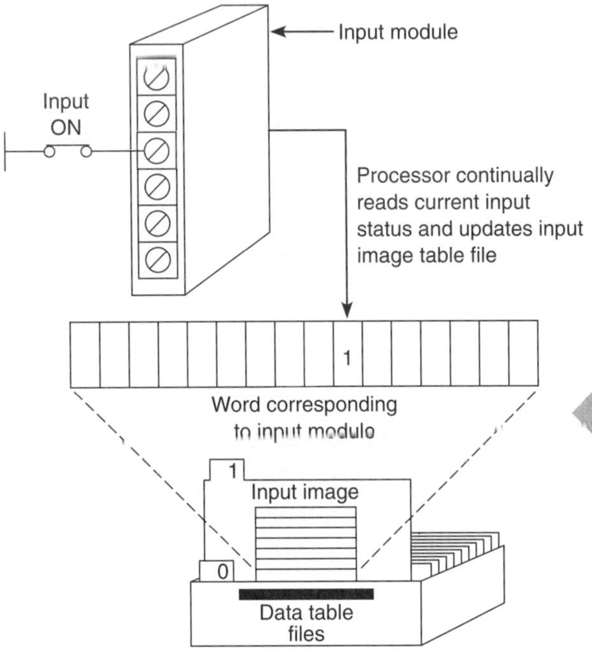

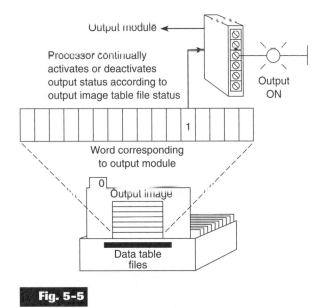

Fig. 5-5

Typical output image table file connection.

terminal to which the output is connected. If the program calls for a specific output to be on, its corresponding bit in the table is set to 1. If the program calls for the output to be off, its corresponding bit in the table is set to 0.

Fig. 5-4

Typical input image table file connection.

The *output image table file* is an array of bits that controls the status of digital output devices, which are connected to output interface circuits. Figure 5-5 shows a typical connection of a light to the output image table file through the output module. The status of this light (ON/OFF) is controlled by the user program and is indicated by the presence of 1's (ON) and 0's (OFF). Each connected output has a bit in the output image table file that corresponds exactly to the

5•2 PROGRAM SCAN

During each operating cycle, the processor reads all the inputs, takes these values, and energizes or de-energizes the outputs according to the user program. This process is known as a *scan*. Figure 5-6 illustrates a single PLC scan, which consists of the *I/O scan* and the *program scan*. Because the inputs can change at any time, the PLC must carry on this process continuously.

The PLC scan time specification indicates how fast the controller can react to changes in inputs. Scan time varies with program content and length. The time required to make a single scan can vary from about 1 ms to 20 ms. If a controller has to react to an input signal that changes states twice during the scan time, it is possible that the PLC will never be able to detect this change. For example, if it takes 8 ms for the CPU to scan

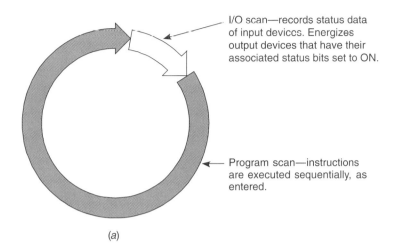

I/O scan—records status data of input devices. Energizes output devices that have their associated status bits set to ON.

Program scan—instructions are executed sequentially, as entered.

(a)

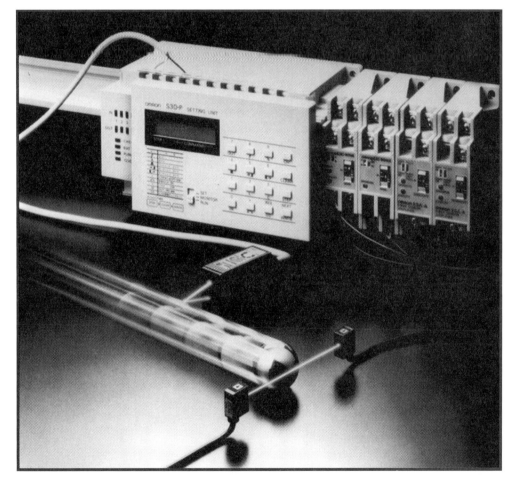

(b)

Fig. 5-6

(a) Single PLC scan. (b) Ideal for high-speed sensing and machine control applications, Omron's model S3D8 Sensor Controller accepts inputs at rates to 3 kHz and responds in 1 millisecond. (Courtesy of Omron Electronics, Inc.)

a program, and an input contact is opening and closing every 4 ms, the program may not respond to the contact changing state. The CPU will detect a change if it occurs during the update of the input image table file, but the CPU will not respond to every change.

The scan is normally a continuous and sequential process of reading the status of inputs, evaluating the control logic, and updating the outputs. Figure 5-7 illustrates this process. When the input device connected to address I:3/6 is closed, the input module circuitry senses a voltage and a 1 (ON) condition is entered into the input image table bit I:3/6. During the program scan, the processor examines bit I:3/6 for a 1 (ON) condition. In this case, since input I:3/6 is 1, the rung is said to be TRUE. The processor then sets the output image table bit O:4/7 to 1. The processor turns on output O:4/7

during the next I/O scan, and the output device (light) wired to this terminal becomes energized. This process is repeated as long as the processor is in the RUN mode. If the input device were to open, a 0 would be placed in the input image table. As a result, the rung would be called FALSE. The processor would then set the output image table bit O:4/7 to 0, causing the output device to turn off.

Each instruction entered into a program requires a certain amount of time for the instruction to be executed. The amount of time required depends on the instruction. For example, it takes less time for a processor to read the status of an input contact than it does to read the accumulated value of a counter.

There are two basic scan patterns that different PLC manufacturers use to accomplish the scan function (Fig. 5-8). Allen-Bradley PLCs use the *horizontal* scan by run method. In this type of system, the processor examines input and output instructions from the first command, top left in the program, horizontally, rung by rung. AEG Modicon PLCs use the *vertical* scan by column method. In this type of system, the processor examines input and output instructions from the top left command entered in the ladder diagram, vertically, column by column and page by page. Pages are executed in sequence. Misunderstanding the way the PLC scans a program can cause programming bugs.

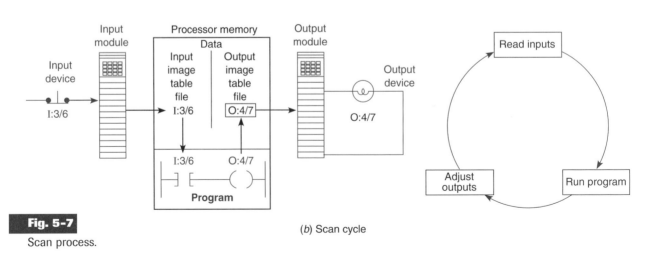

(a) Data flow overview

Fig. 5-7

Scan process.

(b) Scan cycle

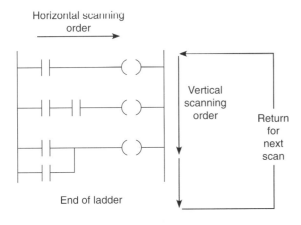

Fig. 5-8

Scanning can be vertical or horizontal.

The time taken to scan the user program is also dependent on the clock frequency of the microprocessor system. The higher the clock frequency, the faster is the scan rate.

5·3 PLC PROGRAMMING LANGUAGES

The term *PLC programming language* refers to the method by which the user communicates information to the PLC. The three most common language structures are *ladder diagram language, Boolean language,* and *function chart.* Although each language structure is similar from one PLC model to another, there are differences between manufacturers in the method of application. However, these differences are usually minor and easy to understand.

Ladder diagram language is by far the most commonly used PLC language. Figure 5-9 shows a comparison of ladder logic programming and Boolean programming. Figure 5-9a shows the original relay ladder diagram drawn as if it were to be hard-wired. Figure 5-9b shows the equivalent logic ladder diagram programmed into the controller. Note that the addressing format shown for input and output devices is generic in nature and varies for different PLC models. Figure 5-9c shows a typical set of generic Boolean state-

ments that could also be used to program the original circuit. This statement refers to the basic AND, OR, and NOT logic gate functions. Also included is the typical Boolean equation for the circuit.

The *function chart* system of programming was originally developed in Europe and is called GRAFCET. It is a method of programming a control system that uses a more structured approach. Function chart programming languages use function blocks (steps and transition units), often controlled by Boolean expressions. A function chart program is a pictorial representation or a special type of flowchart of a sequential control process. It shows the possible paths the process can take and the conditions necessary to go from one block to another.

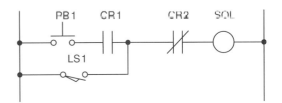

(a) Relay schematic

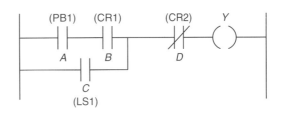

(b) Ladder logic program

START	PB1
AND	CR1
OR	LS1
AND NOT	CR2
OUT	SOL

Boolean equation: $Y = [(AB) + C]\,\overline{D}$

(c)

Fig. 5-9

PLC ladder and Boolean languages.

In principle, function chart programming languages allow the description of the process to become the actual control program. This method aids in understanding the system and localizing problems. A major advantage of function chart programming is that a block of logic can be programmed as a module, and transition logic ensures that only appropriate software modules will operate at any given time. Other advantages include simpler programming, faster scan time, enhanced maintainability, and ease of future enhancements.

Any process or machine, no matter how complex, can be described as a combination of sequential and/or parallel events. The overall program may be fairly complex, but the individual events are often quite simple. With function chart programming instructions, you can divide the program into several steps or stages instead of creating one long ladder program. Figure 5-10 shows the sequence of tasks or steps involved in a simple press process. Each step represents an event that corresponds directly to the machine's sequence of operations.

When you create a program using the function chart system, the steps or stages are programmed individually without concern for how they will affect the rest of the program.

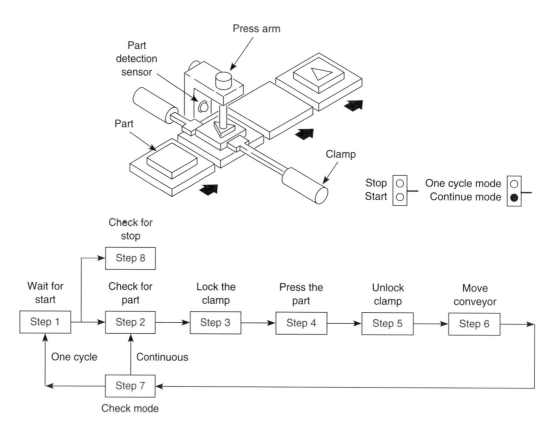

1. The operator presses the start switch to start the machine, or the stop switch to stop the machine.

2. The machine checks for a part. If the part is present, the process continues. If not, the conveyor moves until a part is present.

3. The part is locked in place with a clamp.

4. The press stamps the part.

5. The clamp is unlocked and the finished piece is moved out of the press.

6. The process stops if the machine is in one-cycle mode, or continues if continuous mode is selected.

Fig. 5-10

Steps involved in press process.

Processors that support this type of programming understand which parts of the program are active and scan only the active steps. All other steps are not scanned, thus reducing the program scan time.

Much of the time and effort in writing a ladder logic program is spent programming *interlocks,* which make sure that the process executes in the correct order. A substantial portion of the ladder logic is devoted to interlocking, or making sure certain things *don't* happen. Function chart programming helps eliminate this problem. The processor does not even scan those parts of the program that are inactive, thus reducing the number of complex interlocks required. In addition, it makes control much easier to troubleshoot. When trouble occurs, one can determine quickly which step or stage the process is stuck in and exactly what part of the machine to examine.

The PLC-5 family of Allen-Bradley processors can be programmed using a variation of function chart programming: sequential function chart (SFC) programming. PLC-5 SFC enables the user to program using function chart elements. However, relay ladder logic (rather than Boolean expressions) is used for logic control. This allows the PLC ladder logic program to be structured into sections or modules, which simplifies troubleshooting and maintenance (see Fig. 5-11).

Other programming language variations include state diagrams in which the system is defined as a set of states, and activity from one logic element to another is triggered by a set of Boolean decisions. Other variations, such as traditional programming languages, including BASIC and C, come from the PC industry itself.

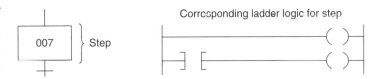

A *step* typically represents an independent machine state. One step of ladder logic runs repeatedly, top to bottom, until a logic condition (transition) lets the processor progress to the next step of the chart. The number (007 in this example) represents the ladder file numbering that contains the ladder logic for that step.

(a) Step

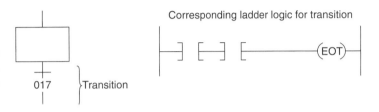

A *transition* represents the logic condition that allows the processor to progress from one step to the next. The transition goes true when it ends with an EOT instruction on a true rung. You draw a transition as a numbered cross below its step.

(b) Transition

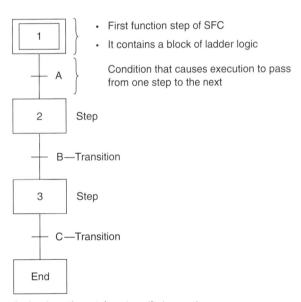

A *simple path* contains steps that execute one at a time in sequence. Each step and transition is a *file.*

(c) Simple path

Fig. 5-11

Sequential Function Chart (SFC) programming

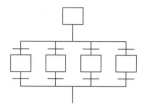

- A *selection branch* contains alternative paths, from which the processor selects one.

- This is equivalent to an OR structure.

- Draw a selection branch as parallel paths connected with *single* horizontal lines.

- Two or more alternative paths where only one is selected.

- The transitions beginning each path are scanned from left to right. The first true transition determines the path taken.

- For example, depending on a build code, one station must either drill or polish.

(*d*) Selection branch

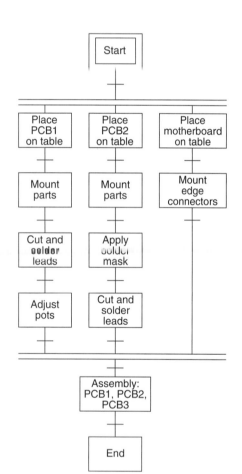

- A *simultaneous branch* runs steps that are in parallel paths simultaneously

- This is equivalent to an AND structure.

- Draw a simultaneous branch as parallel paths connected with *double* horizontal lines.

- All paths are active in the structure.

- Two or more parallel paths must be scanned at least once.

- The processor completes one step in the simultaneous diversion and then goes to the next step.

- A common transition for the paths is *outside* the branch.

- The processor finishes running a simultaneous branch when it has scanned each step in each path at least once and the common transition is true.

- For example, assume that you are mounting parts on printed circuit board PCB1 and soldering on PCB2, and the motherboard is finished and waiting for the other two boards. The only logic scanned is the logic for mounting parts on PCB1 and soldering on PCB2.

(*e*) Simultaneous branch

Fig. 5-11 (continued)

Sequential Function Chart (SFC) programming

5.4 RELAY-TYPE INSTRUCTIONS

The ladder diagram language is basically a *symbolic* set of instructions used to create the controller program. These ladder instruction symbols are arranged to obtain the desired control logic that is to be entered into the memory of the PLC. Because the instruction set is composed of contact symbols, ladder diagram language is also referred to as *contact symbology.*

Representations of contacts and coils are the basic symbols of the logic ladder diagram instruction set. The following three are the fundamental symbols used to translate relay control logic to contact symbolic logic (see Figs. 5-12 through 5-15 on pp. 105-107 for the symbols and examples).

The main function of the ladder logic diagram program is to control outputs based on input conditions. This control is accomplished through the use of what is referred to as a *ladder rung.* In general, a rung consists of a set of input conditions, represented by contact instructions, and an output instruction at the end of the rung, represented by the coil symbol (see Fig. 5-16 on p. 107). Each contact or coil symbol is referenced with an address number that identifies what is being evaluated and what is being controlled. The same contact instruction can be used throughout the program whenever that condition needs to be evaluated. For an output to be activated or energized, at least *one* left-to-right path of contacts must be closed. A complete closed path is referred to as having *logic continuity.* When logic continuity exists in at least one path, the rung condition is said to be TRUE. The rung condition is FALSE if no path has continuity.

During controller operation, the processor determines the ON/OFF state of the bits in the data files, evaluates the rung logic, and changes the state of the outputs according to the logical continuity of rungs.

Symbol

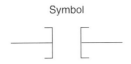

Analogous to the normally open relay contact. For this instruction, we ask the processor to EXAMINE IF (the contact is) CLOSED.

Typically represents any input to the control logic.

The input can be a connected switch or pushbutton, a contact from a connected output, or a contact from an internal output.

Has a bit-level address.

The status bit will be either 1 (ON) or 0 (OFF).

The status bit is examined for an ON condition.

If the status bit is 1 (ON), then the instruction is TRUE.

If the status bit is 0 (OFF), then the instruction is FALSE.

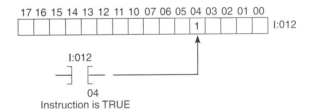

Instruction is TRUE

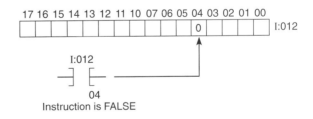

Instruction is FALSE

Fig. 5-12

Examine If Closed (XIC) instruction.

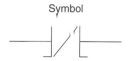

Analogous to the normally closed relay contact. For this instruction, we ask the processor to EXAMINE IF (the contact is) OPEN.

Typically represents any input to the control logic.

The input can be a connected switch or pushbutton, a contact from a connected output, or a contact from an internal output.

Has a bit-level address.

The status bit will be either 1 (ON) or 0 (OFF).

The status bit is examined for an OFF condition.

If the status bit is 0 (OFF), then the instruction is TRUE.

If the status bit is 1 (ON), then the instruction is FALSE.

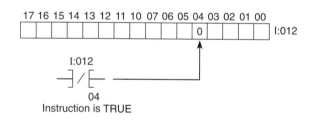

Instruction is TRUE

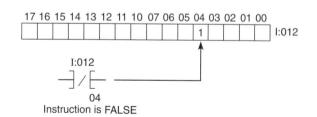

Instruction is FALSE

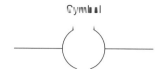

Analogous to the relay coil. The processor makes this instruction true (analogous to energizing a coil) when there is a path of true XIC and XIO instructions in the rung.

Typically represents any output that is controlled by some combination of input logic.

An output can be a connected device or an internal output (internal relay).

If any left-to-right path of input conditions is TRUE, the output is energized (turned ON).

The status bit of the addressed OUTPUT ENERGIZE instruction is set to 1 (ON) when the rung is TRUE.

The status bit of the addressed OUTPUT ENERGIZE instruction is reset to 0 (OFF) when the rung is FALSE.

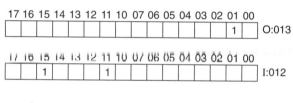

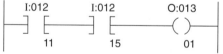

OUTPUT ENERGIZE instruction—TRUE

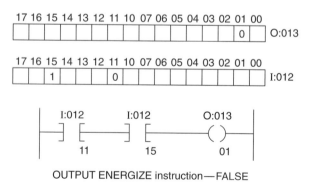

OUTPUT ENERGIZE instruction—FALSE

Fig. 5-13

Examine If Open (XIO) instruction.

Fig. 5-14

Output Energize (OTE) instruction.

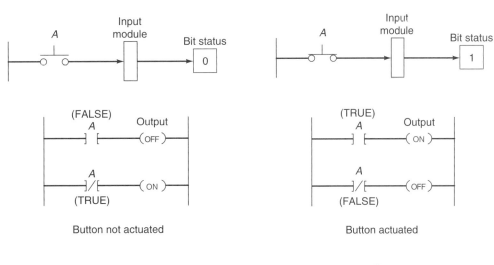

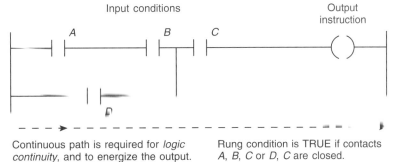

Fig. 5-15

Status bit examples.   Button not actuated   Button actuated

Input conditions   Output instruction

Fig. 5-16

Ladder rung.

Continuous path is required for *logic continuity*, and to energize the output.

Rung condition is TRUE if contacts A, B, C or D, C are closed.

More specifically, input instructions set up the conditions under which the processor will make an output instruction true or false. These conditions are as follows:

◆ When the processor finds a continuous path of true input instructions in a rung, the OUTPUT ENERGIZE (OTE) output instruction will become (or remain) true. We then say that rung conditions are true.

◆ When the processor does *not* find a continuous path of true input instructions in a rung, the OTE input instruction will become (or remain) false. We then say that rung conditions are false.

5•5 INSTRUCTION ADDRESSING

To complete the entry of a relay-type instruction, you must assign an *address* number to it. This number will indicate what PLC input is connected to what input device and what PLC output will drive what output device.

The addressing of real inputs and outputs, as well as internals, depends on the PLC model used. These addresses can be represented in decimal, octal, or hexadecimal depending on the number system used by the PLC. Figure 5-17 on p. 108 shows a typical addressing format for an Allen-Bradley SLC-500 controller. The programming manual of the PLC you are using should be consulted to determine the specific format because the format can vary from model to model, as well as from manufacturer to manufacturer.

The address identifies the function of an instruction and links it to a particular bit in the data table portion of the memory. Figure 5-18 on p. 108 shows the structure of a 16-bit word and its bit values.

The assignment of an I/O address can be included in the I/O connection diagram, as shown in Fig. 5-19 on p. 108. Inputs and outputs are typically represented by squares and diamonds, respectively.

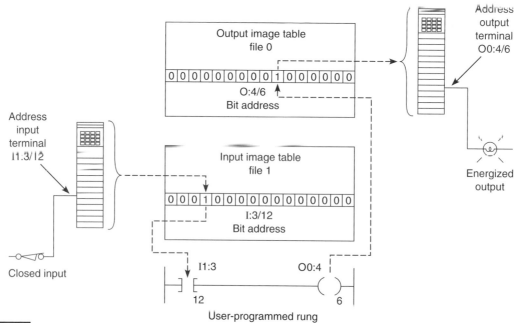

Fig. 5-17

The address identifies a location in the prossessor's data files, where the on/off state of the bit is stored.

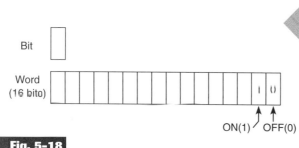

Fig. 5-18

Structure of a 16-bit word.

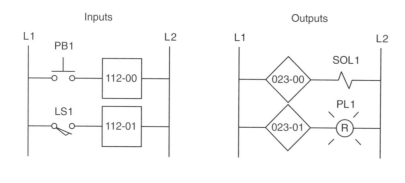

Fig. 5-19

I/O connection diagram.

5•6 BRANCH INSTRUCTIONS

Branch instructions are used to create parallel paths of input condition instructions. This allows more than one combination of input conditions (OR logic) to establish logic continuity in a rung. Figure 5-20 illustrates a simple branching condition. The rung will be TRUE if either instruction A or B is TRUE.

Input branching by formation of parallel branches, can be used in your application program to allow more than one combination of input conditions, as illustrated in Fig. 5-21. If at least one of these parallel branches forms a true logic path, the rung logic is enabled. If none of the parallel branches forms a true logic path, rung logic is not enabled and the output instruction logic will not be true. In the example shown, either A and B, or C provides a true logical path.

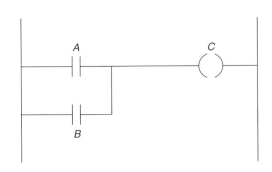

If you want to:	Then:
Add a branch	A. Press **APPEND BRANCH** F3 *Starts to the right of cursor* **or** **INSERT BRANCH** F4 *Starts to the left of cursor* B. Press the function key that corresponds to where you want the branch to start C. Add an instruction

Fig. 5-20

Parallel path (branch) instructions.

On some PLC models, branches can be established at both input and output portions of a rung. With output branching, you can program parallel outputs on a rung to allow a true logic path to control multiple outputs, as illustrated in Fig. 5-22. When there is a true logic path, all parallel outputs become true. In the example shown, either A or B provides a true logical path to all three output instructions: C, D, and E.

Additional input logic instructions (conditions) can be programmed in the output branches to enhance condition control of the outputs, as illustrated in Fig. 5-23. When there is a true logic path, including extra input conditions on an output branch, that branch becomes true. In the example shown, either A and D or B and D provide a true logic path to E.

Input and output branches can be *nested* to avoid redundant instructions and to speed up processor scan time. Figure 5-24 illustrates nested

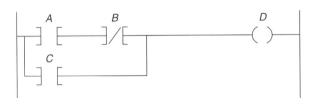

Fig. 5-21

Parallel input branching.

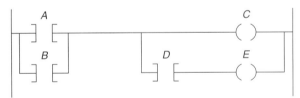

Fig. 5-23

Parallel output branching with conditions.

Fig. 5-22

Parallel output branching.

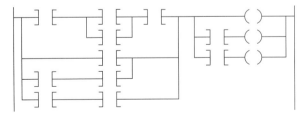

Fig. 5-24

Nested input and output branches.

Basics of PLC Programming

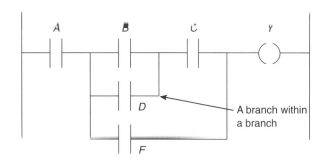

Fig. 5-25

Nested contact program.

input and output branches. A nested branch starts or ends within another branch.

In *some* PLC models, the programming of a branch circuit within a branch circuit or a *nested* branch cannot be done directly. It is possible, however, to program a logically equivalent branching condition. Figure 5-25 shows an example of a circuit that contains a nested contact *D*. To obtain the required logic, the circuit would be programmed as shown in Fig. 5-26. The duplication of contact *C* eliminates the nested contact *D*. Nested branching, can be converted into non-nested branches by repeating instructions to make parallel equivalents.

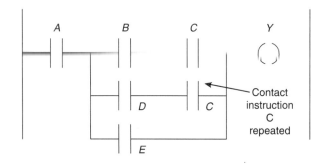

Fig. 5-26

Program required to eliminate nested contact.

For each PLC model, there may be limitations to the number of series contact instructions that can be included in one rung of a ladder diagram, as well as the number of parallel branches. Also, there is an additional limitation with some PLCs: only one output per rung, and the output must be located at the end of the rung. The only limitation on the number of rungs is memory size. Figure 5-27 shows the matrix limitation diagram for a

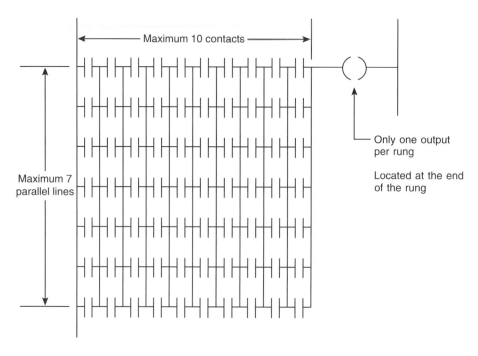

Fig. 5-27

Typical PLC matrix limitation diagram. The exact limitations are dependent on the particular type of PLC used. Programming more than the allowable series elements, parallel branches, or outputs will result in an error message being displayed.

PLC. A maximum of seven parallel lines and ten series contacts per rung is possible.

Another limitation to branch circuit programming is that the PLC will not allow for programming of vertical contacts. A typical example of this is contact *C* of the user program drawn in Fig. 5-28. To obtain the required logic, the circuit would be reprogrammed as shown in Fig. 5-29.

As mentioned, the processor examines the ladder logic rung for power flow from left to right *only* for logic continuity. The processor never allows for flow from right to left. This situation presents a problem for user program circuits similar to that shown in Fig. 5-30. If programmed as shown, contact combination *FDBC* would be ignored. To obtain the required logic, the circuit would be reprogrammed as shown in Fig. 5-31.

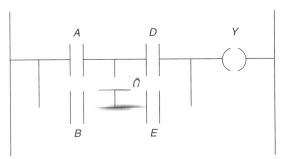

Boolean equation: Y = (AD) + (BCD) + (BE) + (ACE)

Fig. 5-28

Program with vertical contact

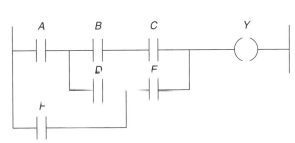

Boolean equation: Y = (ABC) + (ADE) + (FE) + (FDBC)

Fig. 5-30

Original circuit.

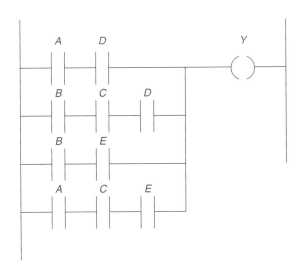

Fig. 5-29

Reprogrammed to eliminate vertical contact.

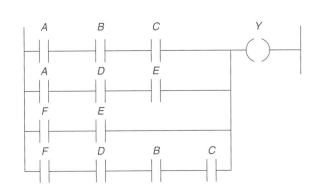

Fig. 5-31

Reprogrammed circuit.

5·7 INTERNAL RELAY INSTRUCTIONS

Most PLCs have an area of the memory allocated for what are known as *internal storage bits.* These storage bits are also called *internal outputs, internal coils, internal control relays,* or just *internals.* The internal output operates just as any output that is controlled by programmed logic; however, the output is used strictly for internal purposes. In other words, the internal output does not directly control an output device.

The advantage of using internal outputs is that there are many situations where an output instruction is required in a program, but no physical connection to a field device is needed. If there are no physical outputs wired to a bit address, the address can be used as an internal storage point. Internal storage bits or points can be programmed by the user to perform relay functions without occupying a physical output. In this way internal outputs can minimize output card requirements whenever practical.

An internal control relay can be used when a circuit requires more series contacts than the rung allows. Figure 5-32 shows a circuit that allows for only seven series contacts when twelve are actually required for the programmed logic. To solve this problem, the contacts are split into two rungs. The first rung contains seven of the required contacts and is programmed to an internal relay. The address of the internal relay would also

be the address of the first EXAMINE IF CLOSED contact on the second rung. The remaining five contacts, followed by the discrete output, are programmed. The advantage of an internal storage bit is that it does not waste space in a physical output.

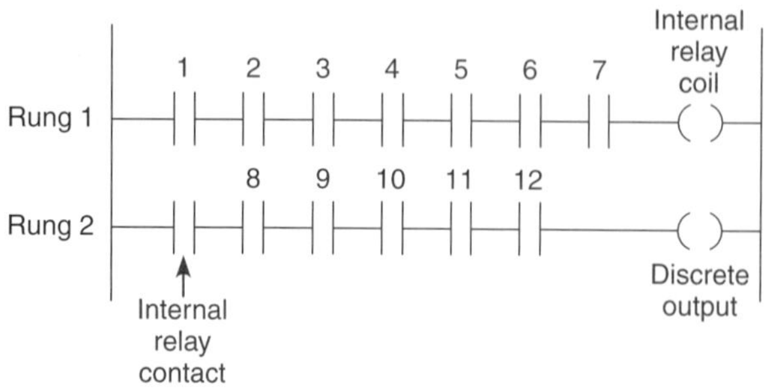

5·8 PROGRAMMING EXAMINE IF CLOSED AND EXAMINE IF OPEN INSTRUCTIONS

A simple program using the EXAMINE IF CLOSED (XIC) instruction is shown in Fig. 5-33. This figure shows a hard-wired circuit and a user program that provides the same results. You will note that *both the NO and the NC* pushbuttons are represented by the EXAMINE IF CLOSED symbol. This is because the normal state of an input (NO or NC) does not matter to the controller. What does matter is that, if contacts need to *close* to energize the output, then the EXAMINE IF CLOSED instruction is used. Since both PB1 and PB2 in Fig. 5-33 must be closed to energize the pilot light, the EXAMINE IF CLOSED instruction is used for both.

A simple program using the EXAMINE IF OPEN (XIO) instruction is shown in Fig. 5-34. Again, both the hard-wired circuit and user program are shown. When the pushbutton is *open* in the hard-wired circuit, relay coil CR is

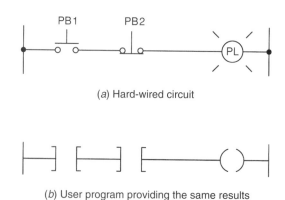

(a) Hard-wired circuit

(b) User program providing the same results

Fig. 5-33

Simple program using the EXAMINE IF CLOSED (XIC) instruction.

Fig. 5-32

Internal control relay.

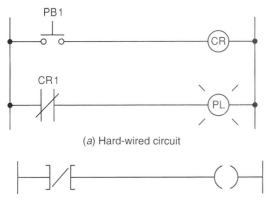

(a) Hard-wired circuit

(b) User program providing the same results

Fig. 5-34

Simple program using the EXAMINE
IF OPEN (XIO) instruction.

de-energized and contacts CR1 close to switch
the pilot light *on.* When the pushbutton is
closed, relay coil CR is energized and contacts
CR1 open to switch the pilot light *off.* The

pushbutton is represented in the user program
by an EXAMINE IF OPEN instruction. This is
because the rung must be TRUE when the
external pushbutton is open, and FALSE
when the pushbutton is closed. Using an
EXAMINE IF OPEN instruction to represent
the pushbutton satisfies these requirements.
The NO or NC mechanical action of the pushbutton is not a consideration. It is important to
remember that the user program is not an electrical circuit but a *logic* circuit. In effect, we
are interested in logic continuity when establishing an output.

Figure 5-35 shows a simple program using
both the XIC and XIO instructions. The logic
states (0 or 1) indicate whether an instruction
is true or false and is the basis of controller
operation. The figure shows the ON/OFF
state of the output as determined by the
changing states of the inputs in the rung.

	The status of the instruction is		
If the data table bit is	XIC EXAMINE IF CLOSED ⊣ ⊢	XIO EXAMINE IF OPEN ⊣/⊢	OTE OUTPUT ENERGIZE ─( )─
Logic 0	False	True	False
Logic 1	True	False	True

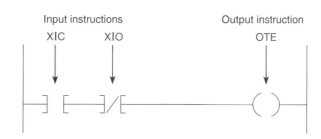

	Inputs		Output		Bit status	
Time	XIC	XIO	OTE	XIC	XIO	OTE
t_1 (initial)	False	True	False	0	0	0
t_2	True	True	Goes true	1	0	1
t_3	True	False	Goes false	1	1	0
t_4	False	False	Remains false	0	1	0

Fig. 5-35
Simple program using both the XIC and XIO instructions.

5.9 ENTERING THE LADDER DIAGRAM

Entering the ladder diagram or actual programming is usually accomplished using a computer keyboard or hand-held programming device. Because hardware and programming techniques vary with each manufacturer, it is necessary to refer to the programming manual for a specific PLC to determine how the instructions are entered.

One method of entering a program is through a hand-held keyboard. Keyboards usually have relay symbol and special function keys, along with numeric keys for addressing. Some also have alphanumeric (letters and numbers) keys for other special programming functions. In hand-held units, the keyboard is small and the keys have multiple functions. Multiple-function keys work like second-function keys on calculators.

A personal computer is most often used today as the programmer. The computer is adapted to the particular PLC model using the relevant programmable controller software (diskette set). A menu system is used instead of dual function keys. Effective menu-guiding on the screen leads the programmer from one programming step to the next (Fig. 5-36a).

With MS-DOS programming software, the ladder diagram is entered by pushing keys on the keyboard in a prescribed sequence. The results are displayed on either the monitor of a computer programmer or with an LED or LCD for a hand-held programmer. Entering the program using Windows based programming software, such as the RSLogix family of ladder logic programming packages (Figure 5-36b), is much quicker than the older DOS programming packages. When entering instructions using Windows based programming, you simply click, drag, and place, the instructions on the desired rungs.

Relay ladder logic is a graphical programming language designed to closely represent the appearance of a wired relay system. It offers considerable advantages for PLC control. Not only is it reasonably intuitive, especially for technicians with relay experience, it is particularly effective in an on-line mode when the PLC is actually performing control. Operation of the logic is apparent from the highlighting of the various relay contacts and coils on screen, which identifies the logic state in real time (Fig. 5-37). An intensified instruction indicates the instruction is true and has logic continuity.

The programming manual for the PLC that you are using should be consulted before attempting any programming. The proper

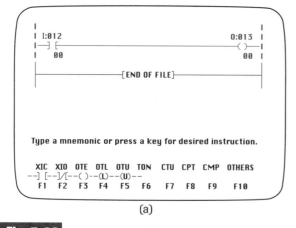

(a)

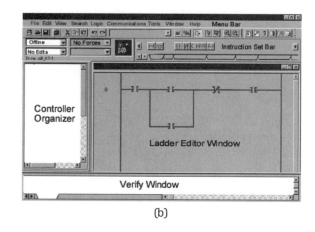

(b)

Fig. 5-36

(a) On-screen menu leads the programmer from one programming step to another. (b) RSLogix Windows based PLC user interface.

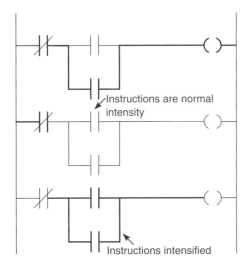

Instructions are normal intensity

Instructions intensified

Monitoring a relay ladder logic diagram.

keystroke sequence is important for entering the program correctly into the controller and is most easily learned through hands-on experience.

For most PLC systems, each EXAMINE IF CLOSED and EXAMINE IF OPEN contact, each output, and each branch START/END instruction requires one word of user memory. When the key for an EXAMINE IF CLOSED contact symbol is pushed and then followed by an address, one full word of user memory is used. Many controllers will display the total number of memory words that have been used on the video display.

5•10 MODES OF OPERATION

The programming device can also be used to select the various processor modes of operation. Again, the number of different operating modes and the method of accessing them varies with the manufacturer. Regardless of PLC model, some common operating modes are CLEAR MEMORY, PROGRAM, TEST, and RUN. A chart describing the purpose of these standard modes of operation is shown in Table 5-1.

Table 5-1

PLC MODES OF OPERATION

Mode	Description
CLEAR MEMORY	Used to erase the contents of the on-board RAM memory.
PROGRAM	Used to enter a new program or update an existing one in the internal RAM memory.
TEST	Used to operate or monitor the user program without energizing any outputs.
RUN	Used to execute the user program. Input devices are monitored and output devices are energized accordingly.

Questions

1. Briefly explain the purpose of the user program portion of a typical PLC memory map.

2. Briefly explain the purpose of the data table portion of a typical PLC memory map.

3.   A. What information is stored in the input image table file?

   B. In what form is this information stored?

4.   A. What information is stored in the output image table file?

   B. In what form is this information stored?

5. Outline the sequence of events involved in a single PLC program scan.

6. Draw the ladder logic program, write the Boolean program, and write the Boolean equation for the relay schematics in Fig. 5-38.

7. Draw the symbol and state the equivalent instruction for each of the following: NO contact, NC contact, and coil.

Fig. 5-38

8.   A. What does an EXAMINE IF CLOSED or EXAMINE IF OPEN instruction represent?

   B. What does an OUTPUT ENERGIZE instruction represent?

   C. The status bit of an EXAMINE IF CLOSED instruction is examined and found to be 0. What does this mean?

   D. The status bit of an EXAMINE IF OPEN instruction is examined and found to be 1. What does this mean?

   E. Under what condition would the status bit of an OUTPUT ENERGIZE instruction be 0?

9.   A. Describe the basic makeup of a ladder logic rung.

   B. How are the contacts and coil of a rung identified?

   C. When is the ladder rung considered TRUE, or as having logic continuity?

10. What two addresses are contained in some five-digit PLC addressing formats?

11. What is the function of an internal control relay?

12. An NO limit switch is to be programmed to control a solenoid. What determines whether an EXAMINE IF CLOSED or EXAMINE IF OPEN contact instruction is used?

13. Briefly describe each of the following modes of operation of PLCs:

a. CLEAR MEMORY   **c.** TEST
b. PROGRAM   **d.** RUN

14. How are data files organized?

15. List 8 different types of data files.

16. Compare the way horizontal and vertical scan patterns examine input and output instructions.

17. Compare the basic concepts of ladder logic, Boolean, and function chart programming.

18. Explain what is meant by a true and false rung condition.

19. Explain the function of input branching.

Problems

1. Assign each of the inputs and outputs shown in the table the correct address based on a typical five-digit addressing.

Inputs			
Device	**Terminal Number**	**Rack Number**	**Module Group Number**
a. Limit switch	1	1	3
b. Pressure switch	2	1	3
c. Pushbutton	3	1	3

Outputs			
Device	**Terminal Number**	**Rack Number**	**Module Group Number**
a. Pilot light	10	1	0
b. Motor starter	11	1	0
c. Solenoid	12	1	0

Basics of PLC Programming

2. Redraw each of the following programs corrected for the problem indicated:

    a. *Problem:* nested programmed contact (Fig. 5-39).

    b. *Problem:* vertical programmed contact (Fig. 5-40).

    c. *Problem:* some logic ignored (Fig. 5-41).

    d. *Problem:* too many series contacts (only four allowed) (Fig. 5-42).

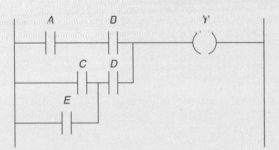

Problem: nested programmed contact

Fig. 5-39

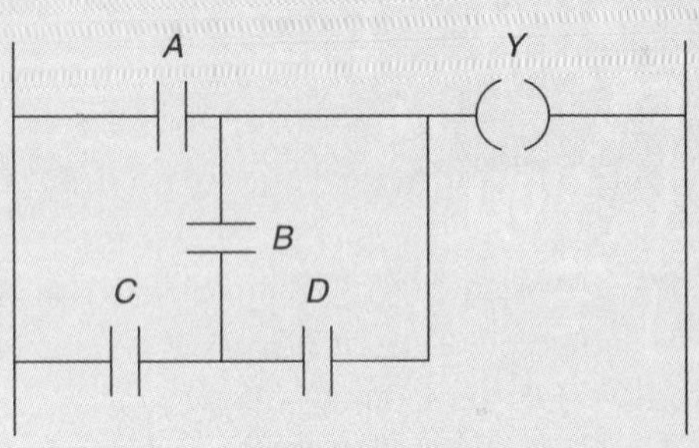

Problem: vertical programmed contact

Fig. 5-40

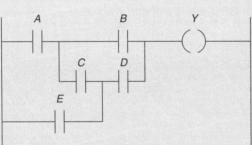

Problem: some logic ignored

Fig. 5-41

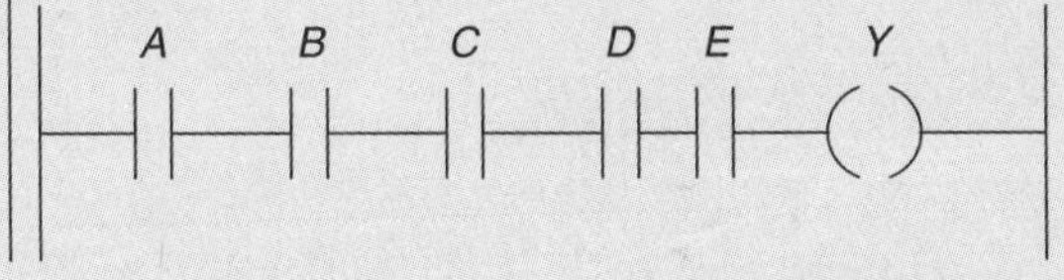

Problem: too many series contacts (only four allowed)

Fig. 5-42

3. Draw the equivalent ladder logic diagram used to implement the hard-wired circuit drawn in Fig. 5-43, wired using:

    a. A limit switch with a single NO contact connected to the PLC input module

    b. A limit switch with a single NC contact connected to the PLC input module

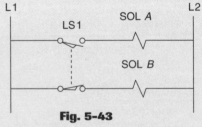

Fig. 5-43

4. Assuming the hard-wired circuit drawn in Fig. 5-44 is to be implemented using a PLC program, identify

    a. All input field devices

    b. All output field devices

    c. All devices that could be programmed using internal relay instructions

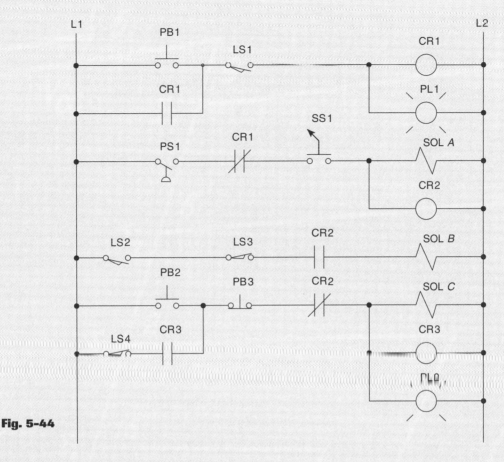

Fig. 5-44

5. What instruction would you select for each of the following input field devices to accomplish the desired task? (State the reason for your answer.)

   a. Turn on a light when a conveyor motor is running in reverse. The input field device is a set of contacts on the conveyor start relay that close when the motor is running forward and open when it is running in reverse.

   b. When a pushbutton is pressed, it operates a solenoid. The input field device is a normally open pushbutton.

   c. Stop a motor from running when a pushbutton is pressed. The input field device is a normally closed pushbutton.

   d. When a limit switch is closed, it triggers an instruction ON. The input field device is a limit switch that stores a 1 in a data table bit when closed.

6. Write the ladder logic program needed to implement each of the following: (assume inputs *A, B,* and *C* are all normally open toggle switches):

   a. When input *A* is closed, turn ON and hold ON outputs *X* and *Y* until *A* opens.

   b. When input *A* is closed and either input *B* or *C* is open, turn ON output *Y;* otherwise, it should be OFF.

   c. When input *A* is closed or open, turn ON output *Y.*

   d. When input *A* is closed, turn ON output *X* and turn OFF output *Y.*

Basics of PLC Programming

6

Developing Fundamental PLC Wiring Diagrams and Ladder Logic Programs

After completing this chapter, you will be able to:

◆ Identify the functions of electromagnetic control relays, contactors, and motor starters

◆ Identify switches commonly found in PLC installations

◆ Explain the operation of sensors commonly found in PLC installations

◆ Explain the operation of output control devices commonly found in PLC installations

◆ Describe the operation of an electromagnetic latching relay and the PLC-programmed LATCH/UNLATCH instruction

◆ Compare sequential and combination control processes

◆ Convert fundamental relay ladder diagrams to PLC ladder logic programs

◆ Write PLC programs directly from a narrative description

For ease of understanding, ladder logic programs can be compared to relay schematics. This chapter gives examples of how traditional relay schematics are converted into PLC ladder logic programs. You will learn more about the wide variety of field devices commonly used in connection with the I/O modules.

6.1 ELECTROMAGNETIC CONTROL RELAYS

As previously stated, the PLC's original purpose was the replacement of electromagnetic relays with a solid-state switching system that could be programmed. Although the PLC has replaced much of the relay control logic, electromagnetic relays are still used as auxiliary devices to switch I/O field devices. The programmable controller is designed to replace the physically small control relays that make logic decisions but are not designed to handle heavy current or high voltage. In addition, an understanding of electromagnetic relay operation and terminology is important for correctly converting relay schematic diagrams to ladder logic diagrams.

An electrical relay is a magnetic switch. It uses electromagnetism to switch contacts. A relay will usually have only one coil but may have any number of different contacts. Figure 6-1 illustrates the operation of a typical control relay. With no current flow through the coil (de-energized), the armature is held away from the core of the coil by spring tension. When the coil is energized, it produces an electromagnetic field. Action of this field, in turn, causes the physical movement of the armature. Movement of the armature causes the contact points of the relay to open or close. The coil and contacts are insulated from each other; therefore, under normal conditions, no electric circuit will exist between them.

The symbol used to represent a control relay is shown in Fig. 6-2. The contacts are represented by a pair of short parallel lines and are identified with the coil by means of the same number and letters (CR1). Both an NO and an NC contact are shown. *Normally open contacts* are defined as those contacts that are *open* when no current flows through the coil but *close* as soon as the coil conducts a current or is energized. *Normally closed contacts* are *closed* when the coil is de-energized and *open* when the coil is energized. Each contact is usually drawn as it would appear with the coil de-energized.

A typical control relay used to control two pilot lights is shown in Fig. 6-3. With the switch *open,* coil CR1 is de-energized. The circuit to the green pilot light is completed through NC contact CR1-2, so this light will be on. At the same time the circuit to the red pilot light is opened through NO contact CR1-1, so this light will be off.

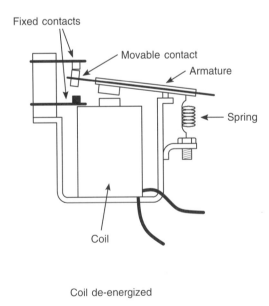
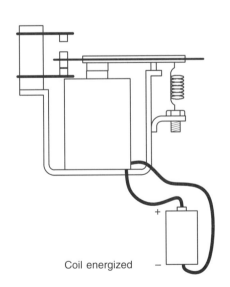

Coil de-energized

Coil energized

Fig. 6-1

Electromagnetic control relay operation.

Coil

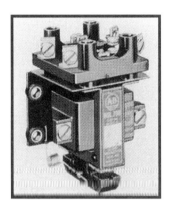

CR1

CR1-1   CR1-2

Normally open (NO) contact   Normally closed (NC) contact

(*a*) Control relay symbol

Fig. 6-2

Control relay.

(*b*) Typical industrial control relay. *(Courtesy of Allen-Bradley Company, Inc.)*

With the switch closed (Fig. 6-4), the coil is energized. The NO contact CR1-1 closes to switch the red pilot light on. At the same time the NC contact CR1-2 opens to switch the green pilot light off.

A relay will usually have only one coil, but it may have any number of normally open and normally closed contacts. Control relays generally do not need to carry heavy currents or high voltages. The contacts are usually rated between 5 and 10 amperes, with the most common rating for the coil voltage being 120 V ac.

6·2 CONTACTORS

A *contactor* is a special type of relay designed to handle heavy power loads that are beyond the capability of control relays. Unlike relays, contactors are designed to make and break electric power circuits

without being damaged. Such loads include lights, heaters, transformers, capacitors, and electric motors for which overload protection is provided separately or not required (Fig. 6-5 on p. 124).

A control relay can pick up a contactor that is built to handle the heavy current and higher voltages. Programmable controllers have I/O capable of operating the contactor, but they do not have the capacity to operate heavy power loads directly. Figure 6-6 on p. 125 illustrates the application of a PLC used in conjunction with a contactor to switch power on and off to a pump. The output module is connected in series with the coil to form a low-current switching circuit. The contacts of the contactor are connected in series with the pump motor to form a high-current switching circuit.

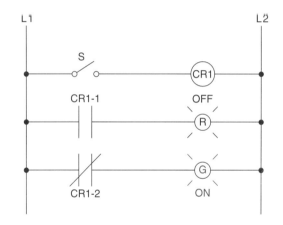

Fig. 6-3

Relay circuit—switch open.

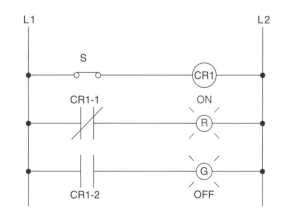

Fig. 6-4

Relay circuit—switch closed.

Developing Fundamental PLC Wiring Diagrams and Ladder Logic Programs

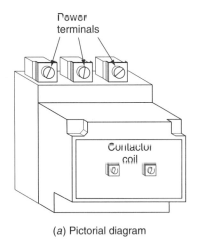

Power terminals

Contactor coil

(a) Pictorial diagram

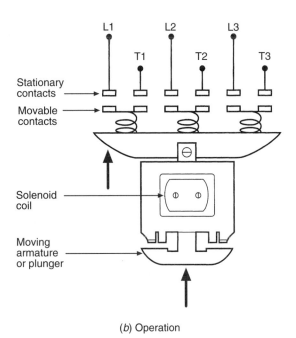

L1   L2   L3

T1   T2   T3

Stationary contacts →

Movable contacts →

Solenoid coil →

Moving armature or plunger →

(b) Operation

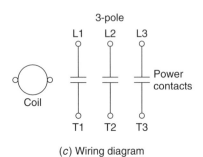

3-pole

L1   L2   L3

Coil

Power contacts

T1   T2   T3

(c) Wiring diagram

Fig. 6-5

Magnetic contactor.

The advantages of using magnetic contactors instead of manually operated control equipment include the following:

◆ Where large currents or high voltages have to be handled, it is difficult to build a suitable manual apparatus. Such an apparatus is large and hard to operate. On the other hand, it is a relatively simple matter to build a magnetic contactor that will handle large currents or high voltages. The manual apparatus must control only the coil of the contactor.

◆ Contactors allow multiple operations to be performed from one operator (one location) and interlocked to prevent false and dangerous operations.

◆ Where the operation must be repeated many times an hour, a distinct savings in effort will result if contactors are used. The operator simply has to push a button and the contactors will automatically initiate the proper sequence of events.

◆ Contactors can be controlled automatically by sensitive pilot devices. Pilot devices of this nature are limited in power and size, and it would be difficult to design them to handle heavy current directly.

◆ High voltage may be handled by the contactor and kept entirely away from the operator, thus increasing the safety of an installation. The operator also will not be in the proximity of high-power arcs, which are always a source of danger from shocks, burns, or perhaps injury to the eyes.

◆ With contactors, the control equipment may be mounted at a remote point. The only space required near the machine will be the space needed for the push-button. It is possible to control one contactor from as many different push-buttons as desired, with the necessity of running only a few light control wires between the stations.

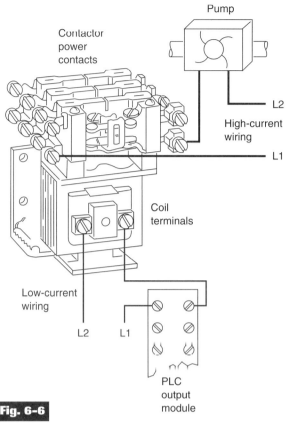

Fig. 6-6

PLC used in conjunction with a contactor to switch power on and off to a pump.

◆ With contactors, automatic and semi-automatic control is possible with equipment such as programmable logic controllers.

6·3 MOTOR STARTERS

A *motor starter* is a relay specially designed to provide power to motors. The magnetic starter (Fig. 6-7) is a contactor with an *overload relay* attached physically and electrically. The overload relay will open the supply voltage to the starter if it detects an overload on a motor. It does this by putting heaters in series with the contactor supplying the voltage to the motor. When these heaters are heated by the current, their heat indirectly heats an element such as a bimetal strip,

which trips a mechanical latch. Tripping this latch opens a set of contacts that are wired in series with the supply to the contactor feeding the motor. The characteristics of the heaters can be matched to the motor so that the motor is protected against overload.

Magnetic motor starters are electromagnetically operated switches that provide a safe method for starting large motor loads. Figure 6-8 on p. 126 shows the wiring diagram for a typical three-phase, magnetically operated, across-the-line ac starter. When the START button is pressed, coil M is energized. When coil M is energized, it closes *all* M contacts. The M contacts in series with the motor close to complete the current path to the motor. These contacts are part of the *power* circuit and must be designed to handle the full load current of the motor. Control contact M (across START button) closes to seal in the coil circuit when the START button is released. This contact is part of the *control* circuit and, as such, is required to handle the small amount of current needed to energize the coil. An overload (OL) relay is provided to protect the motor against current overloads. Normally closed relay contact OL opens automatically when an overload current is sensed, to de-energize the M coil and stop the motor.

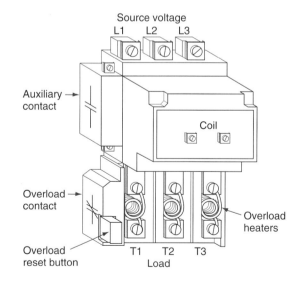

Fig. 6-7

Magnetic starter.

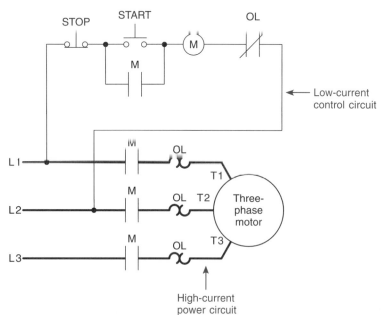

(b) Typical across-the-line AC starter. *(Courtesy of Allen-Bradley Company, Inc.)*

(a)

Across-the-line AC starter.

When a PLC needs to control a large motor, it must work in conjunction with a starter.

Motor starters are available in various standard National Electric Manufacturers Association (NEMA) sizes and ratings. When a PLC needs to control a large motor, it must work in conjunction with a starter. The power requirements to the starter coil must be within the power rating of the output module of the PLC (Fig. 6-9).

6•4 MANUALLY OPERATED SWITCHES

Manually operated switches are controlled by hand. These include toggle switches, pushbutton switches, knife switches, and selector switches.

Pushbutton switches are the most common form of manual control found in industry. Three commonly used pushbutton switches are illustrated by their symbols in Fig. 6-10. The NO pushbutton makes a circuit when it is pressed and returns to its open position when the button is released. The NC pushbutton opens the circuit when it is pressed and returns to the closed position when the button is released. The break-make pushbutton is used for *interlocking* controls. In this switch the top section is NC, while the

Normally open (NO) pushbutton

Normally closed (NC) pushbutton

Break-make pushbutton

Note: The abbreviations NO and NC represent the electrical state of the switch contacts when the switch is not actuated.

(*a*) Pushbutton switches

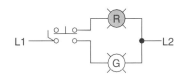

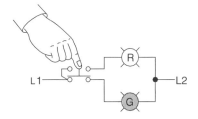

(*b*) Control circuit using a combination break-make pushbutton

(*c*) Typical pushbutton control station. *(Courtesy of Allen-Bradley Company, Inc.)*

Selector switch

Illuminated pushbutton

Momentary contact pushbutton

Fig. 6-10

Various types of pushbutton symbols and switches.

bottom section is NO. When the button is pressed, the bottom contacts are closed as the top contacts open. Also available are *make-before-break* pushbuttons. When the button is pressed with this type of switch, the normally open contacts close before the normally closed contacts open.

The *selector switch* is another common manually operated switch. Selector switch positions are made by turning the operator knob—not pushing it. Selector switches may have two or more selector positions with either maintained contact position or spring return to give momentary contact operation (Fig. 6-11).

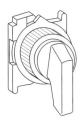

(*a*) Selector switch operator

		Contacts	
Position		A	B
Hand		X	
Off			
Auto			X

(*b*) Three-position selector switch and truth table

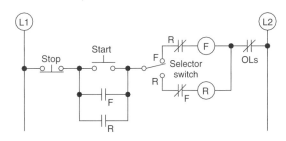

(*c*) Selector switch used in conjunction with a reversing motor starter to select forward or reverse operation of the motor

Fig. 6-11

Selector switch.

Developing Fundamental PLC Wiring Diagrams and Ladder Logic Programs

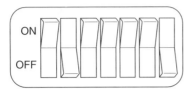

(a) Rocker type

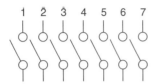

(b) Switching arrangement

Fig. 6-12

DIP switch.

Dual in-line package (DIP) switches are small switch assemblies designed for mounting on printed circuit board modules (Fig. 6-12). The pins or terminals on the bottom of the DIP switch are the same size and spacing as an integrated circuit (IC) chip. The individual switches may be of the toggle, rocker, or slide kind. Switch settings are seldom changed, and the changes occur mainly during installation or configuration of the system.

6•5 MECHANICALLY OPERATED SWITCHES

A *mechanically operated switch* is controlled automatically by factors such as pressure, position, or temperature. The *limit switch,* shown in Fig. 6-13, is a very common industrial control device. Limit switches are designed to operate only when a predetermined limit is reached, and they are usually actuated by contact with an object such as a cam. These devices take the place of a human operator. They are often used in the control circuits of machine processes to govern the starting, stopping, or reversal of motors.

The *temperature switch,* or *thermostat,* shown in Fig. 6-14 is used to sense temperature

Symbols

NO contact

NC contact

(a) Limit switches. *(Photo courtesy of EATON Corporation, Cutler-Hammer Products)*

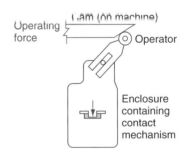

(b) Operation

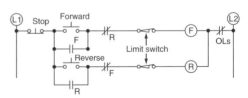

(c) Control circuit for starting and stopping a motor in forward and reverse with limit switches providing over travel protection

Fig. 6-13

Limit switch.

changes. Although there are many types available, they are all actuated by some specific environmental temperature change. Temperature switches open or close when a designated temperature is reached. Industrial applications for these devices include maintaining the desired temperature range of air, gases, liquids, or solids.

The temperature switch is similar to a pressure switch except that a closed, chemically filled bellows system is used. The pressure in the system changes in proportion to the temperature of the bulb. The temperature-responsive medium in the system is a volatile liquid whose vapor pressure increases as the temperature of the bulb rises. Conversely, as the temperature of the bulb falls, the vapor pressure decreases. The pressure change is transmitted to the bellows through a capillary tube operating a precision switch at a predetermined setting.

Symbols

NO contact

NC contact

(a) Temperature switch. *(Photo courtesy of Allen-Bradley Company, Inc.)*

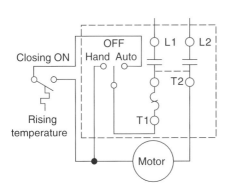

(b) Temperature switch used to automatically control a motor

Symbols

NO contact

NC contact

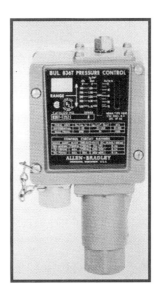

(a) Pressure switch. *(Photo courtesy of Allen-Bradley Company, Inc.)*

Fig. 6-14

Temperature switch.

Fig. 6-15

Pressure switch.

Developing Fundamental PLC Wiring Diagrams and Ladder Logic Programs

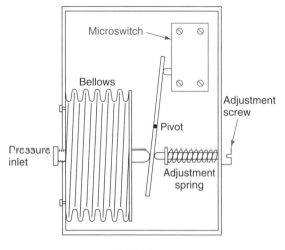

(b) Bellows

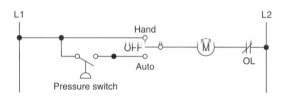

(c) Starter operated by pressure switch

Fig. 6-15 (continued)

Pressure switch.

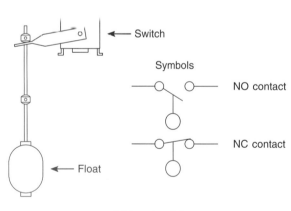

(a) Level switch

(b) Two-wire level switch control of starter

Fig. 6-16

Level switch.

Pressure switches (Fig. 6-15) are used to control the pressure of liquids and gases. Again, although many types are available, they are all basically designed to actuate (open or close) their contacts when a specified pressure is reached. Pressure switches can be pneumatically (air), or hydraulically (liquid) operated switches. Generally a bellows or diaphragm presses up against a small microswitch and causes it to open or close.

Level switches, such as the one illustrated in Fig. 6-16, are used to sense the height of a liquid. The raising or lowering of a float that is mechanically attached to the level switch trips the level switch; the level switch itself is used to control motor-driven pumps that empty or fill tanks. Level switches are also used to open or close piping solenoid valves to control fluids.

6·6 TRANSDUCERS AND SENSORS

A *transducer* is any device that converts energy from one form to another. Transducers may be divided into two classes, *input transducers* and *output transducers* (Fig. 6-17). Electric-input transducers convert nonelectric energy, such as sound or light, into electric energy. Electric-output transducers work in the reverse order. They convert electric energy to forms of nonelectric energy.

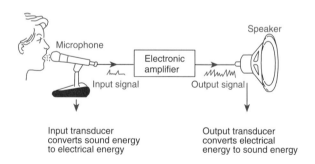

Fig. 6-17

Electric-input and electric-output transducers.

Sensors are transducers for *detecting,* and often *measuring,* the magnitude of something. They convert mechanical, magnetic, thermal, optical, and chemical variations into electric voltages and currents. Sensors are usually categorized by what they measure and play an important role in modern manufacturing process control. They provide the equivalents of our eyes, ears, nose, and tongue to the microprocessor brain of industrial automation systems (Fig. 6-18).

Proximity Sensor

Proximity sensors or switches are pilot devices that detect the presence of an object (usually called the target) without physical contact. They are solid-state electronic devices that are completely encapsulated to protect against excessive vibration, liquids,

chemicals, and corrosive agents found in the industrial environment. Proximity sensors are used when:

◆ The object being detected is too small, lightweight, or soft to operate a mechanical switch.

◆ Rapid response and high switching rates are required, as in counting or ejection control applications.

◆ An object has to be sensed through nonmetallic barriers such as glass, plastic, and paper cartons.

◆ Hostile environments demand improved sealing properties, preventing proper operation of mechanical switches.

◆ Long life and reliable service are required.

◆ A fast-automatic control system requires a bounce-free input signal.

An *inductive proximity sensor* is actuated by a metal object. Inductive proximity sensors are commonly used in the machine tool

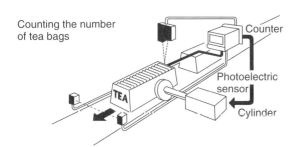

Counting the number of tea bags

Counter

Photoelectric sensor

TEA

Cylinder

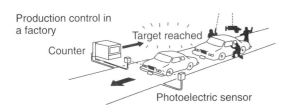

Production control in a factory

Target reached

Counter

Photoelectric sensor

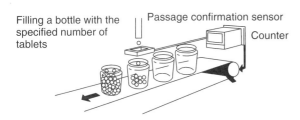

Filling a bottle with the specified number of tablets

Passage confirmation sensor

Counter

(a) Sensors used in manufacturing process control.
(Courtesy of Keyence Corp. of America)

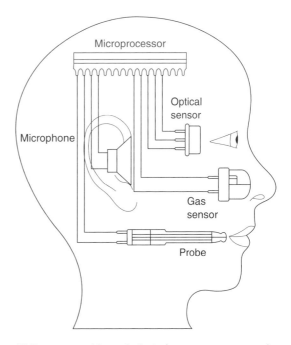

Microprocessor

Optical sensor

Microphone

Gas sensor

Probe

(b) Sensors provide equivalent of eyes, ears, nose, and tongue to microprocessor brain

Fig. 6-18

Sensors.

industry. A typical application is shown in Fig. 6-19*b*. Proximity sensors (A′ and B′) detect targets A and B moving in the directions indicated by the arrows. When A reaches A′, the machine reverses its motion; the machine reverses again when B reaches B′.

In principle, an *inductive sensor consists of* a coil, oscillator, detector circuit, and solid-state output (Fig. 6-20). When energy is supplied, the oscillator operates to generate a high-frequency field. At this moment, there must not be any conductive material in the high-frequency field. When a metal object enters the high-frequency field, eddy currents are induced in the surface of the target. These currents result in a loss of energy in the oscillator circuit, which in turn causes a smaller amplitude of oscillation. The detector circuit recognizes a specific change in amplitude and generates a signal that will turn the solid-state output on or off. When the metal object leaves the sensing area, the oscillator regenerates, allowing the sensor to return to its normal state.

The method of connecting and exciting a proximity sensor varies with the type of sensor and its application (Fig. 6-21). With a *current-sourcing output,* or *PNP* transistor, the load is connected between the sensor and ground. Current flows from the sensor through the load to ground (open emitter). With a *current-sinking output,* or *NPN* transistor, the load is connected between the positive supply and sensor. Current flows from the load through the sensor to ground (open collector).

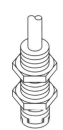

(*a*) Barrel-type physical appearance

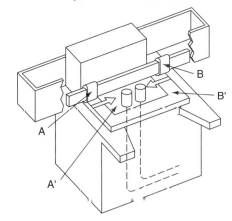

(*b*) Inductive proximity sensor—typical machine tool application

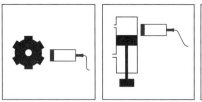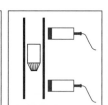

(*c*) Proximity switch applications. *(Courtesy of Rechmer Electronics Industries, Inc.)*

Fig. 6-19

Proximity sensor.

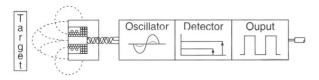

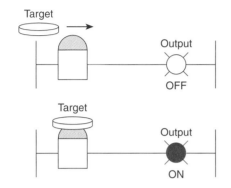

(a) Block diagram

(b) Operation—as the target moves into the sensing area, the sensor switches the output ON

Fig. 6-20

Inductive proximity sensor.

Hysteresis is the distance between the operating point when the target approaches the proximity sensor face and the release point when the target is moving away from the sensor face (Fig. 6-22 on p. 134). The object must be closer to turn the sensor on rather than to turn it off. If the target is moving toward the sensor, it will have to move to a closer point. Once the sensor turns on, it will remain on until the target moves to the release point. Hysteresis is needed to keep proximity sensors from chattering when subjected to shock and vibration, slow-moving targets, or minor disturbances such as electrical noise and temperature drift.

As a result of solid-state switching of the output, a small leakage current flows through the sensor, even when the output is turned off. Similarly, when the sensor is on, a small voltage drop is lost across its output terminals. To operate properly, a proximity sensor should be powered continuously. Figure 6-23 on p. 134 illustrates the use of a bleeder resistor connected to allow enough current for the sensor to operate but not enough to turn on the input of the PLC.

A *capacitive proximity sensor* is a sensing device actuated by conductive and nonconductive materials. The operation of *capacitive sensors* is also based on the principle of an oscillator. Instead of a coil, however, the active face of a capacitive sensor is formed by two metallic electrodes—rather like an "opened" capacitor. The electrodes (Fig. 6-24a on p. 134) are placed in the feedback

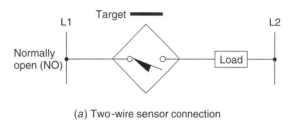

(a) Two-wire sensor connection

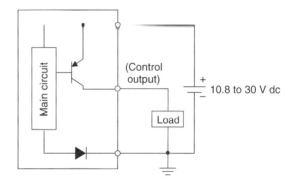

(b) Current-sourcing output (PNP)

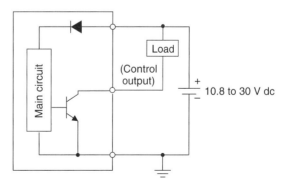

(c) Current-sinking output (NPN)

Fig. 6-21

Proximity sensor connections.

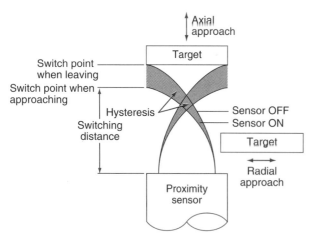

Axial approach

Target

Switch point when leaving

Switch point when approaching

Hysteresis

Sensor OFF
Sensor ON

Switching distance

Target

Radial approach

Proximity sensor

(*a*) Direction and distance are important sensing considerations

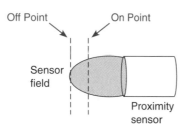

Off Point   On Point

Sensor field

Proximity sensor

(*b*) The difference between the ON point and OFF point is called hysteresis

Fig. 6-22
Hysteresis.

loop of a high-frequency oscillator that is inactive with "*no* target present." As the target approaches the face of the sensor, it enters the electrostatic field formed by the electrodes. This approach causes an increase in the coupling capacitance, and the circuit begins to oscillate. The amplitude of these oscillations is measured by an evaluating circuit that generates a signal to turn the solid-state output on or off.

A typical application is shown in Fig. 6-24*b*. Liquid filling a glass or plastic container can be monitored from the outside of the container with a capacitive proximity sensor. In some applications, the empty container is detected by a second sensor, which starts the flow of liquid. The flow is shut off when the level reaches the upper sensor.

To actuate inductive sensors, we need a conductive material. Capacitive sensors

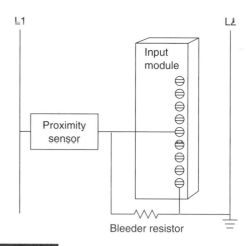

L1   L2

Input module

Proximity sensor

Bleeder resistor

Fig. 6-23
Connection of a proximity sensor to the input module of a PLC.

may be actuated by both conductive materials and by nonconductive materials such as wood, plastics, liquids, sugar, flour, and wheat. Along with this advantage of the capacitive sensor (compared to the inductive sensor) comes some disadvantages. For example, inductive proximity switches may be actuated only by a metal and are insensitive to humidity, dust, dirt, and the

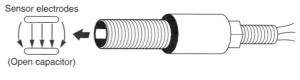

Sensor electrodes

(Open capacitor)

(*a*) Operation

(*b*) Typical application

Fig. 6-24
Capacitive proximity sensor.

like. Capacitive proximity switches, however, can be actuated by any dirt in their environment. For general applications, the capacitive proximity switches are not really an alternative but a *supplement* to the inductive proximity switches. They are a supplement when there is no metal available for the actuation (e.g., for woodworking machines and for determining the exact level of liquids or powders).

Magnetic Switches

A *magnetic switch* (also called a *reed switch*) contact is composed of two flat contact tabs that are hermetically sealed (airtight) in a glass tube filled with protective gas. As a permanent magnet approaches, the ends of the overlapped contact tab attract one another and come into contact. As the permanent magnet is moved further away, the contact tab ends are demagnetized and return to their original positions (Fig. 6-25). The magnetic reed switch is almost inertia free. Because the contacts are sealed, they are unaffected by dust, humidity, and fumes; thus, their life expectancy is quite high.

A permanent magnet is the most common actuator for a reed switch. Permanent magnet actuation can be arranged in several ways, dependent upon the switching requirement. The most commonly used arrangements are proximity motion, rotation, and the shielding method (Fig. 6-26 on p. 136). The device can also be actuated by a dc electromagnet. When operated by an electromagnet, it is known as a *reed relay*. Reed relays are faster, and more reliable, and they produce less arcing than conventional electromechanical switches. However, the current-handling capabilities of the reed relay are limited.

Light Sensors

The *photovoltaic cell,* or *solar cell,* is a common light-sensor device that converts light energy directly into electric energy (Fig. 6-27

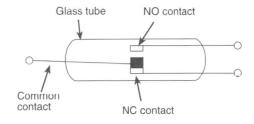

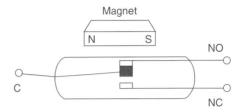

Fig. 6-25
Magnetic switch (reed switch).

on p. 136). Modern silicon solar cells are basically PN junctions with a transparent P-layer. Shining light on the transparent P layer causes a movement of electrons between P- and N-sections, thus producing a small dc *voltage.* The typical output voltage is about 0.5 V per cell in full sunlight.

The *photoconductive cell* (also called a *photoresistive cell*) is another popular type of light transducer (Fig. 6-28 on p. 136). Light energy falling on a photoconductive cell will cause a change in the *resistance* of the cell. One of the more popular types is the *cadmium sulphide photocell.* When the surface of this device is dark, the resistance of the device is high. When brightly lit, its resistance drops to an extremely low value.

Most industrial photoelectric sensors use a light-emitting diode (LED) for the light source and a phototransistor to sense the presence or absence of light (Fig. 6-29 on p. 136). In operation, light from the LED falls on the input of the phototransistor, and the amount of conduction through the transistor changes. *Analog* outputs provide an output proportional to the quantity of light seen by the photodetector.

LEDs are chosen for the light source in photoelectric sensors because they are small, sturdy, and very efficient, and they can be turned on

Developing Fundamental PLC Wiring Diagrams and Ladder Logic Programs

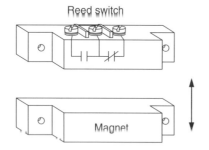

Reed switch

(a) Proximity motion—movement of the switch or magnet will activate the switch

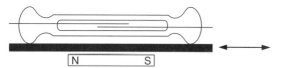

(b) Rotary motion—switch is activated twice for every complete revolution

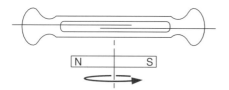

(c) Shielding—ferromagnetic (iron-based) shield short-circuits the magnetic field holding the contacts; switch is activated by removal of the shield

Fig. 6-26

Reed switch activation.

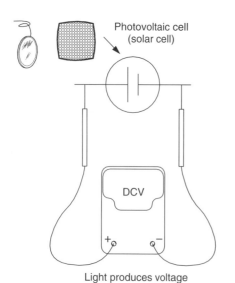

Photovoltaic cell (solar cell)

DCV

+   −

Light produces voltage

Fig. 6-27

Photovoltaic or solar cell.

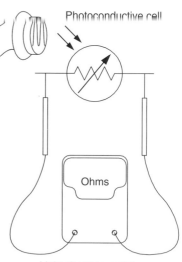

Photoconductive cell

Ohms

Light changes resistance

Fig. 6-28

Photoconductive cell.

and off at extremely high speeds. The LEDs in sensors are switched on and off continually. The on-time is extremely small compared to the off-time. LEDs are pulsed on and off for two reasons: so that the sensor is unaffected by ambient light, and to increase the life of the LED. The pulsed light is sensed by the phototransistor. The phototransistor essentially sorts out all ambient light and looks for the pulsed light. The light sources chosen are typically invisible to the human eye. The wavelengths are chosen so that the sensors are

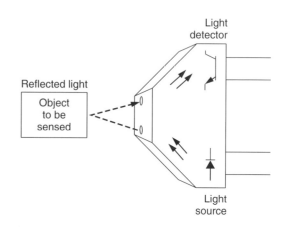

Light detector

Reflected light

Object to be sensed

Light source

Fig. 6-29

Photoelectric sensor operation.

unaffected by other lighting in the plant. The use of different wavelengths allows some sensors, called color mark sensors, to differentiate between colors.

There are two main types of photoelectric sensors for sensing objects. Each emits a light beam (visible, infrared, or laser) from its light-emitting element (Fig. 6-30). A *reflective-type photoelectric sensor* is used to detect the light beam reflected from the target. A *through-beam photoelectric sensor* is used to measure the change in light quantity caused by the target's crossing the optical axis. Typical photoelectric sensing modes include:

◆ **Light Sensing**
The output is energized (on) when the sensor receives the modulated beam.

◆ **Dark Sensing**
The output is energized (on) when the sensor does not receive the modulated beam.

Figure 6-31 on p. 138 shows typical photoelectric sensor applications. Features of this type of sensor include:

◆ **Noncontact Detection**
Noncontact detection eliminates damage either to the target or sensor head, ensuring long service life and maintenance-free operation. You cannot use contact devices in a clean room due to particle generation.

◆ **Detection of Targets of Almost Any Material**
Detection is based on the quantity of light received, or the change in the quantity of light received, or the change in the quantity of reflected light. This method allows detection of targets of diverse materials such as glass, metal, plastics, wood, and liquid.

◆ **Long Detecting Distance**
The reflective-type photoelectric sensor has a detecting distance of 1 m, and the through-beam type has a detecting distance of 10 m.

◆ **High Response Speed**
The photoelectric sensor is capable of a response speed as high as 50 μs (1/20,000 s).

◆ **Color Discrimination**
The sensor has the ability to detect light from an object based on the reflectance and absorption of its color, thus permitting color detection and discrimination.

◆ **Highly Accurate Detection**
A unique optical system and a precision electronic circuit allow highly accurate positioning and detection of minute objects.

Bar code technology is widely implemented in industry and is rapidly increasing its broad range of applications. It is easy to use, can be used to enter data much more quickly than manual methods, and is highly accurate. A bar code system consists of three

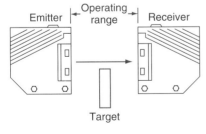

Diffused-reflective

Emitter/receiver | Operating range | Target

Retro-reflective

Emitter/ receiver | Operating range | Reflector

(*a*) Reflective type

Emitter | Operating range | Receiver | Target

(*b*) Through-beam type

Fig. 6-30
Types of photoelectric sensors.

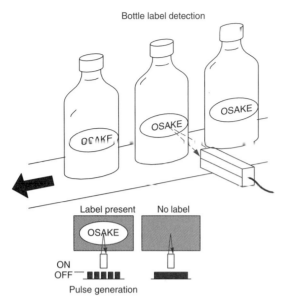

Bottle label detection

Label present   No label

ON
OFF

Pulse generation

(a) A reflective-type photoelectric sensor used for the detection of the presence or absence of a label

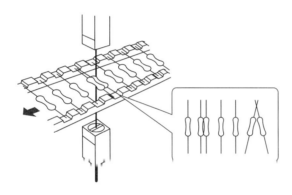

(b) Through-beam optical sensor heads are positioned above and below the resistors traveling on a production line. A variation on the line changes the quantity of the laser beam, thus signaling a defect.

Fig. 6-31

Photoelectric sensor applications.

basic elements: the bar code symbol, a scanner, and a decoder.

The *bar code symbol* contains up to 30 characters encoded in a machine-readable form. The characters are usually printed above or below the bar code so data can be entered manually if a symbol cannot be read by the machine. The blank space on either side of the bar code symbol, called the quiet zone,

along with the start and stop characters lets the scanner know where data begin and end (Fig. 6-32).

Bar code   Quiet zone

100 74983 12345 9
Human readable characters

Fig. 6-32

Bar code symbol.

There are several different kinds of bar codes. In each one, a number, letter, or other character is formed by a certain number of bars and spaces. In the United States, the Universal Product Code (UPC) is the standard bar code symbol for retail food packaging. The UPC symbol (Fig. 6-33) contains all of the encoded information in one symbol. It is strictly a numeric code containing the:

◆ UPC type (1 character)

◆ UPC manufacturer, or vendor, ID number (5 characters)

◆ UPC item number (5 characters)

◆ Check digit (1 character) used to mathematically check the accuracy of the read

0 74983 12345 2
                           Check digit
                        Item number
              Manufacturer number

Fig. 6-33

UPC bar code symbol.

Bar code scanners are the eyes of the data collection system. A light source within the scanner illuminates the bar code symbol; those bars absorb light, and spaces reflect light. A photodetector collects this light in the form of an electronic-signal pattern representing the printed symbol. The *decoder* receives the signal from the scanner and converts these data into the character data representation of the symbol's code. Although the scanner and decoder operate as a team, they can be integrated or separate, depending on the application (Fig. 6-34).

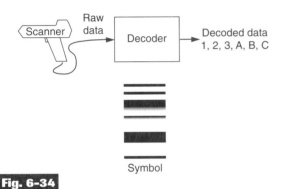

Fig. 6-34

Bar code scanner and decoder.

Bar code modules are available for PLCs. A typical application might involve a bar code module reading the bar code on boxes as they move along a conveyor line. The PLC is then used to divert the boxes to the appropriate product lines (Fig. 6-35).

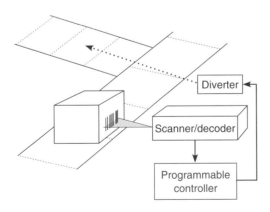

The PLC checks the part number and diverts the product accordingly

Fig. 6-35

Typical PLC bar-code application.

Ultrasonic Sensors

An *ultrasonic sensor* operates by sending sound waves toward the target and measuring the time it takes for the pulses to bounce back. The time taken for this echo to return to the sensor is directly proportional to the distance or height of the object because sound has a constant velocity. In Fig. 6-36, the returning echo signal is electronically converted to a *4-mA to 20-mA output,* which supplies the monitored flow rate to external control devices. Solids, fluids, granular objects, and textiles can be detected by an ultrasonic sensor. The sonic reflectivity of liquid surfaces is the same as solid objects. Textiles and foams absorb the sonic waves and reduce the sensing range.

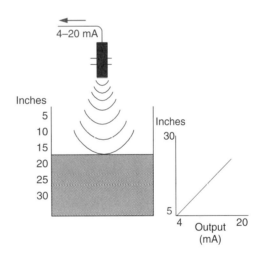

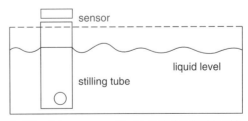

A stilling tube may need to be added to liquids with a choppy surface in order to smooth out the surface for an accurate measurement

Fig. 6-36

Ultrasonic sensor.

Strain/Weight Sensors

A *strain gauge transducer* converts a mechanical strain into an electric signal. Strain gauges are based on the principle that the resistance of a conductor varies with length and cross-sectional area (Fig. 6-37). The force applied to the gauge causes the gauge to bend. This bending action also distorts the physical size of the gauge, which in turn changes its *resistance.* This resistance change is fed to a bridge circuit that detects small changes in the gauge's resistance. *Strain gauge load cells* are usually made with steel and sensitive strain gauges. As the load cell is loaded, the metal elongates or compresses very slightly. The strain gauge detects this movement and translates it to a *varying voltage* signal. Many sizes and shapes of load cells are available and they range in sensitivity from grams to millions of pounds.

Temperature Sensors

There are four basic types of temperature sensors commonly used today: *thermocouple, resistance temperature detector (RTD), thermistor,* and *IC sensor.* Figure 6-38 compares the important features of these devices.

The thermocouple is the most commonly used device for temperature measurement in industrial applications. A *thermocouple* consists essentially of a pair of dissimilar conductors welded or fused together at one end to form the "hot," or measuring, junction, with the free ends available for connection to the "cold," or reference, junction. A temperature difference between the measuring and reference junctions must exist for this device to function as a thermocouple. When this temperature difference occurs, a small dc voltage is generated. Because of their ruggedness and wide temperature range, thermocouples are used in industry to monitor and control oven and furnace temperatures (Fig. 6-39).

Flow Measurement

Many industrial processes depend on accurate measurement of fluid flow. Although there are several ways to measure fluid flow, the usual approach is to convert the kinetic

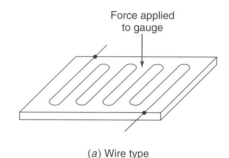

(a) Wire type

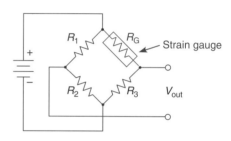

(b) Bridge measuring circuit

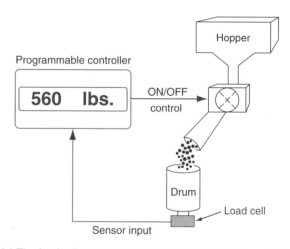

(c) The *load cell* provides sensor input to the controller, which displays the weight and controls the hopper chute

Fig. 6-37

Strain gauge.

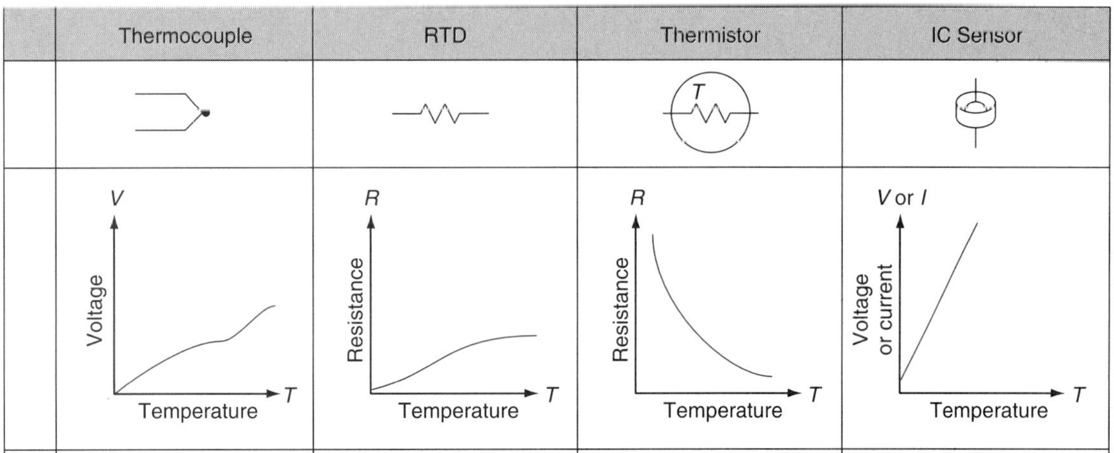

	Thermocouple	RTD	Thermistor	IC Sensor
Advantages	• Self-powered • Simple • Rugged • Inexpensive • Wide variety • Wide temperature range	• Most stable • Most accurate • More linear than thermocouple	• High output • Fast • Two-wire ohms measurement	• Most linear • Highest output • Inexpensive
Disadvantages	• Nonlinear • Low voltage • Reference required • Least stable • Least sensitive	• Expensive • Power supply required • Small ΔR • Low absolute resistance • Self-heating	• Nonlinear • Limited temperature range • Fragile • Power supply required • Self-heating	• $T < 200°C$ • Power supply required • Slow • Self-heating • Limited configurations

Fig. 6-38

Common temperature sensors.

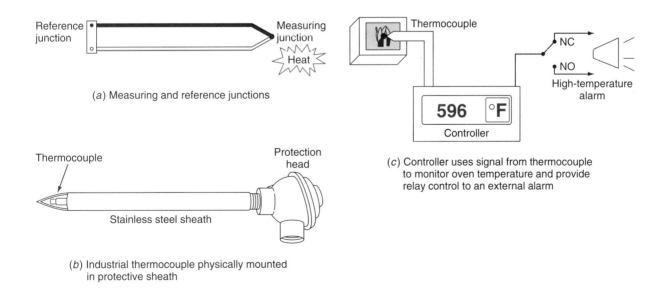

(a) Measuring and reference junctions

(b) Industrial thermocouple physically mounted in protective sheath

(c) Controller uses signal from thermocouple to monitor oven temperature and provide relay control to an external alarm

Fig. 6-39

Thermocouple.

energy that the fluid has into some other measurable form. This conversion can be as simple as connecting a paddle to a potentiometer or as complex as connecting rotating vanes to a pulse-sensing system or tachometer. With the *turbine flowmeter* shown in Fig. 6-40*a*, the turbine blades turn at a rate proportional to the fluid velocity and are magnetized to induce voltage pulses in the coil. Figure 6-40*b* shows an *electronic magnetic flowmeter*, which can be used with electrically conducting fluids and offers *no* restriction to flow. A coil in the unit sets up a magnetic field. If a conductive liquid flows through this magnetic field, a voltage is induced and is sensed by two electrodes.

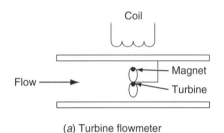

(a) Turbine flowmeter

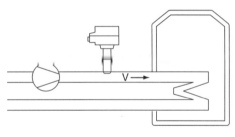

(b) Electronic magnetic flowmeter

Fig. 6-40

Flow measurement.

Velocity/RPM Sensors

The output voltage of a generator varies with the speed at which the generator is driven. A *tachometer* normally refers to a small permanent magnet dc generator. When the generator is rotated, it produces a dc voltage directly proportional to speed. Tachometers coupled to motors are commonly used in motor speed control applications to provide a feedback voltage to the controller that is proportional to motor speed (Fig. 6-41).

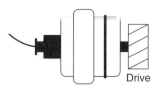

(a) Tachometer is normally a small permanent magnet dc generator

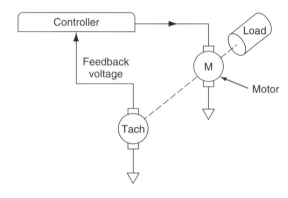

(b) Tachometer provides feedback voltage to the controller that is proportional to motor speed

Fig. 6-41

Tachometer.

The rotating speed of a shaft is often measured using a *magnetic (inductive) pickup sensor.* A magnet is attached to the shaft. A small coil of wire held near the magnet receives a pulse each time the magnet passes. By measuring the frequency of the pulses, the shaft speed can be determined. The voltage output of a typical pickup coil is quite small and requires amplification to be measured (Fig. 6-42).

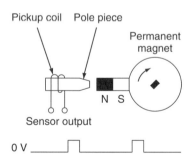

Fig. 6-42

Magnetic pickup sensor.

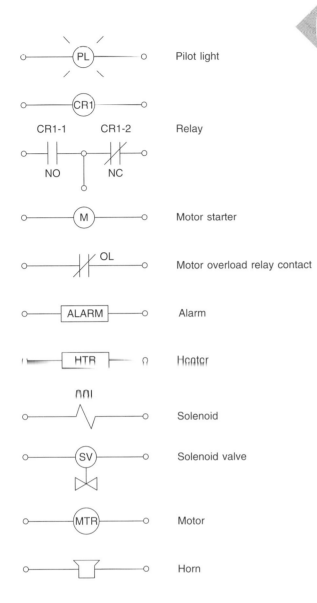

Pilot light

Relay

CR1-1   CR1-2

NO   NC

Motor starter

OL

Motor overload relay contact

ALARM

Alarm

HTR

Heater

Solenoid

SV

Solenoid valve

MTR

Motor

Horn

Fig. 6-43

Symbols for output control devices.

6•7 OUTPUT CONTROL DEVICES

A variety of output control devices can be operated by the controller output module to control traditional industrial processes. These devices include pilot lights, control relays, motor starters, alarms, heaters, solenoids, solenoid valves, small motors, and horns. Electrical symbols are used to represent these devices both on relay schematics and PLC output connection diagrams. For this reason, recognition of the symbols used is important. Figure 6-43 shows common electrical symbols used for various output devices. While these symbols are generally accepted by industry personnel, some differences among manufacturers do exist.

In the electrical sense, an actuator is any device that converts an electrical signal into mechanical movement. The principal types of actuators are relays, solenoids, and motors.

A *solenoid* is a device used to convert an electrical signal or electrical current into linear mechanical motion. As shown in Fig. 6-44, the solenoid is made up of a coil with a movable iron core. When the coil is energized, the core (or armature, as it is sometimes called) is pulled inside the coil. The amount of pulling or pushing force produced by the solenoid is determined by the number of turns of copper wire and the amount of current flowing through the coil.

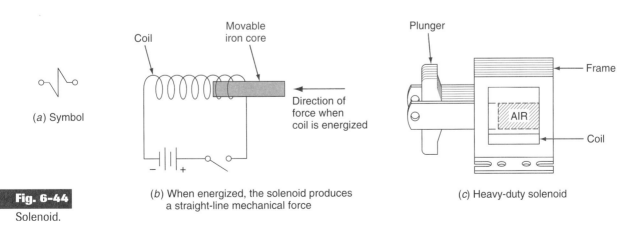

(a) Symbol

Coil

Movable iron core

Direction of force when coil is energized

− +

(b) When energized, the solenoid produces a straight-line mechanical force

Plunger

Frame

AIR

Coil

(c) Heavy-duty solenoid

Fig. 6-44

Solenoid.

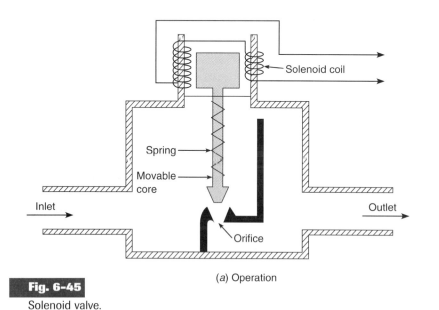

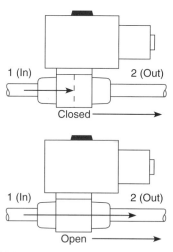

Fig. 6-45
Solenoid valve.

(a) Operation

Valve must be installed with direction of flow in accordance with markings

(b) Solenoid valve installation

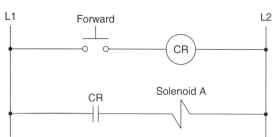

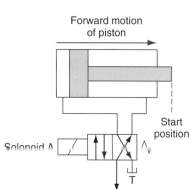

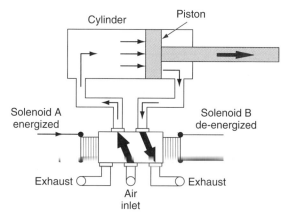

(b) Pneumatic cylinder extends when solenoid A is energized and solenoid B is de-energized

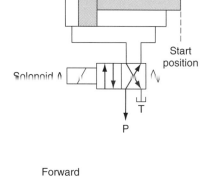

(a) A two-position, single-solenoid valve used to operate a double acting cylinder. When the pushbutton is depressed, the valve spool in the solenoid valve is shifted so that the fluid is directed to the back side of the cylinder piston, causing it to move forward. When the pushbutton is released, the valve spool returns to its original, de-energized position so that the fluid is directed to the front side of the cylinder piston, causing it to move in reverse.

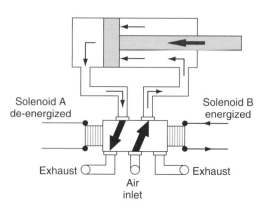

(c) Energizing solenoid B reverses the pressure and exhaust valve positions as shown; cylinder retracts

Fig. 6-46
Typical directional solenoid control valve applications.

A *solenoid valve* is a combination of two basic, functional units:

◆ A solenoid (electromagnet), with its core or plunger

◆ A valve body containing an orifice in which a disc or plug is positioned to restrict or allow flow

Flow through an orifice is stopped or allowed by the movement of the core and depends on whether the solenoid is energized or de-energized. When the coil is energized, the core is drawn into the solenoid coil to *open* the valve. The spring returns the valve to its original *closed* position when the current ceases. Solenoid valves are available to control *hydraulics* (oil fluid), *pneumatics* (air), or water flow (Fig. 6-45).

Directional solenoid valves (commonly called spool valves) start, stop, and control the direction of the flow path. These valves direct the flow by opening and closing flow paths in definite valve positions. They are classified by their number of connections and valve positions. Figure 6-46 illustrates typical directional solenoid control valve applications.

A *stepper motor* converts electrical pulses applied to it into discrete rotor movements called *steps*. A one-degree-per-step motor will require 360 pulses to move through one revolution. Microstep motors, with thousands of steps per revolution, are also available. The rating of the stepper motor is generally given in steps per revolution of the motor. They are generally low speed and low torque, and they provide precise position control of movement.

Figure 6-47 illustrates a typical PLC stepper motor control system. The stepper motor module provides pulse trains to a stepper motor translator, which enables control of a stepper motor. The PLC is free to do other tasks once it communicates with the stepper motor module. This module will send the command to the translator and will not accept a different command until that

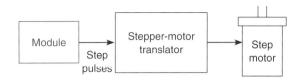

(a) Typical PLC stepper motor control system

(b) The motor will move one step for each pulse received by the driver. The computer provides the desired number of pulses at a specified or programmed rate, which translates into distance and speed.

Fig. 6-47

Stepper motor.

command has been executed. The commands for the module are determined by the control program in the PLC.

6•8 SEAL-IN CIRCUITS

Seal-in, or *hold-in, circuits* are very common in both relay logic and PLC logic. Essentially, a seal-in circuit is a method of maintaining current flow after a momentary switch has been pressed and released. In these types of circuits, the seal-in contact is usually in parallel with the momentary device.

The motor stop/start circuit shown in Fig. 6-48 on p. 146 is a typical example of a seal-in circuit. The hardwired circuit consists of a normally closed stop button in series with a normally open start button. The seal-in auxiliary contact of the starter is connected in parallel with the start button to keep the starter coil energized when the start button is released. When this circuit is programmed into a PLC, both the start and stop buttons are examined for a closed condition because both buttons must be closed to cause the motor starter to operate.

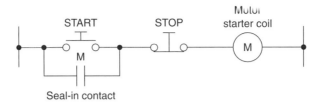

(a) Hard-wired circuit

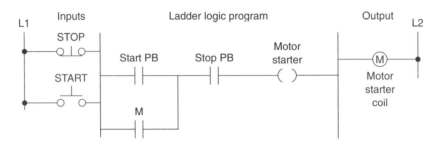

(b) Programmed circuit

Fig. 6-48

Seal-in circuit.

6.9 LATCHING RELAYS

Electromagnetic latching relays are designed to hold the relay closed after power has been removed from the coil. Latching relays are used where it is necessary for contacts to stay open and/or closed, even though the coil is energized only momentarily. Figure 6-49 shows a latching relay that uses two coils. The *latch* coil is momentarily energized to set the latch and hold the relay in the latched position. The *unlatch* or release coil is momentarily energized to disengage the mechanical latch and return the relay to the unlatched position.

Figure 6-50 shows the schematic diagram for an electromagnetic latching-type relay. The contact is shown with the relay in the *unlatched* position. In this state the circuit to the pilot light is open and so the light is off. When the ON button is *momentarily* actuated, the latch coil is energized to set the relay to its latched position. The contacts close, completing the circuit to the pilot light, and so the light is switched on.

Note that the relay coil does *not* have to be continuously energized to hold the contacts closed and keep the light on. The only way to switch the lamp off is to actuate the OFF button, which will energize the unlatch coil and return the contacts to their open, unlatched state. In cases of power loss, the relay will remain in its original latched or unlatched state when power is restored.

An electromagnetic latching relay function can be programmed on a PLC to work like its real-world counterparts. The use of the OUTPUT LATCH and OUTPUT UNLATCH coil instruction is illustrated in the ladder program of Fig. 6-51. Both the latch (L) and the unlatch (U) coil have the *same* address (O:013/10). When the ON button is momentarily actuated, the latch rung becomes true and the latch status bit (10) is set to 1, and so the output is switched on. This status bit *will remain on* when logical continuity of the latch rung is lost. When the unlatch rung becomes true (OFF button actuated), the status bit (10) is reset back to 0 and so the light is switched off.

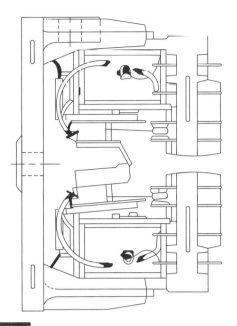

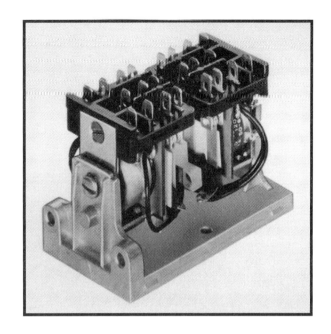

Output latch is an output instruction with a bit-level address. When the instruction is true, it sets a bit in the output image file. It is a retentive instruction because the bit remains set when the *latch* instruction goes false. In most applications it is used with an *unlatch* instruction. The *output latch* instruction is also an output instruction with a bit-level address. When the instruction is true, it resets a bit in the output image file. It, too, is a retentive instruction because the bit remains reset when the instruction goes false.

Instruction	Symbol	Mnemonic
Output latch	–(L)–	OTL
Output unlatch	–(U)–	OTU

$$(L) \quad \frac{XXX}{\text{Latch coil}}$$

$$(U) \quad \frac{XXX}{\text{Unlatch coil}} \quad \text{Same address}$$

(a) Latch and unlatch coils have the same address

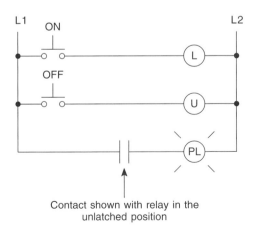

Contact shown with relay in the unlatched position

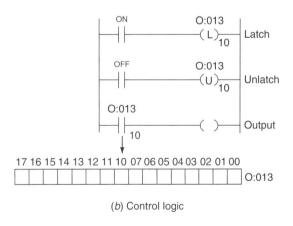

(b) Control logic

Fig. 6-50

Schematic of electromagnetic latching relay.

Fig. 6-51

OUTPUT LATCH and OUTPUT UNLATCH instructions.

Developing Fundamental PLC Wiring Diagrams and Ladder Logic Programs

147

The program of Fig. 6-52 illustrates an application of the output latch and output unlatch instructions. The program is designed to control the level of water in a storage tank by turning a discharge pump on or off. The operation is as follows:

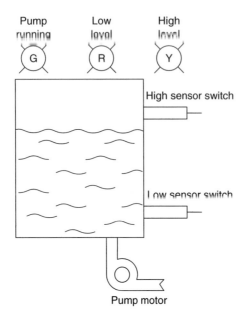

♦ **OFF Position**
The water pump will *stop* if it is running and will *not* start if it is stopped.

♦ **Manual Mode**
The pump will start if the water in the tank is at any level except low.

♦ **Automatic Mode**
- If the level of water in the tank *reaches a high point*, the water pump will *start* so that water can be removed from the tank, thus lowering the level.
- When the water level *reaches a low point*, the pump will *stop*.

♦ **Status Indicating Lights**
- Water pump running light (green)
- Low water level status light (red)
- High water level status light (yellow)

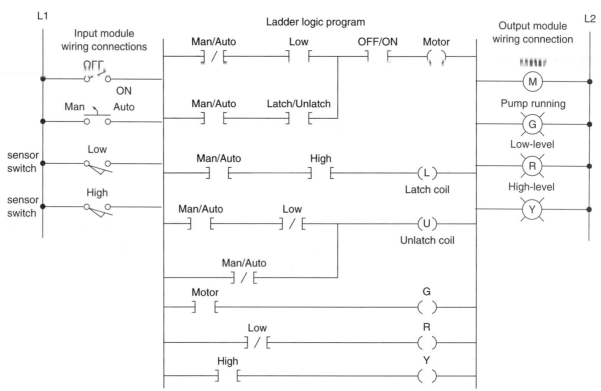

Fig. 6-52
Program designed to control the level of water in a storage tank.

6.10 CONVERTING RELAY SCHEMATICS INTO PLC LADDER PROGRAMS

The best approach to developing a PLC program from a relay schematic is to understand first the operation of each relay ladder rung. As each relay ladder rung is understood, an equivalent PLC rung can be generated. This process will require access to the relay schematic, documentation of the various input and output devices used, and possibly a process flow diagram of the operation.

Most industrial processes require the completion of several operations to produce the required output. Manufacturing, machining, assembling, packaging, finishing, or transporting of products requires the precise coordination of tasks. The majority of industrial control processors use *sequential controls*. Sequential controls are required for processes that demand that certain operations be performed in a specific order. Figure 6-53 illustrates part of a soda-bottling process. In the filling and capping operations, the tasks are (1) fill bottle and (2) press on cap. These tasks must be performed in the proper order. Obviously we could not fill the bottle after the cap is pressed on. This process, therefore, requires sequential control.

Combination controls require that certain operations be performed without regard to the order in which they are performed. Figure 6-54 illustrates another part of the same soda-bottling process. Here, the tasks are (1) place label 1 on bottle and (2) place label 2 on bottle. The order in which the tasks are performed does not really matter. In fact, however, many industrial processes

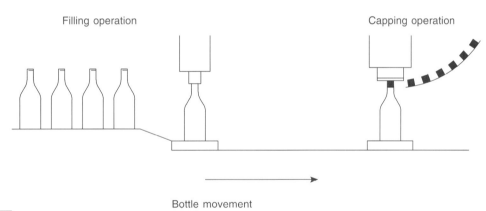

Fig. 6-53

Sequential control process.

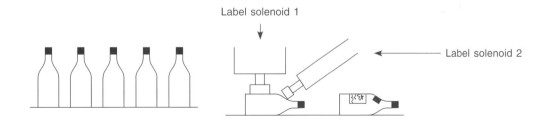

Fig. 6-54

Combination control process.

that are not inherently sequential in nature are performed in a sequential manner for the most efficient order of operations.

The converting of a simple sequential process can be examined with reference to the simple task illustrated in Fig. 6-55. Shown is a process flow diagram along with the relay ladder schematic of its electrical control circuit. The sequential task is as follows:

1. START button is pressed.
2. Table motor is started.

3. Package moves to the position of the limit switch and stops.

Other auxiliary features include:

1. An emergency STOP button that will stop the table, for any reason, before the package reaches the limit switch position
2. A red pilot light to indicate the table is stopped
3. A green pilot light to indicate the table is running

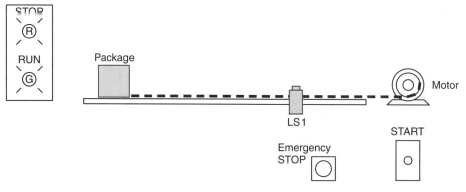

(a) Process flow diagram

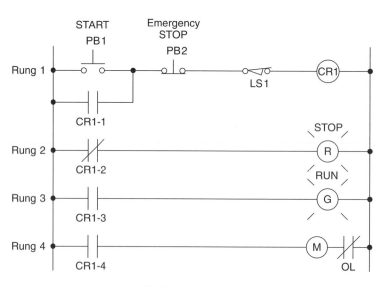

(b) Relay schematic

Fig. 6-55

Sequential process.

A summary of the control task for the process illustrated in Fig. 6-55 could be written as follows:

1. START button is actuated; CR1 is energized if emergency STOP button and limit switch are not actuated.
2. Contact CR1-1 closes, sealing in CR1 even if the START button is released.
3. Contact CR1-2 opens, switching the red pilot light from ON to OFF.
4. Contact CR1-3 closes, switching the green pilot light from OFF to ON.
5. Contact CR1-4 closes to energize the motor starter coil, starting the motor and moving the package toward the limit switch.
6. Limit switch is actuated, de-energizing relay coil CR1.
7. Contact CR1-1 opens, opening the seal-in circuit
8. Contact CR1-2 closes, switching the red pilot light from OFF to ON.
9. Contact CR1-3 opens, switching the green pilot light from ON to OFF.
10. Contact CR1-4 opens, de-energizing the motor starter coil to stop the motor and end the sequence.

At this point, it is wise to make an I/O address chart for the circuit. Each input and output device should be represented along with its address. These addresses will indicate what PLC input is connected to what input device and what PLC output will drive what output device. The address code, of course, will depend on the PLC model used. Figure 6-56a shows a typical I/O address chart that uses SLC-500 addressing for the process. Note that the electromagnetic control relay CR1 is *not* needed because its function is replaced by an *internal* PLC control relay. Symbols are added to the inputs and outputs to relate the program better to real-world field devices.

The four rungs of the relay schematic of Fig. 6-55b can be converted to four rungs of PLC language, as illustrated in Fig. 6-56b. In converting these rungs, the operation of

each rung must be understood. The PB1, PB2, and LS1 references are all programmed using the EXAMINE IF CLOSED (⊣ ⊢) instruction to produce the desired logic control action. Also internal relay is used to replace control relay CR1. To obtain the desired control logic, all internal relay contacts are programmed using the PLC contact instruction that matches their normal state (NO or NC). Whereas the original hard-wired relay CR1 required *four* different contacts, the internal relay uses only *one* contact, which can be examined for an ON or OFF condition as many times as you like. The use of these internal relay equivalents is one of the things that makes the PLC unique.

There is more than one correct way to implement the ladder logic for a given control process. In some cases one arrangement may be more efficient in terms of the amount of memory used and the time required to scan the program. Figure 6-57 on p. 152 illustrates how to arrange instructions for optimum performance.

Field Device	Logical Address	Symbolic Address
Start Button	I:3/0	PB1
Emergency Stop Button	I:3/1	PB2
Limit Switch	I:3/2	LS1
Motor Starter Coil	O:4/1	M
Red - Stop Pilot Light	O:4/2	PL1
Green - Run Pilot Light	O:4/3	PL2

(a) • A logical Address is a specific arrangement of numbers, letters, and punctuation which are used to identify a location in the Data Table.

• A Symbolic Address is a real name or code the programmer can substitute for a logical address because it relates physically to the application. It is a physical name convention for a location in the Data Table.

Fig. 6-56

Process program.

Ladder logic program

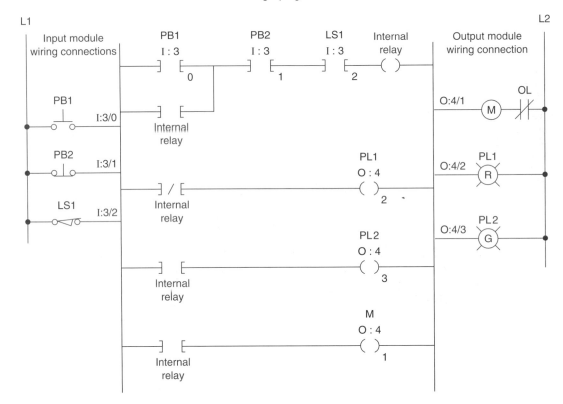

Fig. 6-56 (continued)

Process program.

(b) Control logic

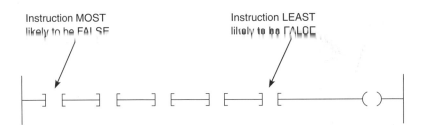

(a) Sequence series instructions from the most likely to be FALSE (far left) to least likely to be FALSE (far right). Once the processor sees a FALSE input instruction in series, it executes the remaining instructions FALSE, even if they are TRUE.

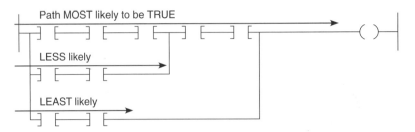

(b) If your rung contains parallel branches, place the path that is most often TRUE on the top. The processor will not look at the others unless the top path is FALSE.

Fig. 6-57

Arranging instructions for optimum performance.

6·11 WRITING A LADDER LOGIC PROGRAM DIRECTLY FROM A NARRATIVE DESCRIPTION

In most cases, it is possible to prepare a ladder diagram directly from the narrative description of a control process. Some of the steps in planning a program are as follows:

◆ Define the process to be controlled.

◆ Draw a sketch of the process, including all sensors and manual controls needed to carry out the control sequence.

◆ List the sequence of operational steps in as much detail as possible.

◇ Write the ladder logic program to be used as a basis for the PLC program

◇ Consider different scenarios where the process sequence may go astray and make adjustments as needed.

◇ Consider the safety of operating personnel and make adjustments as needed.

EXAMPLE 6-1

A simple drilling operation requires the drill press to turn on only if there is a part present and the operator has one hand on each of the start switches. This precaution will ensure that the operator's hands are not in the way of the drill.

Solution

◆ Figure 6-58a is an illustration of the process.

◆ The sequence of operation is as follows:
  Switches 1 and 2 and the part sensor must be activated to make the drill motor operate.

◆ Figure 6-58b shows the ladder logic required for the process.

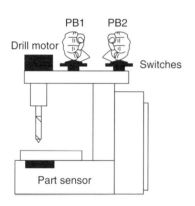

(a) Sketch of the process

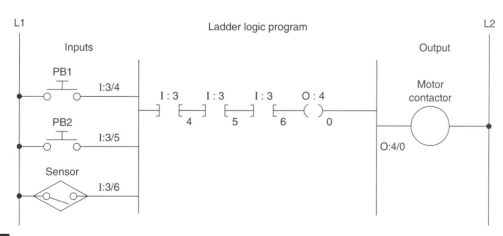

Fig. 6-58
Drilling operation process.

(b) Control logic

EXAMPLE 6-2

A motorized overhead garage door is to be operated automatically to preset open and closed positions. The field devices include one of each of the following.

◆ Reversing *motor contactor* for the up and down directions

◆ Normally *closed down limit switch* to sense when the door is fully closed

◆ Normally closed *up limit switch* to sense when the door is fully opened

◆ Normally open *door up button* for the up direction

◆ Normally open *door down button* for the down direction

◆ Normally closed *door stop button* for stopping the door

◆ Red *door ajar light* to signal when the door is partially open

◆ Green *door open light* to signal when the door is fully open

◆ Yellow *door closed light* to signal when the door is fully closed

Solution:

◆ The sequence of operation is as follows:
  - When the up button is pushed, the up motor contactor energizes and the door travels upward until the up limit switch is actuated.
  - When the down button is pushed, the down motor contactor energizes and the door travels down until the down limit switch is activated.
  - When the stop button is pushed, the motor stops. The motor must be stopped before it can change direction.

◆ Figure 6-59 shows the ladder logic required for the process.

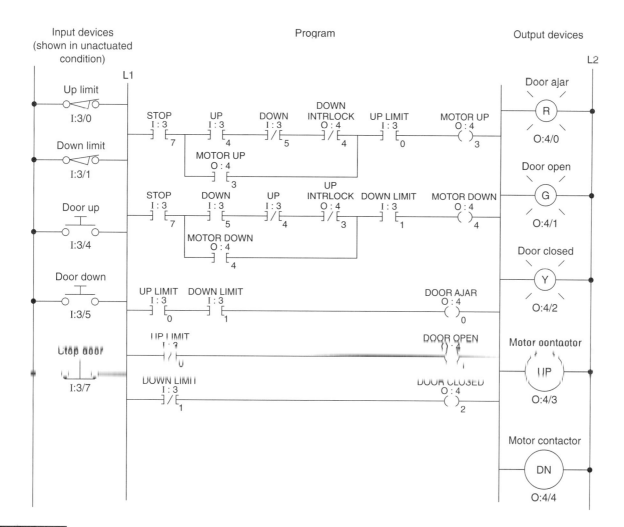

Fig. 6-59

Motorized door PLC program.

EXAMPLE 6-3

A continuous filling operation requires boxes moving on a conveyor to be automatically positioned and filled.

Solution:

◆ Figure 6-60*a* is an illustration of the process.

◆ The sequence of operation is as follows:
- Start the conveyor when the START button is momentarily pressed.
- Stop the conveyor when the STOP button is momentarily pressed.
- Energize the RUN status light when the process is operating.
- Energize the STANDBY status light when the process is stopped.
- Stop the conveyor and energize the STANDBY light when the right edge of the box is first sensed by the photosensor.
- With the box in position and the conveyor stopped, open the solenoid valve and allow the box to fill. Filling should stop when the LEVEL sensor goes true.
- Energize the FULL light when the box is full. The FULL light should remain energized until the box is moved clear of the photosensor.

◆ Figure 6-60*b* shows the ladder logic required for the process.

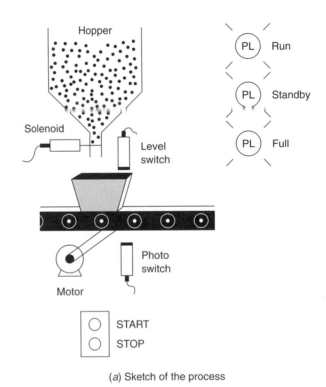

(*a*) Sketch of the process

Fig. 6-60

Continuous filling operation.

Ladder logic program

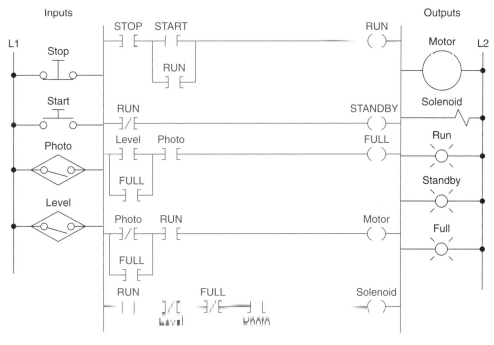

(b) Control logic

Fig. 6-60 (continued)

Continuous filling operation.

Questions

1. **A.** Explain the basic operating principle of an electromagnetic control relay.

    B. Explain the terms *normally open contact* and *normally closed contact* as they apply to this relay.

2. In what way is the construction of a contactor different from that of a control relay?

3. What is the main difference between a contactor and a magnetic motor starter?

4. **A.** Draw the schematic for an across-the-line ac starter.

    B. With reference to this schematic, explain the function of each of the following parts:

    1. Main contact M
    2. Control contact M
    3. Starter coil M
    4. OL relay coils
    5. OL relay contact

    C. Explain the difference between the current requirements for the control circuit and power circuit of the starter.

5. **A.** Compare the method of operation of each of the following types of switches.

    1. Manually operated switch
    2. Mechanically operated switch
    3. Proximity switch

    B. What do the abbreviations NO and NC represent when used to describe switch contacts?

    C. Draw the electrical symbol used to represent each of the following switches:

    1. NO pushbutton
    2. NC pushbutton
    3. Break-make pushbutton
    4. Single-pole selector switch
    5. NO limit switch
    6. NC temperature switch
    7. NO pressure switch
    8. NC level switch
    9. NO proximity switch

6. Compare the methods used to actuate inductive and capacitive proximity sensors.

7. Compare the operation of a photovoltaic cell with that of a photo-conductive cell.

8. What do most industrial photoelectric sensors use for the light source and light sensing device?

9. Compare the operation of the reflective-type and through-beam type photoelectric sensor.

10. Explain how bar code scanners operate to read bar code symbols.

11. Explain how an ultrasonic sensor operates.

12. Explain the principle of operation of a strain gauge.

13. Explain the principle of operation of a thermocouple.

14. Compare the operation of a turbine and electronic magnetic flowmeter.

15. State two types of speed sensors used and explain the basic operation of each.

16. Draw the electrical symbol used to represent each of the following PLC output control devices:

a. Pilot light   **f.** Heater
b. Relay   **g.** Solenoid
c. Motor starter coil   **h.** Solenoid valve
d. OL relay contact   **i.** Motor
c. Alarm   **j.** Horn

17. Explain the function of each of the following actuators:

a. Solenoid   **c.** Directional solenoid valve
b. Solenoid valve   **d.** Stepper motor

18. What is a seal-in circuit?

19. **A.** Draw the schematic of a simple electromagnetic latching relay circuit wired to operate a pilot light.

B. With reference to this circuit, explain how the pilot light is switched on and off.

C. In this circuit, assume that the pilot light was on and power to the circuit is lost. When the power is restored, will the light be on or off? Why?

20. Explain the difference between a sequential and a combination control process.

21. In what ways can different programming arrangements for a given control process be more efficient?

Problems

1. Design and draw the schematic for a conventional hard-wired relay circuit that will perform each of the following circuit functions when an NC pushbutton is pressed:

   - Switch a pilot light on
   - De-energize a solenoid
   - Start a motor running
   - Sound a horn

2. Design and draw the schematic for a conventional hard-wired circuit that will perform the following circuit functions using two break-make pushbuttons:

   - Turns on light L1 when pushbutton PB1 is pressed.
   - Turns on light L2 when pushbutton PB2 is pressed.
   - Electrically interlock the pushbuttons so that L1 and L2 cannot both be turned on at the same time.

3. Study the ladder logic program in Fig. 6-61 and answer the questions that follow:

   a. Under what condition will the latch rung 1 be TRUE?

   b. Under what conditions will the unlatch rung 2 be TRUE?

   c. Under what condition will rung 3 be TRUE?

   d. When PL1 is on, the relay is in what state (LATCHED or UNLATCHED)?

   e. When PL2 is on, the relay is in what state (LATCHED or UNLATCHED)?

   f. If ac power is removed and then restored to the circuit, what pilot light will automatically come on when the power is restored?

   g. Assume the relay is in its LATCHED state and all three inputs are FALSE. What input change(s) must occur for the relay to switch into its UNLATCHED state?

   h. If the EXAMINE IF CLOSED instructions at addresses 001, 002, and 003 are all TRUE, what state will the relay remain in (LATCHED or UNLATCHED)?

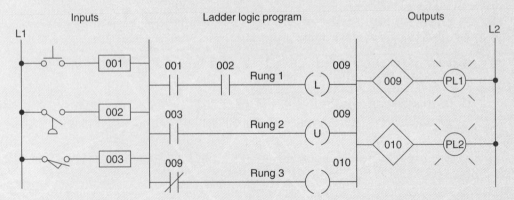

Fig. 6-61

4. Design a PLC program and prepare a typical I/O connection diagram and ladder logic program that will correctly execute the hard-wired control circuit in Fig. 6-62.

5. Design a PLC program and prepare a typical I/O connection diagram and ladder logic program that will correctly execute the hard-wired control circuit in Fig. 6-63.

6. Design a PLC program and prepare a typical I/O connection diagram and ladder logic program that will correctly execute the hard-wired control circuit in Fig. 6-64.

7. Design a PLC program and prepare a typical I/O connection diagram and ladder logic program for the following motor control specifications:

- A motor must be started and stopped from any one of three START/STOP pushbutton stations.
- Each START/STOP station contains one NO START button and one NC STOP button.
- Motor OL contacts are to be hard-wired.

8. Design a PLC program and prepare a typical I/O connection diagram and ladder logic program for the following motor control specifications:

- Three starters are to be wired so that each starter is operated from its own START/STOP pushbutton station.

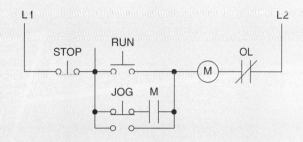

Assume: STOP is wired using an NO pushbutton.
RUN is wired using an NO pushbutton.
JOG is wired using *one* set of NO contacts.
OL is hard-wired.

Fig. 6-62

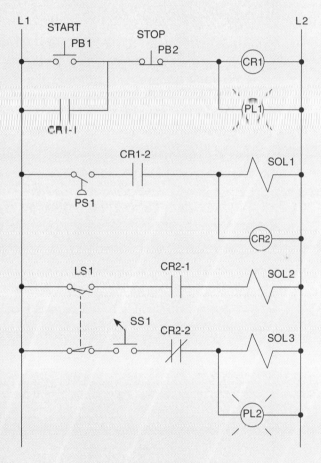

Assume: PB1 and PS1 are wired NO.

PB2 is wired NC.

LS1 is wired using *one* set of NC contacts.

Fig. 6-63

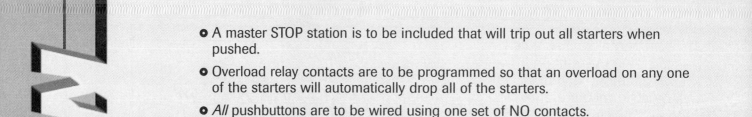

- A master STOP station is to be included that will trip out all starters when pushed.

- Overload relay contacts are to be programmed so that an overload on any one of the starters will automatically drop all of the starters.

- *All* pushbuttons are to be wired using one set of NO contacts.

9. A temperature control system consists of four thermostats controlling three heating units. The thermostat contacts are set to close at 50°, 60°, 70° and 80°F, respectively. The PLC ladder logic program is to be designed so that, at a temperature below 50°F, three heaters are to be ON. Between 50° to 60°F, two heaters are to be ON. For 60° to 70°F, one heater is to be ON. Above 80°F, there is a safety shut-off for all three heaters in case one stays on due to a malfunction. A master switch is to be used to turn the system ON and OFF. Prepare a typical PLC program for this control process.

10. A pump is to be used to fill two storage tanks. The pump is manually started by the operator from a start/stop station. When the first tank is full, the control logic must be able to automatically stop flow to the first tank and direct flow to the second tank through the use of sensors and electric solenoid valves. When the second tank is full, the pump must shut down automatically. Indicator lamps are to be included to signal when each tank is full.

   a. Draw a sketch of the process.
   b. Prepare a typical PLC program for this control process.

11. Write the optimum ladder logic rung for each of the following scenarios and arrange the instructions for optimum performance:

   a. If limit switches LS1 or LS2 or LS3 are on, or if LS5 and LS7 are on, turn on; otherwise, turn off. (Commonly, if LS5 and LS7 are on, the other conditions rarely occur.)

   b. Turn on an output when switches SW6, SW7, and SW8 are all on, or when SW55 is on. (SW55 is an indication of an alarm state, so it is rarely on; SW7 is on most often, then SW8, then SW6.)

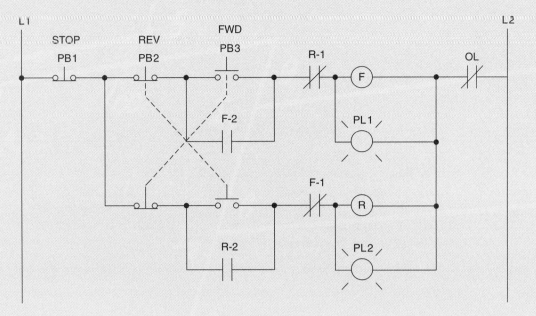

Assume: PB1 is wired NC.

PB2 and PB3 are wired using
one set of contacts

OL is hard-wired.

Fig. 6-64

7

Programming Timers

After completing this chapter, you will be able to:

◆ Describe the operation of pneumatic on-delay and off-delay timers

◆ Describe PLC timer instruction and differentiate between a nonretentive and retentive timer

◆ Convert fundamental timer relay schematic diagrams to PLC ladder logic programs

◆ Analyze and interpret typical PLC timer ladder logic programs

◆ Program the control of outputs using the timer instruction control bits

The most commonly used PLC instruction, after coils and contacts, is the timer. This chapter deals with how timers time intervals and the way in which they can control outputs. The basic PLC on-delay timer function, as well as other timing functions derived from it, will be discussed. Typical industrial timing tasks are also discussed.

7·1 MECHANICAL TIMING RELAY

There are very few industrial control systems that do not need at least one or two timed functions. Mechanical timing relays are used to delay the opening or closing of contacts for circuit control. The operation of a mechanical timing relay is similar to that of a control relay, except that certain of its contacts are designed to operate at a preset time interval, after the coil is energized or de-energized.

Figure 7-1 shows the construction of an on-delay pneumatic (air) timer. The time-delay function depends on the transfer of air through a restricted orifice. The time-delay period is adjusted by positioning the needle valve to vary the amount of orifice restriction. When the coil is energized, the timed contacts are prevented from opening or closing. However, when the coil is de-energized, the timed contacts return instantaneously to their normal state. This particular pneumatic timer has nontimed contacts in addition to timing contacts. These nontimed contacts are controlled directly by the timer coil, as in a general-purpose control relay.

Mechanical timing relays provide time delay through two arrangements. The first arrangement, *on delay* (previously illustrated), provides time delay when the relay is *energized.* The second arrangement, *off delay,* provides time delay when the relay is *de-energized.* Figure 7-2 illustrates the standard relay diagram symbols used for timed contacts.

The circuits of Figs. 7-3, 7-4, 7-5, and 7-6 on pp. 168-169 are designed to illustrate the basic timed-contact functions. In each circuit, the time-delay setting of the timing relay is assumed to be 10 s.

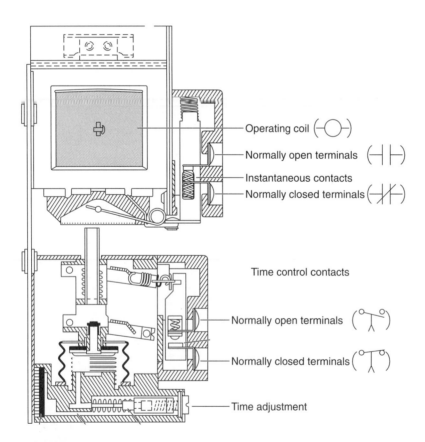

Operating coil
Normally open terminals
Instantaneous contacts
Normally closed terminals

Time control contacts

Normally open terminals

Normally closed terminals

Time adjustment

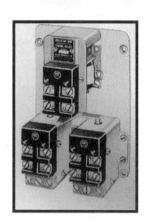

Fig. 7-1

Pneumatic on-delay timer. *(Courtesy of Allen-Bradley Company, Inc.)*

On-delay symbols

Normally open, timed
closed contact (NOTC).

Contact is open when
relay coil is de-energized.

When relay is energized,
there is a time delay in
closing.

Normally closed, timed
open contact (NCTO).

Contact is closed when
relay coil is de-energized.

When relay is energized,
there is a time delay in
opening.

Off-delay symbols

Normally open, timed
open contacts (NOTO).

Contact is normally
open when relay coil
is de-energized.

When relay coil is
energized, contact
closes instantly.

When relay coil is
de-energized, there is
a time delay before the
contact opens.

Normally closed, timed
closed contact (NCTC).

Contact is normally
closed when relay coil
is de-energized.

When relay coil is
energized, contact
opens instantly.

When relay coil is
de-energized, there is
a time delay before the
contact closes.

Fig. 7-2

Timed contact symbols.

7•2 TIMER INSTRUCTIONS

PLC timers are output instructions that provide the same functions as mechanical timing relays. They are used to activate or
de-activate a device after a preset interval of
time. The timer and counter instructions are
the second oldest pair of PLC instructions
beside the standard relay instruction discussed in the previous unit. While earlier,
first-generation PLC systems did not include
these instructions, they are found on all

PLCs manufactured today. The number of
timers that can be programmed depends on
the model of PLC you are using. However,
the availability usually far exceeds the
requirement.

The advantage of PLC timers is that their
settings can be altered easily, or the number
of them used in a circuit can be increased or
decreased, by programming changes without wiring changes. Timer addresses are
usually specified by the programmable controller manufacturer and are located in a
specific area of the data organization table.
Another advantage of the PLC timer is that
its timer accuracy and repeatability are
extremely high since it is based on solid-
state technology.

In general, there are three different timers:
the on-delay timer (TON), off-delay timer
(TOF), and retentive timer on (RTO). The
most common is the on-delay timer, which
is the basic function. There are also many
other timing configurations, all of which can
be derived from one or more of the basic
time delay functions.

There are several quantities associated with
the timer instruction:

◆ The *preset time* represents the time
   duration for the timing circuit. For
   example, if a time delay of 10 s is
   required, the timer will have a preset
   of 10 s.

◆ The *accumulated time* represents the
   amount of time that has elapsed from
   the moment the timing coil became
   energized.

◆ Once the timing rung has continuity, the
   timer counts in time-based intervals and
   times until the preset value and accumulated value are equal or, depending
   on the type of controller, up to the maximum time interval of the timer. The
   intervals that the timers time out at are
   generally referred to as *time bases* of
   the timer. Each timer will have a time

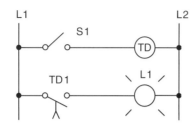

Sequence of operation:
S1 open, TD de-energized, TD1 open, L1 off.

S1 closes, TD energizes, timing period starts, TD1 is still open, L1 is still off.

After 10 s, TD1 closes, L1 is switched on.

S1 is opened, TD de-energizes, TD1 opens instantly, L1 is switched off.

(a)

(b)

Fig. 7-3

On-delay timer circuit (NOTC contact). (a) Operation. (b) Timing diagram.

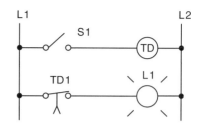

Sequence of operation:
S1 open, TD de-energized, TD1 closed, L1 on.

S1 closes, TD energizes, timing period starts, TD1 is still closed, L1 is still on.

After 10 s, TD1 opens, L1 is switched off.

S1 is opened, TD de-energizes, TD1 closes instantly, L1 is switched on.

(a)

(b)

Fig. 7-4

On-delay timer circuit (NCTO contact). (a) Operation. (b) Timing diagram.

base. Timers can typically be programmed with several different time bases: 1 s, 0.1 s, and 0.01 s are typical time bases. If a programmer entered 0.1 for the time base and 50 for the number of delay increments, the timer would have a 5 s delay (50 × 0.1 s = 5 s).

The timers in a programmable controller are operated by an internally generated clock that originates in the processor module. The majority of programmable controllers have time bases of 1 s and 0.1 s; some have a time base of 0.01 s. The 0.01 s (10-ms) timer is valuable when the PLC is controlling high-speed events or when it is necessary to generate short duration pulses. The 10 ms timer may cause problems when operating in

conjunction with extremely long user programs and scan times. To overcome this problem, the 10 ms timers or other devices that could be affected by a long scan time can be inserted more than once in the program. The additional rungs will ensure that devices are scanned by the processor in a time less than the increment time of the device.

There are two methods used to represent a timer within a PLC's logic ladder program. The first depicts the timer instruction as a relay coil similar to that illustrated in Fig. 7-7 on p. 170. The timer is assigned an address as well as being identified as a timer. Also included as part of the timer instruction is the time base of the timer, the timer's preset value or time-delay period, and the accumulated value or current

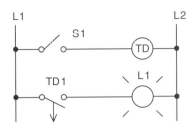

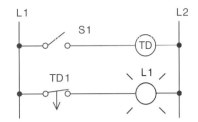

Sequence of operation:
S1 open, TD de-energized, TD1 open, L1 off.

S1 closes, TD energizes, TD1 closes instantly,
L1 is switched on.

S1 is opened, TD de-energizes, timing period starts,
TD1 is still closed, L1 is still on.

After 10 s, TD1 opens, L1 is switched off.

(a)

Sequence of operation:
S1 open, TD de-energized, TD1 closed, L1 on.

S1 closes, TD energizes, TD1 opens instantly,
L1 is switched off.

S1 is opened, TD de-energizes, timing period starts,
TD1 is still open, L1 is still off.

After 10 s, TD1 closes, L1 is switched on.

(a)

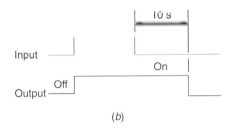

(b)

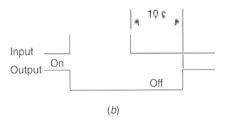

(b)

Fig. 7-5
Off-delay timer circuit (NOTO contact). (a) Operation.
(b) Timing diagram.

Fig. 7-6
Off-delay timer circuit (NCTC contact). (a) Operation.
(b) Timing diagram.

time-delay period for the timer. When the timer rung has logic continuity, the timer begins counting time-based intervals and times until the accumulated value equals the preset value. When the accumulated time equals the preset time, the output is energized and the timed output contact associated with the output is closed. The timed contact can be used as many times as you wish throughout the program as an NO or NC contact.

The second timer format is referred to as a *block format*. Figure 7-8 on p. 170 illustrates a generic block format for a retentive timer that requires two input lines. The timer block has two input conditions associated with it, namely, the *control* and *reset*. The control line controls the actual timing operation of

the timer. Whenever this line is true or power is supplied to this input, the timer will time. Removal of power from the control line input halts the further timing of the timer.

The reset line resets the timer's accumulated value to zero. Some manufacturers require that *both* the control and reset lines be true for the timer to time; removal of power from the reset input resets the timer to zero. Other manufacturers' PLCs require power flow for the control input *only* and no power flow on the reset input for the timer to operate. For this type of timer operation, the timer is reset whenever the reset input is true.

The timer instruction block contains information pertaining to the operation of the

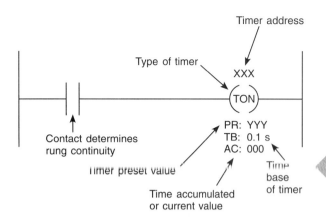

timer, including the preset time, the time base of the timer, and the current or accumulated time. All block-formatted timers provide at least one output signal from the timer. When a single output is provided, it is used to signal the completion of the timing cycle.

7·3 ON-DELAY TIMER INSTRUCTION

Timers are output instructions that you can condition with input instructions such as EXAMINE IF CLOSED and EXAMINE IF

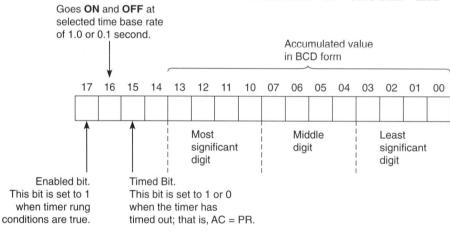

(b)

Fig. 7-7

Coil-formatted timer instruction. (a) Generic instruction.
(b) Allen-Bradley PLC-2 timer accumulated value word.

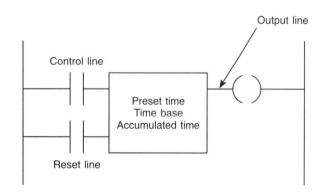

Fig. 7-8

Block-formatted timer instruction.

OPEN. They time intervals as determined by your application program logic. The *on-delay timer* operates so that, when the rung containing the timer is true, the timer time-out period commences. At the end of the timer time-out period, an output is made active, as shown in Fig. 7-9. The timed output becomes true sometime after the timer rung becomes true; hence, the timer is said to have an on delay.

Allen-Bradley PLC-5 and SLC-500 controller timer elements each take three data table words: the control word, preset word, and accumulated word. The *control word* uses three control bits:

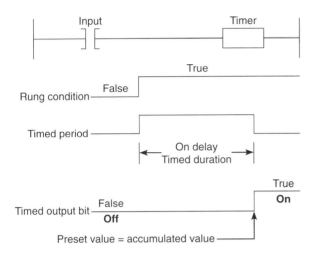

Fig. 7-9

On-delay timer sequence.

Enable (EN) bit

The *enable bit* is true (has a status of 1) whenever the timer instruction is true. When the timer instruction is false, the enable bit is false (has a status of 0).

Timer-timing (TT) bit

The *timer-timing bit* is true whenever the accumulated value of the timer is changing, which means the timer is timing. When the timer is not timing, the accumulated value is not changing, so the timer-timing bit is false.

Done-bit (DN) bit

The *done bit* changes state whenever the accumulated value reaches the preset value. Its state depends on the type of timer being used.

The *preset value (PRE) word* is the set point of the timer, that is, the value up to which the timer will time. The preset word has a range of 0 through 32,767 and is stored in binary form. The preset will not store a negative number.

The *accumulated value (ACC) word* is the value that increments as the timer is timing. The accumulated value will stop incrementing when its value reaches the preset value.

The timer instruction also requires that you enter a *time base*, which is either 1.0 s or 0.01 s. The actual preset time interval is the time base multiplied by the value stored in the timer's preset word. The actual accumulated time interval is the time base multiplied by the value stored in the timer's accumulated word.

Figure 7-10 shows an example of the on-delay timer instruction used as part of the Allen-Bradley PLC-5 and SLC-500 controller instruction sets. The information to be entered includes:

Timer number

This number must come from the timer file. In the example shown, the timer number is T4:0, which represents timer file 4, timer 0 in that file. There may be up to 1000 timers in each timer file, numbered from 0 through 999. The timer address must be unique for this timer and may not be used for any other timer.

Time base

The time base (which is always expressed in seconds) may be either 1.0 s or 0.01 s. In the example shown, the time base is 1.0 s.

Preset value

In the example shown, the preset value is 15. The timer preset value can range from 0 through 32,767.

Accumulated value

In the example shown, the accumulated value is 0. The timer's accumulated

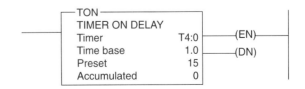

Fig. 7-10

On-delay timer instruction.

value normally is entered as 0, although it is possible to enter a value from 0 through 32,767. Regardless of the value preloaded, the timer will reset to zero whenever it is reset.

The on-delay timer (TON) is the most commonly used timer. Figure 7-11 shows a PLC program that uses an on-delay timer. The timer is activated by closing the switch. The preset time for this timer is 10 s, at which time output D will be energized. When the switch is closed, the timer begins counting and counts until the accumulated time equals the preset value; the output is then energized. If the switch is opened before the timer is timed out, the accumulated time is automatically reset to zero. This timer configuration is termed *nonretentive* because loss of power flow to the timer causes the timer instruction to reset. This timing operation is that of an on-delay timer because output D is switched on 10 s after the switch has been actuated from the off to the on position.

In Fig. 7-11b, the timing diagram first shows the timer timing to 4 s and then going false. The timer resets, and both the timer-timing bit and the enable bit go false. The accumulated value also resets to 0. Input A then goes

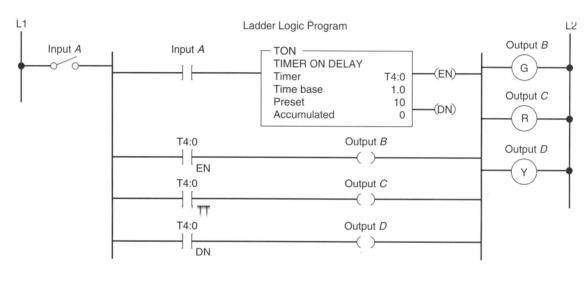

(a) Ladder diagram

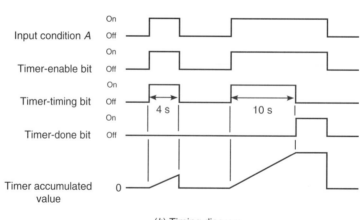

(b) Timing diagram

Fig. 7-11

On-delay timer.

Timer element

15 14 13 12 11 10 9 8 7 6 5 4 3 2 1 0	Word
EN TT DN | Internal use	0
Preset value PRE	1
Accumulated value ACC	2

Addressable bits

EN = Bit 15 enable
TT = Bit 14 timer timing
DN = Bit 13 done

Addressable words

PRE = Preset value
ACC = Accumulated value

(c) Timers are 3-word elements. Word 0 is the control word, word 1 stores the preset value, and word 2 stores the accumulated value (Allen-Bradley PLC-5 and SLC-500 format).

Fig. 7-11 (continued)

On-delay timer.

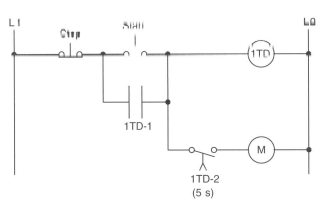

(a) Relay ladder schematic diagram

true again and remains true in excess of 10 s. When the accumulated value reaches 10 s, the done bit (DN) goes from false to true and the timer-timing bit (TT) goes from true to false. When input *A* goes false, the timer instruction goes false and also resets, at which time the control bits are all reset and the accumulated value resets to 0.

Timers may or may not have an instantaneous output (also known as the enable bit) signal associated with them. If an instantaneous output signal is required from a timer and it is not provided as part of the timer instruction, an equivalent instantaneous contact instruction can be programmed using an internally referenced relay coil.

Figure 7-12 shows an application of this technique. According to the relay ladder schematic diagram, coil M is to be energized 5 s after the start pushbutton is pressed. Contact 1TD-1 is the instantaneous contact, and contact 1TD-2 is the timed contact. The ladder logic program shows that a contact instruction referenced to an internal relay is now used to operate the timer. The instantaneous contact is referenced to the internal relay coil, while the time-delay contact is referenced to the timer output coil.

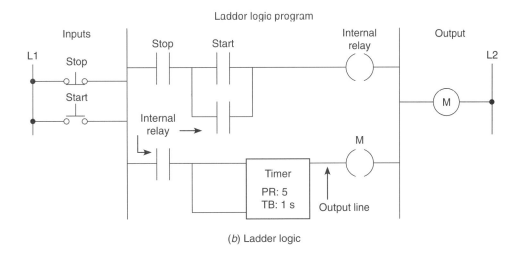

(b) Ladder logic

Fig. 7-12

On-delay timer with instantaneous output programming.

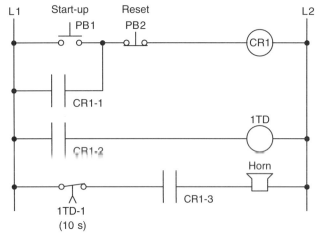

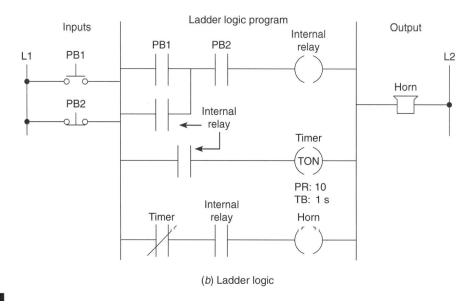

Figure 7-13 shows an application for an on-delay timer that uses an NCTO contact. This circuit is used as a warning signal when moving equipment, such as a conveyor motor, is about to be started. According to the relay ladder schematic diagram, coil CR1 is energized when the start pushbutton PB1 is momentarily actuated. As a result, contact CR1-1 closes to seal in CR1, contact CR1-2 closes to energize timer coil 1TD, and contact CR1-3 closes to sound the horn. After a 10-s time-delay period, timer contact 1TD-1 opens to automatically switch the horn off. The ladder logic program

(a) Relay ladder schematic diagram

(b) Ladder logic

Fig. 7-13

Starting-up warning signal circuit.

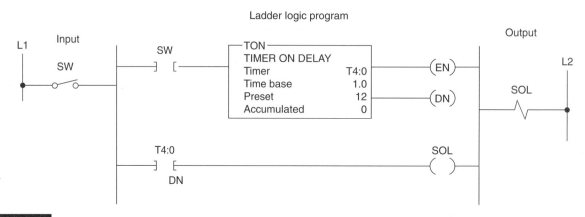

Fig. 7-14

Solenoid valve timed closed.

shows how the circuit could be programmed using a PLC with a coil-formatted timer instruction.

Figure 7-14 shows an application for an on-delay timer that uses a block-formatted timer instruction. This program calls for the solenoid valve to be energized if the switch is closed for 12 s.

Timers are often used as part of automatic sequential control systems. Figure 7-15 shows how a series of motors can be started automatically with only one start/stop control station. According to the relay ladder schematic, lube-oil pump motor starter coil M1 is energized when the start pushbutton PB2 is momentarily actuated. As a result, M1-1 control contact closes to seal in M1, and the lube-oil pump motor starts. When the lube-oil pump builds up sufficient oil pressure, the lube-oil pressure switch PS1 closes. This in turn energizes coil M2 to start the main drive motor and energizes coil 1TD to begin the time-delay period. After the preset time-delay period of 15 s, 1TD-1 contact closes to energize coil M3 and start the feed motor. The ladder logic program shows how the circuit could be programmed using a PLC.

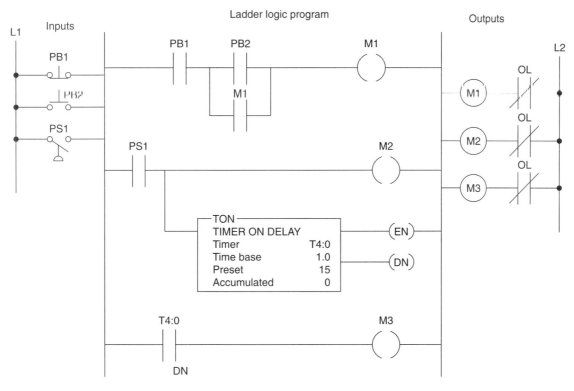

(a) Relay ladder schematic diagram

(b) Ladder logic

Fig. 7-15

Automatic sequential control system.

7·4 OFF-DELAY TIMER INSTRUCTION

The *off-delay timer (TOF)* operation will keep the output energized for a time period after the rung containing the timer has gone false. Figure 7-16 illustrates the generic programming of an off-delay timer that uses the coil-formatted timer instruction. If logic continuity is *lost,* the timer begins counting time-based intervals until the accumulated time equals the programmed preset value. When the switch connected to input 001 is first closed, timed output 009 is set to 1 immediately and the lamp is switched on. If this switch is now opened, logic continuity is lost and the timer begins counting. After 15 s, when the accumulated time equals the preset time, the output is reset to 0 and the lamp switches off. If logic continuity is gained before the timer is timed out, the accumulated time is reset to zero. For this reason, this timer is also classified as nonretentive.

Figure 7-17 illustrates the use of the block-formatted off-delay timer instruction. In this application, closing the switch immediately turns on motors M1, M2, and M3. When the switch is opened, motors M1, M2, and M3 turn off at 5-s intervals.

Figure 7-18 shows how a relay circuit with a pneumatic off-delay timer could be programmed using a PLC. According to the relay schematic diagram, when power is first

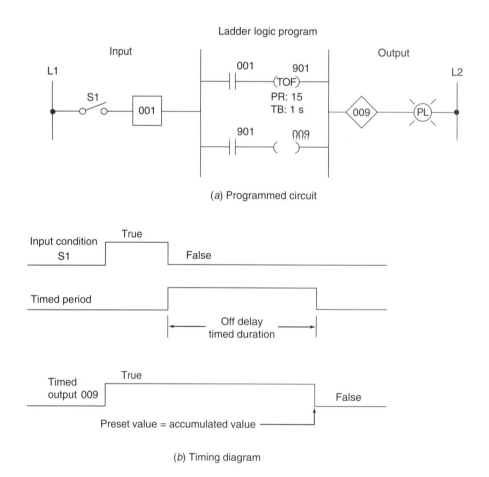

(*a*) Programmed circuit

(*b*) Timing diagram

Fig. 7-16

Off-delay programmed timer.

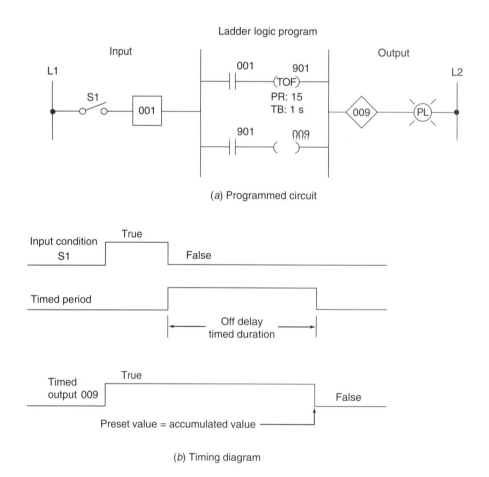

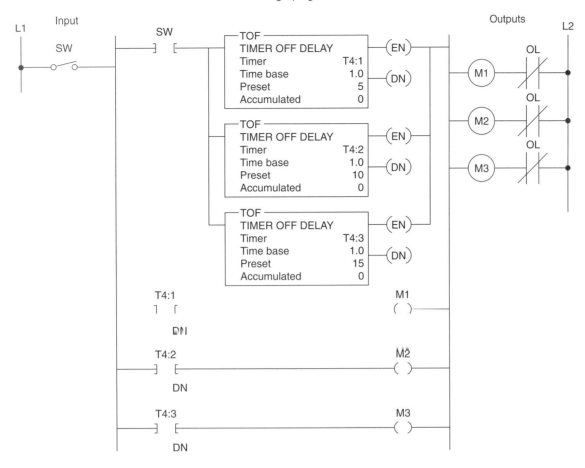

Fig. 7-17

Off-delay timer instructions programmed
to switch motors off at 5-s intervals.

applied (limit switch LS1 open), motor
starter coil M1 is energized and the green
pilot light is on. At the same time, motor
starter coil M2 is de-energized, and the red
pilot light is off.

When limit switch LS1 closes, off-delay timer
coil TD1 energizes. As a result, timed contact
TD1-1 opens to de-energize motor starter coil
M1, timed contact TD1-2 closes to energize
motor starter coil M2, instantaneous contact
TD1-3 opens to switch the green light off, and
instantaneous contact TD1-4 closes to switch
the red light on. The circuit remains in this
state as long as limit switch LS1 is closed.

When limit switch LS1 is opened, the off-
delay timer coil TD1 de-energizes. As a

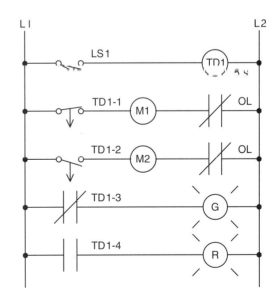

(a) Relay schematic diagram

Fig. 7-18

Programming a pneumatic off-delay timer circuit.

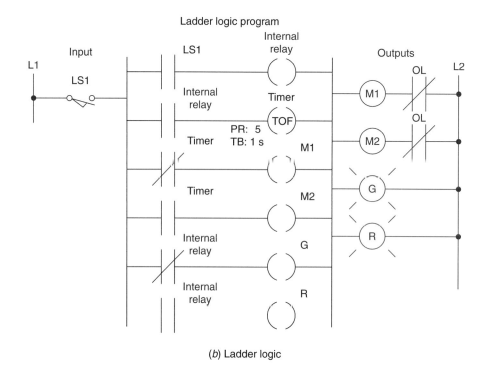

Ladder logic program

(b) Ladder logic

Fig. 7-18 (continued)

Programming a pneumatic off-delay timer circuit.

result, the time-delay period is started, instantaneous contact TD1-3 closes to switch the green light on, and instantaneous contact TD1-4 opens to switch the red light off. After a 5-s time-delay period, timed contact TD1-1 closes to energize motor starter M1, and timed contact TD1-2 opens to de-energize motor starter M2. An internal relay and timer instruction are used to implement the program.

Figure 7-19 shows a program that uses both the on-delay and the off-delay timer instruction. The process involves pumping fluid from tank A to tank B. The operation of the process can be described as follows:

◆ When the start button is pushed, the pump starts. The button can then be released and the pump continues to operate.

◆ When the stop button is pushed, the pump stops.

◆ Before starting, PS1 must be closed.

◆ PS2 and PS3 must be closed 5 s after the pump starts. If either PS2 or PS3 opens, the pump will shut off and will not be able to start again for another 14 s.

7·5 RETENTIVE TIMER

A *retentive timer* accumulates time whenever the device receives power, and maintains the current time should power be removed from the device. Once the device accumulates time equal to its preset value, the contacts of the device change state. Loss of power to the device after reaching its preset value does not affect the state of the contacts. The retentive timer must be *intentionally reset* with a separate signal for the accumulated time to be reset and for the contacts of the device to return to their shelf state.

Figure 7-20 illustrates the action of a motor-driven, electromechanical retentive timer

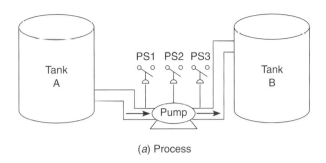

(a) Process

Ladder logic program

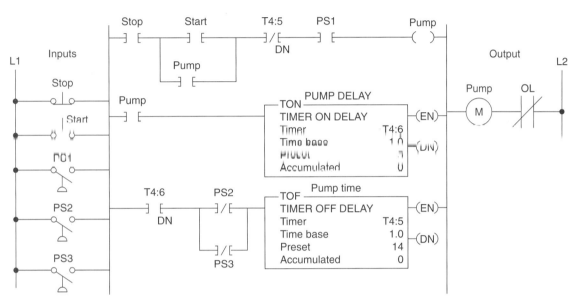

(b) Ladder logic

Fig. 7-19

Fluid pumping process.

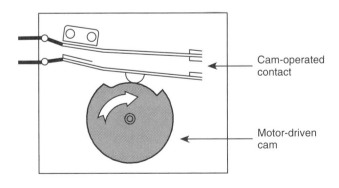

Cam-operated contact

Motor-driven cam

Fig. 7-20

Electromechanical retentive timer.

used in some appliances. The shaft-mounted cam is driven by a motor. Once power is applied, the motor starts turning the shaft and cam. The positioning of the lobes of the cam and the gear reduction of the motor determine the time it takes for the motor to turn the cam far enough to activate the contacts. If power is removed from the motor, the shaft stops but *does not reset.*

The PLC-programmed RETENTIVE ON-DELAY timer (RTO) operates in the same way as the nonretentive on-delay timer (TON), with one major exception. This is a retentive timer reset (RTR) instruction. Unlike the TON, the RTO will hold its accumulated value when the timer rung goes false and will continue timing where it left off when the timer rung goes true again. This timer must be

Programming Timers

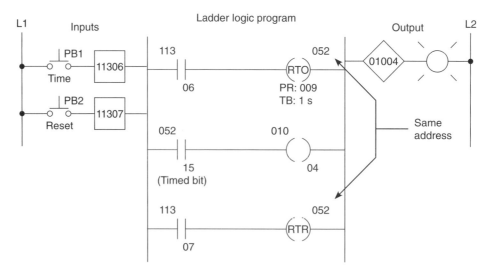

(a) Programmed logic

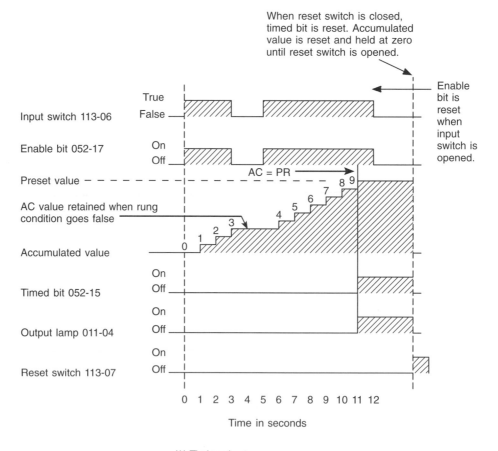

(b) Timing chart

Fig. 7-21

Retentive on-delay timer program and timing chart.

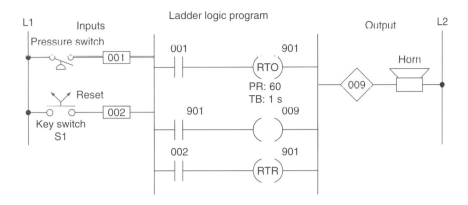

Fig. 7-22

Retentive on-delay alarm program.

accompanied by a timer reset instruction to reset the accumulated value of the timer to zero. The RTR instruction is the *only* automatic means of resetting the accumulated value of a retentive timer. The RTR instruction must be addressed to the same word as the RTO instruction. Normally, if any RTR rung path has logic continuity, then the accumulated value of the referenced timer is reset to zero.

Figure 7-21 shows a typical PLC program for an RTO along with a timing chart for the circuit. The timer will start to time when the NO pushbutton PB1 referenced to input 11306 is closed. If the pushbutton is opened after 3 s, the timer accumulated value stays at 003. When the pushbutton is closed again, the timer picks up the time at 3 s and continues timing. When the accumulated value equals the preset value 000, bit 15 of word 052 is set to 1 and output 01004 and the light are both on.

Since the retentive timer does not reset to zero when the timer is de-energized, reset rung 3 must be used to reset the timer. This rung consists of a NO pushbutton PB2 referenced to input 11307 and an RTR instruction. The RTR instruction is given the same address (052) as the RTO. When pushbutton PB2 closes, RTR resets the accumulated time to zero and changes timed bit 15 of word 052 to 0, turning that light off.

The program drawn in Fig. 7-22 illustrates a practical application for an RTO. The purpose of the RTO timer is to detect whenever

a piping system has sustained a *cumulative* overpressure condition of 60 s. At that point, a horn is sounded automatically to call attention to the malfunction. When they are alerted, maintenance personnel can silence the alarm by switching the key switch S1 to the reset (contact closed) position. After the problem has been corrected, the alarm system can be reactivated by switching the key switch to the on (contact open) position.

Figure 7-23 on p. 182 shows a practical application that uses the on-delay, off-delay, and retentive on-delay instructions in the same program. In this industrial application, there is a machine with a large steel shaft supported by babbitted bearings. This shaft is coupled to a large electric motor. The bearings need lubrication, which is supplied by an oil pump driven by a small electric motor. The sequence of operation is as follows:

◆ To start the machine, the operator turns SW on.

◆ Before the *motor* shaft starts to turn, the bearings are supplied with oil by the *pump* for 10 s.

◆ The bearings also receive oil when the machine is running.

◆ When the operator turns SW off to stop the machine, the oil pump continues to supply oil for 15 s.

◆ A retentive timer is used to track the total running time of the pump. When

Ladder logic program

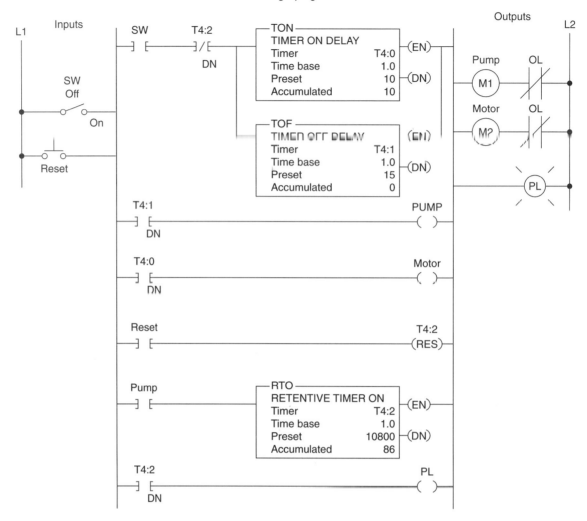

Fig. 7-23

Bearing lubrication program.

the total running time is 3 h, the motor is shut down and a pilot light is turned on to indicate that the filter and oil need to be changed.

◆ A reset button is provided to reset the process after the filter and oil have been changed.

A retentive off-delay timer is programmed in the same manner as an RTO. Both maintain their accumulated time value even if logic continuity is lost before the timer is timed out or if power is lost. These retentive timers do *not* have to be timed out completely to be reset. Rather, such a timer can be reset at any time during

its operation. It should be noted that the reset input to the timer will override the control input of the timer, to reset it even though the control input to the timer has logic continuity.

7·6 CASCADING TIMERS

The programming of two or more timers together is called *cascading*. Timers can be interconnected, or cascaded, to satisfy any required control logic. Figure 7-24 shows how three motors can be started automatically in

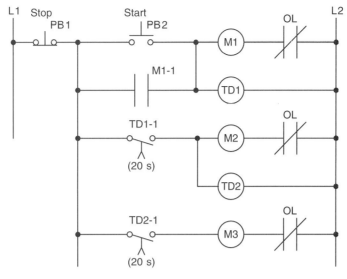

(a) Relay schematic diagram

Ladder logic program

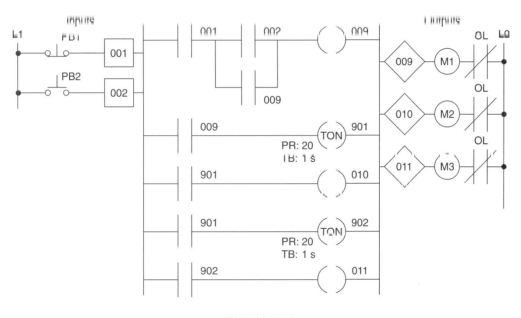

(b) Ladder logic

Fig. 7-24

Sequential time-delayed motor-starting circuit.

sequence with a 20-s time delay between each motor startup. According to the relay schematic diagram, motor starter coil M1 is energized when the start pushbutton PB2 is momentarily actuated. As a result, motor 1 starts, contact M1-1 closes to seal in M1, and timer coil TD1 is energized to begin the first time-delay period. After the preset time period of 20 s, TD1-1 contact closes to energize motor starter coil M2. As a result, motor 2 starts and timer coil TD2 is energized to begin the second time-delay period. After the preset time period of 20 s, TD2-1 contact closes to energize motor starter coil M3, and so motor 3 starts. The ladder logic program shows how the circuit could be programmed using a PLC. Note that two internal timers are used and the output of the first timer is used to control the input logic to the second timer.

Two timers can be interconnected to form an oscillator circuit. The oscillator logic is basically a timing circuit programmed to generate periodic output pulses of any duration. Figure 7-25 shows the program for an annunciator flasher circuit. Two internal timers form the oscillator circuit, which generates a timed, pulsed output. The oscillator circuit output is programmed in series with the alarm condition. If the alarm condition (temperature, pressure, or limit switch) is true, the appropriate output indicating light will flash. Note that any number of alarm conditions could be programmed using the same flasher circuit.

At times you may require a time-delay period longer than the maximum preset time allowed for the single timer instruction of the PLC being used. When this is the case, the problem can be solved by simply cascading timers, as illustrated in Fig. 7-26. The type of timer programmed for this example is a TON. The total time delay required is 1200 s (999 + 201 s, or 20 min). The first timer is programmed for its maximum preset time of 999 s and begins timing when field switch S1 is closed. When it completes its time-delay period, 999 s later, internally referenced contacts 901 will close. This action in turn will activate the second timer, which is preset for the remaining 201 s of the total 1200-s time delay. Once the second timer reaches its preset time, internally referenced contact 902 closes to turn the light on and indicate the completion of the full 20-min time delay. Opening of field switch S1 at any time will reset both timers and switch the light off.

A typical application for PLC timers is the control of traffic lights. The ladder logic circuit of Fig. 7-27 illustrates a simulated control of a set of traffic lights in one direction

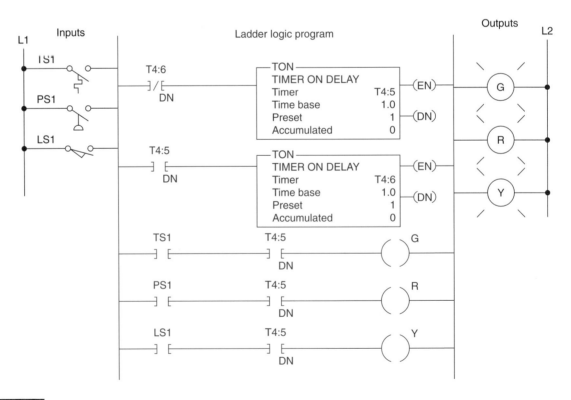

Fig. 7-25

Annunciator flasher program.

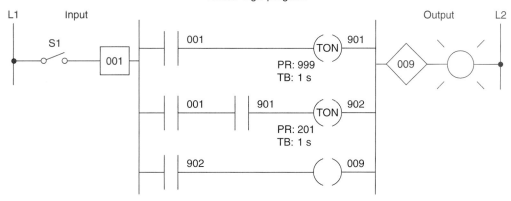

Fig. 7-26

Cascading of timers for longer time delays.

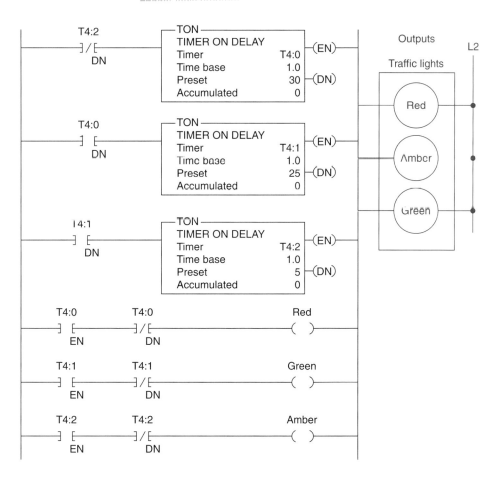

Fig. 7-27

Control of traffic lights in one direction.

Ladder logic program

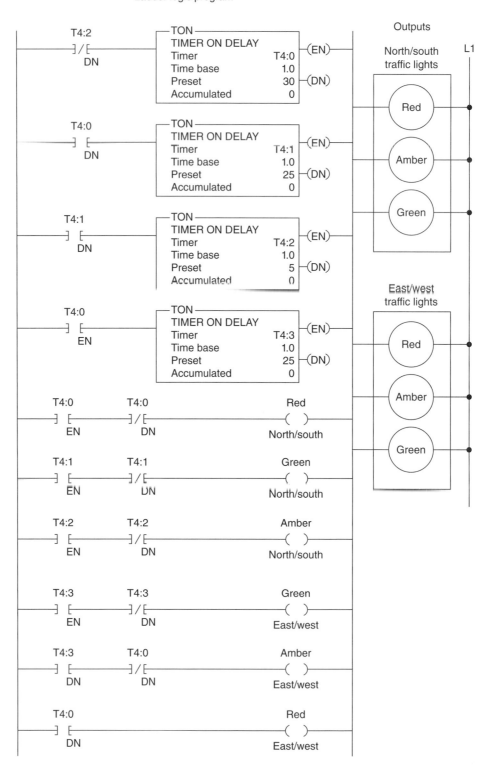

(a) Ladder logic

Fig. 7-28

Control of traffic lights in two directions.

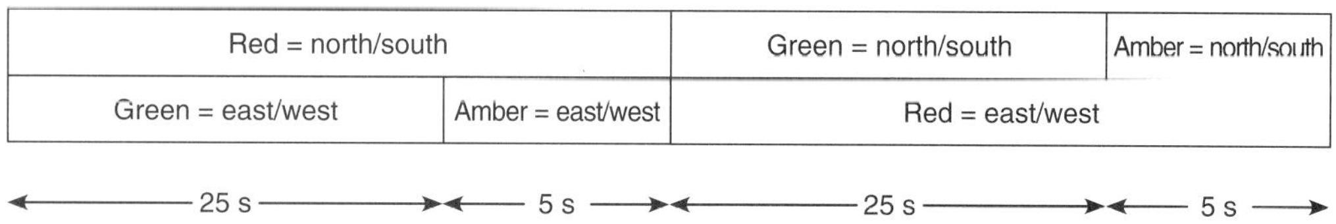

Red = north/south		Green = north/south	Amber = north/south
Green = east/west	Amber = east/west	Red = east/west	

◄─────── 25 s ───────►◄─── 5 s ───►◄─────── 25 s ───────►◄─── 5 s ───►

(b) Timing chart

Fig. 7-28 (continued)

Control of traffic lights in two directions.

only. Transition from red to green to amber is accomplished by a cascading timer circuit. The sequence of operation is:

Red	30 s on
Green	25 s on
Amber	5 s on

The sequence then repeats itself. Figure 7-28 shows the original traffic light program modified to include three more lights that control traffic flow in the other direction.

Questions

1. Explain the difference between the timed and instantaneous contacts of a pneumatic timer.

2. Draw the symbol and explain the operation of each of the following timed contacts of a pneumatic timer:

    a. On-delay timer—NOTC contact
    b. On-delay timer—NCTO contact
    c. Off-delay timer—NOTO contact
    d. Off-delay timer—NCTC contact

3. State five pieces of information usually associated with a PLC timer instruction.

4. When is the output of a programmed timer energized?

5. What are the two methods commonly used to represent a timer within a PLC's ladder logic program?

6. **A.** Explain the difference between the operation of a nonretentive and a retentive timer.

    B. Explain how the accumulated count of programmed retentive and non-retentive timers is reset to zero.

7. State three advantages of using programmed PLC timers.

8. **A.** Name three different types of PLC timers.

    B. Which of the three is most commonly used?

9. Explain what each of the following quantities associated with a PLC timer instruction represents:

    a. Preset time
    b. Accumulated time
    c. Time base

10. **A.** When is the enable bit of a timer instruction true?

    B. When is the timer-timing bit of a timer instruction true?

    C. When does the done bit of a timer change state?

11. State the method used to reset the accumulated time of each of the following:

    a. TON timer
    b. TOF timer
    c. RTO timer

Problems

1. A. With reference to the relay schematic diagram in Fig. 7-29, state the status of each light (on or off) after each of the following sequential events:

   1. Power is first applied and switch S1 is open.
   2. Switch S1 has just closed.
   3. Switch S1 has been closed for 5 s.
   4. Switch S1 has just opened.
   5. Switch S1 has been opened for 5 s.

B. Design a PLC program and prepare a typical I/O connection diagram and ladder logic program that will execute this hard-wired control circuit correctly.

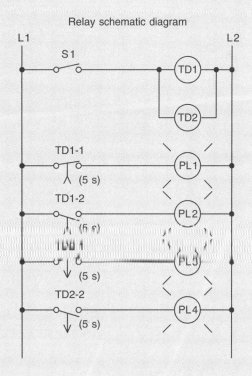

Relay schematic diagram

Fig. 7-29

2. Design a PLC program and prepare a typical I/O connection diagram and ladder logic program (Fig. 7-30) that will execute this hard-wired control circuit correctly.

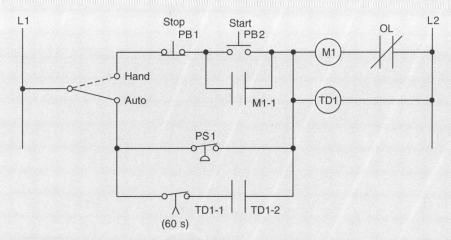

Fig. 7-30

3. Study the ladder logic program in Fig. 7-31 and answer the questions that follow:

a. What type of timer has been programmed?

b. What is the length of the time-delay period?

c. What is the value of the accumulated time when power is first applied?

d. Which address is an internal relay instruction?

e. When does the timer start timing?

f. When does the timer stop timing and reset itself?

g. When input 001 is first turned on, which rungs are true and which are false?

h. When input 001 is first turned on, state the status (on or off) of each output.

i. When the timer's accumulated value equals the preset value, which rungs are true and which are false?

j. When the timer's accumulated value equals the preset value, state the status (on or off) of each output.

k. Suppose that rung 1 is true for 5 s and then power is lost. What will the accumulated value of the counter be when power is restored?

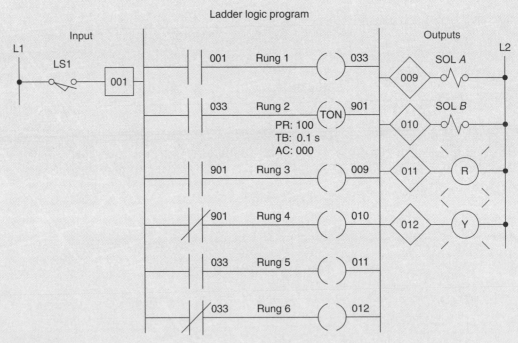

Fig. 7-31

4. Study the ladder logic program in Fig. 7-32 and answer the questions that follow:

    a. What type of timer has been programmed?

    b. What is the length of the time-delay period?

    c. When does the timer start timing?

    d. When is the timer reset?

    e. When will rung 3 be true?

    f. When will rung 5 be true?

    g. When will output 012 be energized?

    h. Assume your accumulated time value is up to 020 and power to your system is lost. What will your accumulated time value be when power is restored?

    i. What happens if inputs 001 and 002 are both true at the same time?

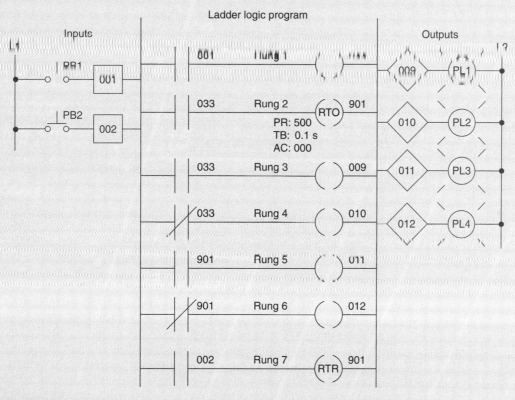

Fig. 7-32

5. Study the ladder logic program in Fig. 7-33 and answer the questions that follow:

  a. What is the purpose of interconnecting the two timers?

  b. How much time must elapse before output 009 is energized?

  c. What two conditions must be satisfied for timer 902 to start timing?

  d. Assume that output 009 is on and power to the system is lost. When power is restored, what will the status of this output be?

  e. When input 002 is on, what will happen?

  f. When input 001 is on, how much accumulated time must elapse before rung 2 will be true?

  g. Assume that the two timers are to be programmed as nonretentive timers. What changes would have to be made in the ladder logic program?

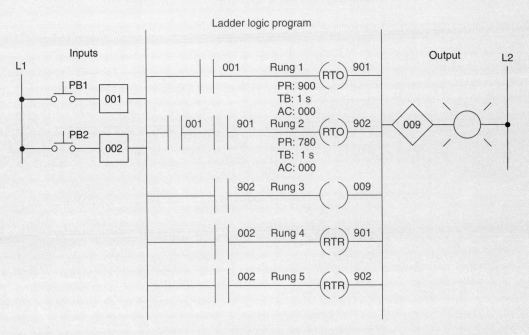

Fig. 7-33

6. You have a machine that cycles on and off during its operation. You need to keep a record of its total run time for maintenance purposes. Which timer would accomplish this?

7. Write a ladder logic program that will turn on a light, PL, 15 s after switch S1 has been turned on.

8. Study the on-delay timer ladder logic program in Fig. 7-34 and, from the conditions stated, determine whether the timer is reset, timing, timed out, or if the conditions stated are not possible.

    a. The input is true and EN is 1, TT is 1, and DN is 0.
    b. The input is true and EN is 1, TT is 1, and DN is 1.
    c. The input is false and EN is 0, TT is 0, and DN is 0.
    d. The input is true and EN is 1, TT is 0, and DN is 1.

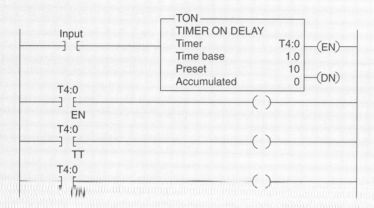

Fig. 7-34

9. Study the off-delay timer ladder logic program in Fig. 7-35 and, from the conditions stated, determine whether the timer is reset, timing, timed out, or if the conditions stated are not possible.

    a. The input is true, and EN is 0, TT is 0, and DN is 1.

    b. The input is true, and EN is 1, TT is 1, and DN is 1.

    c. The input is true, and EN is 1, TT is 0, and DN is 1.

    d. The input is false, and EN is 0, TT is 1, and DN is 1.

    e. The input is false, and EN is 0, TT is 0, and DN is 0.

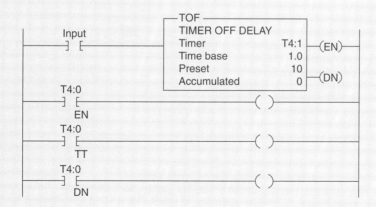

Fig. 7-35

Programming Timers

10. Write a program for an "anti-tie down circuit" that will disallow a punch press solenoid from operating unless both hands are on the 2 palm start buttons. Both buttons must be pressed at the same time within ½ s. The circuit also will not allow the operator to tie down one of the buttons and operate only the press with just one button. (Hint: Once either of the buttons is pressed, begin timing ½ s. Then if both buttons are not pressed, prevent the press solenoid from operating.)

11. Modify the traffic control program of Fig. 7-28 so that there is a 2-s period when both directions will have their red lights illuminated.

12. Write a program to implement the process illustrated in Fig. 7-36. The sequence of operation is to be as follows:

- Normally open start and normally closed stop pushbuttons are used to start and stop the process.
- When the start button is pressed, solenoid *A* energizes to start filling the tank.
- As the tank fills, the empty level sensor switch closes.
- When the tank is full, the full level sensor switch closes.
- Solenoid *A* is de-energized.
- The agitate motor starts automatically and runs for 3 min to mix the liquid.
- When the agitate motor stops, solenoid *B* is energized to empty the tank.
- When the tank is completely empty, the empty sensor switch opens to de-energize solenoid *B*.
- The start button is pressed to repeat the sequence.

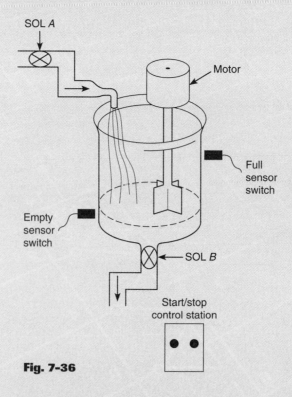

Fig. 7-36

13. When the lights are turned off in a building, an exit door light is to remain on for an additional 2 min, and the parking lot lights are to remain on for an additional 3 min after the door light goes out. Write a program to implement this process.

14. Write a program to simulate the operation of a sequential taillight system. The light system consists of three separate lights on each side of the car. Each set of lights will be activated separately, by either the left or right turn signal switch. There is to be a 1-s delay between the activation of each light, and a 1-s period when all the lights are off. Ensure that, with both switches on, the system will not operate. Use the least number of timers possible. The sequence of operation should be as follows:

- The switch is operated.
- Light 1 is illuminated.
- Light 2 is illuminated 1 s later.
- Light 3 is illuminated 1 s later.
- Light 3 is illuminated for 1 s.
- All lights are off for 1 s.
- The system repeats while the switch is on.

Programming Counters

After completing this chapter, you will be able to:

◆ List and describe the functions of PLC counter instructions

◆ Describe the operating principle of a transitional, or one-shot, contact

◆ Analyze and interpret typical PLC counter ladder logic programs

◆ Apply the PLC counter function and associated circuitry to control systems

◆ Apply combinations of counters and timers to control systems

Most PLCs include both up-counters and down-counters, which function similarly. Counter instructions and their function in ladder logic are explained in this chapter. Typical examples of PLC counters controlling processes studied include the following: straight counting in a process, two counters used to give the sum of two counts, and two counters used to give the difference between two counts.

Programmed counters can serve the same function as mechanical counters. Figure 8-1 shows the construction of a simple mechanical counter. Every time the actuating lever is moved over, the counter adds one number, while the actuating lever returns automatically to its original position. Resetting to zero is done with a pushbutton located on the side of the unit.

Electronic counters, (Fig. 8-2) can count up, count down, or be combined to count up and down. While the majority of counters used in industry are up-counters, numerous applications require the implementation of down-counters or of combination up/down-counters. Every PLC model offers some form of counter instruction as part of its instruction set. Figure 8-3 illustrates typical counter applications.

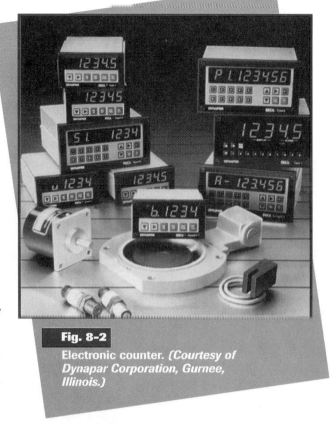

Fig. 8-2

Electronic counter. *(Courtesy of Dynapar Corporation, Gurnee, Illinois.)*

Counters are similar to timers, except that they do not operate on an internal clock but are dependent on external or program sources for counting. As was the case with the timer instruction, there are two methods used to represent a counter within a PLC's ladder logic program. Like the timer, a coil programming format, similar to that illustrated in Fig. 8-4, is used by many manufacturers. The counter is assigned an address as well as being identified as a counter. Also included as part of the counter instruction is the counter's *preset value* and the current *accumulated count* for the counter. The up-counter increments its accumulated value by 1 each time the counter rung makes a false-to-true transition. When the accumulated count equals the preset count, the output is energized, and the counter output is closed. The counter contact can be used as many times as you wish throughout the program as an NO or NC contact.

A COUNTER RESET instruction, which permits the counter to be reset, is also used in conjunction with the counter instruction. Up-counters are always reset to zero. Down-counters may be reset to zero or to some preset value. Some manufacturers include the reset function as a part of the general counter instruction, while others dedicate a separate instruction for resetting of the

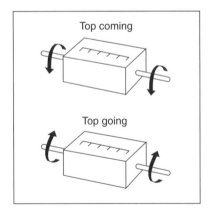

Top coming

Top going

Fig. 8-1

Mechanical counter.

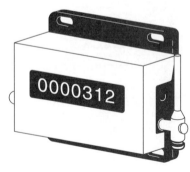

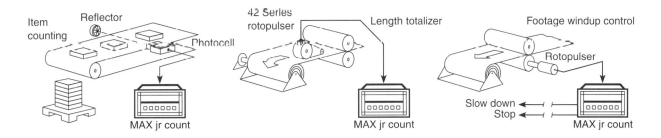

Fig. 8-3

Counter applications. *(Courtesy of Dynapar Corporation, Gurnee, Illinois.)*

counter. Figure 8-5 on p. 200 shows a generic coil-formatted counter instruction with a separate instruction for resetting of the counter. When programmed, the counter reset coil (CTR) is given the *same* reference

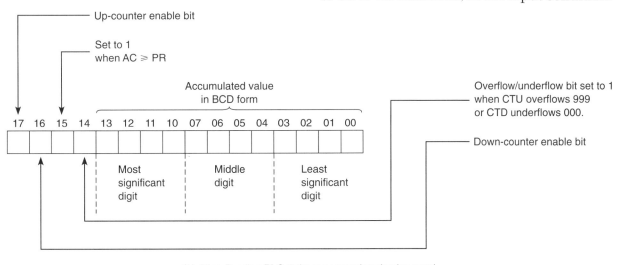

(a) Generic instruction

(b) Allen-Bradley PLC-2 timer accumulated value word (bit addressing is in octal)

address as the counter (CTU) that it is to reset. The reset instruction is activated whenever the CTR rung condition is true.

The second counter format is referred to as a *block format.* Figure 8-6 on p. 200 illustrates a generic block-formatted counter. The instruction block indicates the type of counter (up or down), along with the counter's preset value and accumulated or current value. The counter has two input conditions associated with it; namely the count and reset. All PLC counters operate, or count, on the leading edge of the input signal. The counter will either increment or decrement whenever the count input transfers from an off state to an on state. The counter will *not* operate on the trailing edge, or on-to-off transition, of the input condition.

Fig. 8-4

Coil-formatted counter instruction.

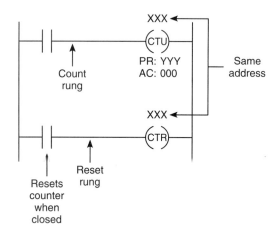

Fig. 8-5

Coil-formatted counter and reset instructions.

value. The *up-counter* is incremented by 1 each time the rung containing the counter is energized. The *down-counter* decrements by 1 each time the rung containing the counter is energized. These rung transitions can result from events occurring in the program, such as parts traveling past a sensor or actuating a limit switch. The preset value of a programmable controller counter can be set by the operator or can be loaded into a memory location as a result of a program decision. Figure 8-7 illustrates the counting sequence of an up-counter and a down-counter. The value indicated by the counter is termed the *accumulated value*. The counter will increment or decrement, depending on the type of counter, until the accumulated value of the counter is equal to or greater than the preset value, at which time an output will be produced. A counter reset is always provided to cause the counter accumulated value to be reset to a predetermined value.

Some manufacturers require the reset rung or line to be true to reset the counter, while others require it to be false to reset the counter. For this reason, it is wise to consult the PLC's operator's manual before attempting any programming of counter circuits.

Most PLC counters are normally retentive; that is, whatever count was contained in the counter at the time of a processor shutdown will be restored to the counter on power-up. The counter may be reset, however, if the reset condition is activated at the time of power restoration.

PLC counters can be designed to count up to a preset value or to count down to a preset

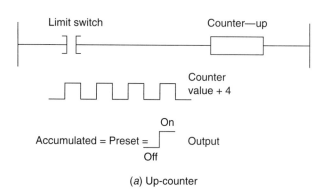

(a) Up-counter

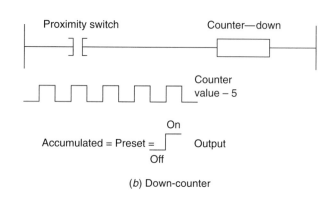

(b) Down-counter

Fig. 8-7

Counter counting sequence.

Fig. 8-6

Block-formatted counter instruction.

8·2 UP-COUNTER

The count-up counter is an output instruction whose function is to increment its accumulated value on false-to-true transitions of its instruction. It thus can be used to count false-to-true transitions of an input instruction and then trigger an event after a required number of counts or transitions. The up-counter output instruction will increment by 1 each time the counted event occurs. Figure 8-8 shows the program and timing diagram for a simple up-counter. This control application

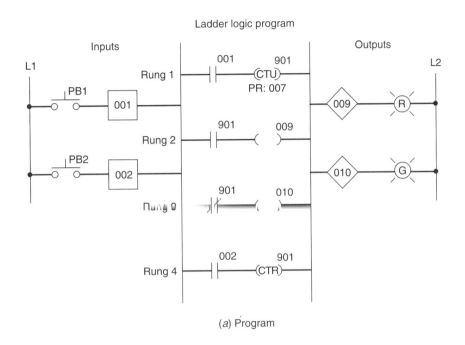

Ladder logic program

(a) Program

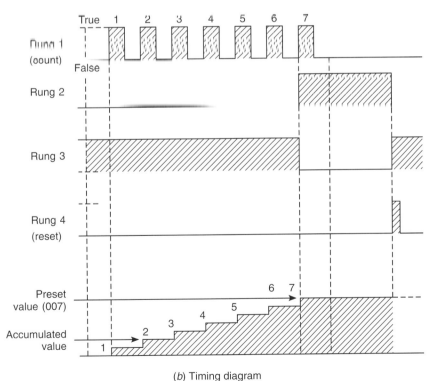

(b) Timing diagram

Fig. 8-8

Simple up-counter program.

is designed to turn the red pilot light on and the green pilot light off after an accumulated count of 7. Operating pushbutton PB1 provides the off-to-on transition pulses that are counted by the counter. The preset value of the counter is set for 007. Each false-to-true transition of rung 1 increases the counter's accumulated value by 1. After 7 pulses, or counts, when the preset counter value equals the accumulated counter value, output 901 is energized. As a result, rung 2 becomes true, energizing output 009 to switch the red pilot light on. At the same time, rung 3 becomes false, de-energizing output 010 to switch the green pilot light off. The counter is reset by closing pushbutton PB2, which makes rung 4 true and resets the accumulated count to zero. Counting can resume when rung 4 goes false again.

Allen-Bradley PLC-5 and SLC-500 controller counter elements each take three data table words: the control word, preset word, and accumulated word. The *control word* uses five control bits consisting of the following:

◆**Count-Up (CU) Enable Bit**
The count-up enable bit is used with the count-up counter and is true whenever the *count-up counter* instruction is true. If the *count-up counter* instruction is false, the (CU) bit is false.

◆**Count-Down (CD) Enable Bit**
The count-down enable bit is used with the count-down counter and is true whenever the *count-down counter* instruction is true. If the *count-down counter* instruction is false, the (CD) bit is false.

◆**Done Bit (DN)**
The done bit is true whenever the accumulated value is equal to or greater than the preset value of the counter, for either the count-up or the count-down counter.

◆**Overflow Bit (OV)**
The overflow bit is true whenever the counter counts past its maximum value,

which is 32,767. On the next count, the counter will wrap around to −32,768 and will continue counting from there toward 0 on successive false-to-true transitions of the count-up counter.

◆**Underflow Bit (UN)**
The underflow bit will go true when the counter counts below −32,768. The counter will wrap around to +32,767 and continue counting down toward 0 on successive false-to-true rung transitions of the count-down counter.

The *preset value (PRE) word* specifies the value that the counter must count to before it changes the state of the done bit. The preset value is the set point of the counter and ranges from −32,768 through +32,767. The number is stored in binary form, with any negative numbers being stored in 2's-complement binary.

The *accumulated value (ACC) word* is the current count based on the number of times the rung goes from false to true. The accumulated value either increments with a false-to-true transition of the *count-up counter* instruction or decrements with a false-to-true transition of the *count-down counter* instruction. It has the same range as the preset: −32,768 through +32,767. The accumulated value will continue to count past the preset value instead of stopping at the preset like a timer does.

Figure 8-9 shows an example of the count-up counter instruction used as part of the Allen-Bradley PLC-5 and SLC-500 controller instruction set. The information to be entered includes:

◆**Counter Number**
This number must come from the counter file. In the example shown, the counter number is C5:0, which represents counter file 5, counter 0 in that file. There may be up to 1000 counters, numbered from 0 through 999, in each counter file. The address for this counter should not be used for any other count-up counter.

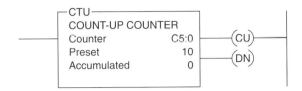

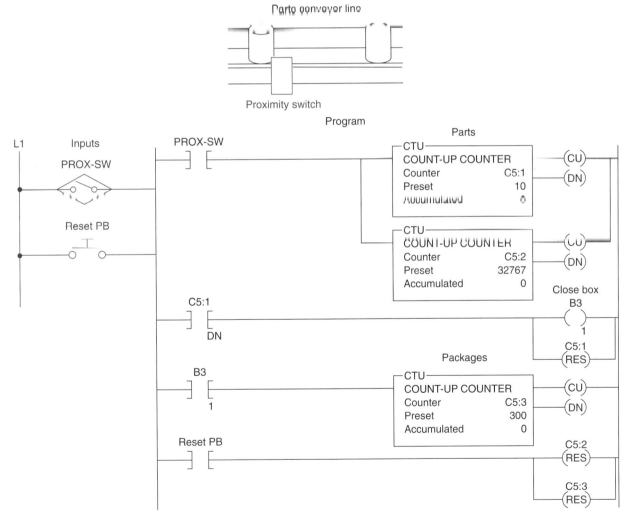

Fig. 8-9

Example of the count-up counter instruction.

◆**Preset Value**

The preset value can range from −32,768 to +32,767. In the example shown, the preset value is 10.

◆**Accumulated Value**

The accumulated value can also range from −32,768 through +32,767.

Typically, as in this example, the value entered in the accumulated word is 0. Regardless of what value is entered, the reset instruction will reset the accumulated value to 0.

Figure 8-10 shows a PLC parts-counting program that uses three up-counters. Counter C5:2 counts the total number of parts coming off an assembly line for final packaging. Each package must contain 10 parts. When 10 parts are detected, counter C5:1 sets bit B3/1 to initiate the box closing sequence. Counter C5:3 counts the total number of packages filled in a day. (The maximum number of packages per day is 300.) A push-button is used to restart the total part and package count from zero daily.

Fig. 8-10

Parts counting program.

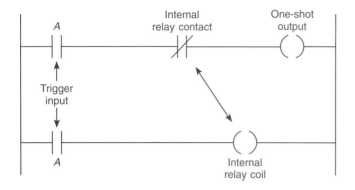

Fig. 8-11

One-shot, or transitional, contact program.

Figure 8-11 shows the program for a *one-shot, or transitional, contact circuit* that is often used to automatically clear or reset a counter. The program is designed to generate an output pulse that, when triggered, goes on for the duration of one program scan and then goes off. The one-shot can be triggered from a momentary signal, or one that comes on and stays on for some time. Whichever signal is used, the one-shot is triggered by the leading edge (off-to-on) transition of the input signal. It stays on for one scan and goes off. It stays off until the trigger goes off, and then comes on again. The one-shot is perfect for resetting both counters and timers since it stays on for one scan only.

Some PLCs provide transitional contacts or one-shot instruction in addition to the standard NO and NC contact instructions. The transitional contact (Fig. 8-12a) is programmed to provide a one-shot pulse when the referenced trigger signal makes a positive (off-to-on) transition. This contact will close for exactly one program scan whenever the trigger signal goes from off to on. The contact will allow logic continuity for one scan and then open, even though the triggering signal may stay on. The on-to-off transitional contact (Fig. 8-12b) provides the same operation as the off-to-on transitional contact instruction, except that it allows logic continuity for a single scan whenever the trigger signal goes from an on to an off state.

The conveyor motor PLC program of Fig. 8-13 illustrates the application of an up-counter along with a programmed one-shot reset circuit. The counter counts the number of cases coming off the conveyor. When the total number of cases reaches 50, the conveyor motor stops automatically. The trucks being loaded will take a total of only 50 cases of this particular product; however, the count can be changed for different product lines. A proximity switch is used to sense the passage of cases.

The sequential task is as follows:

1. The start button is pressed to start the conveyor motor.
2. Cases move past the proximity switch and increment the counter's accumulated value.
3. After a count of 50, the conveyor motor stops automatically and the counter's accumulated value is reset to zero.
4. The conveyor motor can be stopped and started manually at any time without loss of the accumulated count.

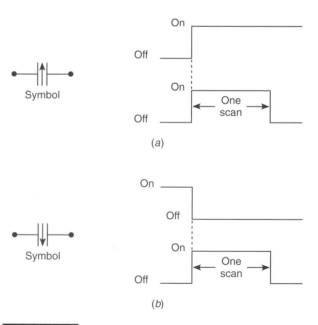

Fig. 8-12

The two types of transitional contact. (a) off-to-on transitional contact. (b) on-to-off transitional contact.

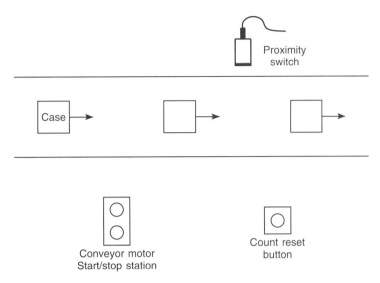

(a) Process flow diagram

Ladder logic program

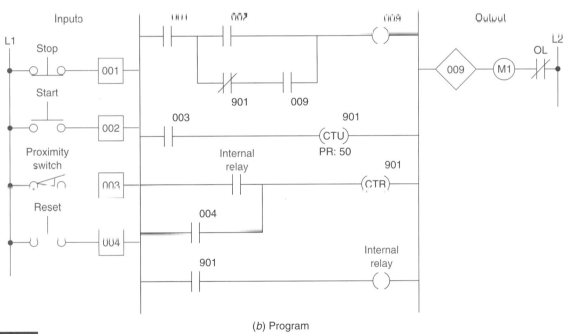

(b) Program

Fig. 8-13

Conveyor motor program.

5. The accumulated count of the counter can be reset manually at any time by means of the count reset button.

The Allen-Bradley one-shot rising (OSR) instruction shown in Fig. 8-14 on p. 206 is a retentive input instruction that triggers an event to occur once. The OSR instruction is used when an event must start based on the change of state of the rung from false to true, not on the resulting status. The address assigned to the OSR instruction is *not* the one-shot address referenced by your program. This address allows the OSR instruction to *remember* its previous rung state. The output instruction(s) that follow the OSR instruction can be referenced by your program as the "one-shot."

The alarm monitor PLC program of Fig. 8-15 illustrates the application of an up-counter used in conjunction with the timed oscillator circuit studied in the previous unit. The operation of the alarm monitor is as follows:

1. The alarm is triggered by the closing of liquid level switch LS1.
2. The light will flash whenever the alarm condition is triggered and has not been acknowledged, even if the alarm condition clears in the meantime.
3. The alarm is acknowledged by closing selector switch SS1.
4. The light will operate in the steady on mode when the alarm trigger condition still exists but has been acknowledged.

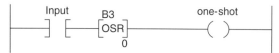

(a) When the input instruction goes from false to true, the OSR instruction conditions the rung so that the output goes true for one program scan. The output goes false and remains false for successive scans until the input makes another false-to-true transition. The addressed OSR bit is set (1) as long as rung conditions preceding the OSR instruction are true; the bit is reset (0) when rung conditions preceding the OSR instruction are false.

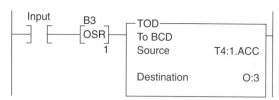

(b) Applications include freezing rapidly displayed LED values. In this case, the accumulated value of a timer is converted to BCD and moved to an output word where an LED display is connected. When the timer is running, the accumulated value changes rapidly. This value can be frozen and displayed for each false-to-true transition of the input condition of the rung.

Fig. 8-14

Allen-Bradley one-shot rising (OSR) instruction.

8.3 DOWN-COUNTER

The down-counter output instruction will count down or decrement by 1 each time the counted event occurs. Each time the down-count event occurs, the accumulated value is decremented. Normally the down-counter is used in conjunction with the up-counter to form an up/down-counter. Figure 8-16 shows the program and timing diagram for a generic, block-formatted up/down-counter. Separate count-up and count-down inputs are provided. Assuming the preset value of the counter is 3 and the accumulated count is zero, pulsing the count-up input (PB1) three times will switch the output light from off to on. This particular PLC counter keeps track of the number of counts received above the preset value. As a result, three additional pulses of the count-up input (PB1) produce an accumulated value of 6 but no change in the output. If the count-down input (PB2) is now pulsed four times, the accumulated count is reduced to 2 (6−4). As a result, the accumulated count drops below the preset count and the output light switches from on to off. Pulsing the reset input (PB3) at any time will reset the accumulated count to zero and turn the output light off.

Not all counter instructions count in the same manner. Some up-counters count only to their preset values, and additional counts are ignored. Other up-counters keep track of the number of counts received above the counter's preset value. Conversely, some down-counters will simply count down to zero and no further. Other down-counters may count below zero and begin counting down from the largest preset value that can be set for the PLC's counter instruction. For example, a PLC up/down-counter that has a maximum counter preset limit of 999 may count up as follows: 997, 998, 999, 000, 001, 002, etc. The same counter would count down in the following manner: 002, 001, 000, 999, 998, 997, etc.

Ladder logic program

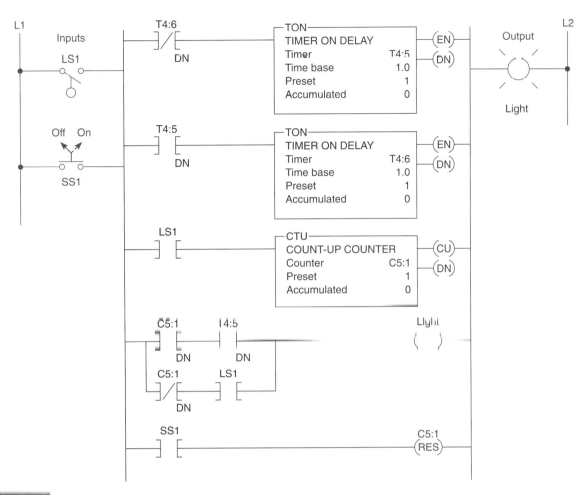

Fig. 8-15

Alarm monitor program.

Ladder logic program

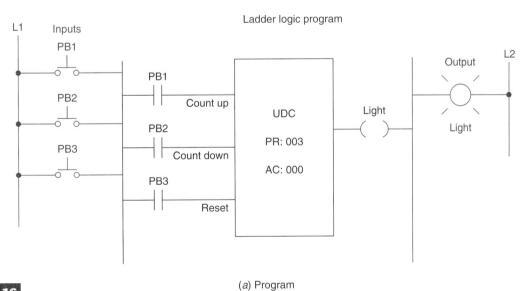

(a) Program

Fig. 8-16

Generic up/down-counter program.

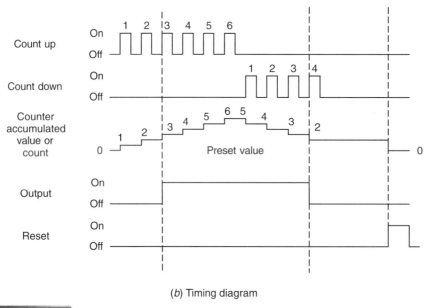

(b) Timing diagram

Fig. 8-16 (continued)

Generic up/down-counter program.

A typical application for an up/down-counter could be to keep count of the cars that enter and leave a parking garage. As a car enters, it triggers the up-counter output instruction and increments the accumulated count by 1. Conversely, as a car leaves, it triggers the down-counter output instruction and decrements the accumulated count by 1. Since both the up- and down-counters have the same address, the accumulated value will be the same in both. Whenever the accumulated value equals the preset value, the counter output is energized to light up the Lot Full sign. Figure 8-17 shows a typical

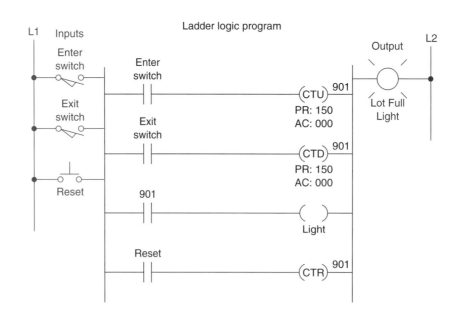

Fig. 8-17

Parking garage counter.

PLC program that could be used to implement the circuit. A coil-formatted type of counter instruction is used, and a reset button has been provided to reset the accumulated count.

Figure 8-18 shows an example of the count-down counter instruction used as part of the Allen-Bradley PLC-5 and SCL-500 controller instruction set. The information to be entered into the instruction is the same as for the count-up counter instruction.

Figure 8-19 shows an up/down-counter program that will increase the counter's accumulated value when pushbutton PB1 is pressed and will decrease the counter's accumulated value when pushbutton PB2 is pressed. Note that the same address is given to both the *up-counter* instruction and the *reset* instruction. All three instructions will be looking at the *same address* in the counter file. When input *A* goes from false to true, one count is added to the accumulated value. When input *B* goes from false to true, one count is subtracted from the accumulated value. The operation of the program can be summarized as follows:

◆ When the CTU instruction is true, C5:2/CU will be true, causing output *A* to be true.

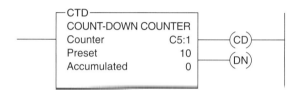

Fig. 8-18

Example of the count-down counter instruction.

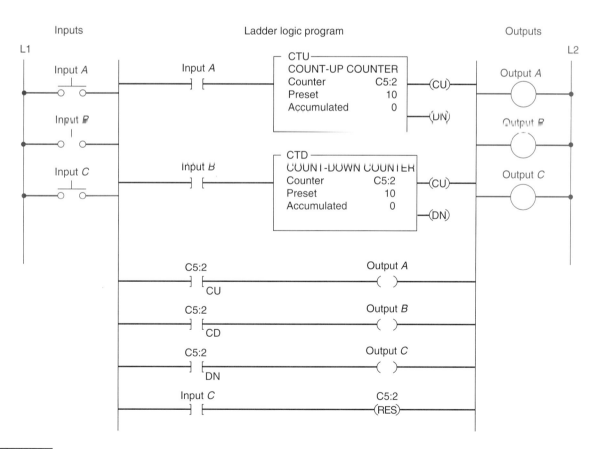

Fig. 8-19

Up/down-counter program.

- When the CTD instruction is true, C5:2/CD will be true, causing output B to be true.

- When the accumulated value is greater than or equal to the preset value, C5:2/DN will be true, causing output C to be true.

- Input C going true will cause both counters to reset. When reset by the *reset* instruction, the accumulated

value will be reset to 0 and the done bit will be reset.

Figure 8-20 illustrates the operation of the up/down-counter program used to provide continuous monitoring of items in-process. An in-feed photoelectric sensor counts raw parts going into the system, and an out-feed photoelectric sensor counts finished parts leaving the machine. The number of parts between the in-feed and out-feed is indicated by the accumulated count of the counter. Counts applied to the up-input are added, and counts applied to the down-input are subtracted. The operation of the program can be summarized as follows:

- Before start-up, the system is completely empty of parts, and the counter is reset manually to zero.

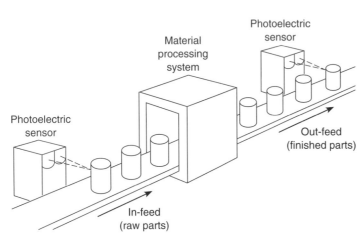

(a) Process

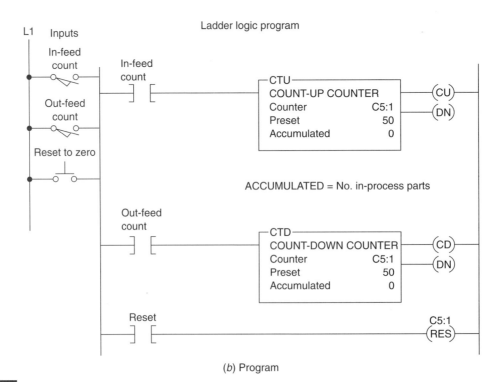

(b) Program

Fig. 8-20

Up/down-counter used in an in-process monitoring system.

◆ When the operation begins, raw parts move through the in-feed sensor, with each part generating an up count.

◆ After processing, finished parts appearing at the out-feed sensor generate down counts, so the accumulated count of the counter continuously indicates the number of in-process parts.

The counter preset value is irrelevant in this application. It does not matter whether the counter outputs are on or off. The output on-off logic is not used. We have arbitrarily set the counter's preset values to 50.

8.4 CASCADING COUNTERS

The maximum count in most controllers is 9999 per counter instruction. Depending on the application, it may be necessary to count events that exceed the maximum number allowable per counter instruction. One way of accomplishing this is by interconnecting, or cascading, two counters. The program of Fig. 8-21 illustrates the application of the technique. In this program the output of the first counter is programmed into the input of the second counter. The status bits of both counters are programmed in series to produce an output. These two counters allow twice as many counts to be measured.

Another method of cascading counters is sometimes used when an extremely large number of counts must be stored. For example, if we require a counter to count up to 250,000, it is possible to achieve this by using only two counters. Figure 8-22 on p. 212 shows how the two counters would be programmed for this purpose. Counter C5:1 has a preset value of 500 and counter C5:2 has a preset value of 500. Whenever counter C5:1 reaches 500, its done bit resets counter C5:1 and increments counter C5:2 by 1. When the done bit of counter C5:1 has turned on and off 500 times, the output light becomes energized. Therefore, the output light turns on after 500 × 500, or 250,000, transitions of the count input.

Some control systems incorporate a 24-h clock to display the time of day or for the

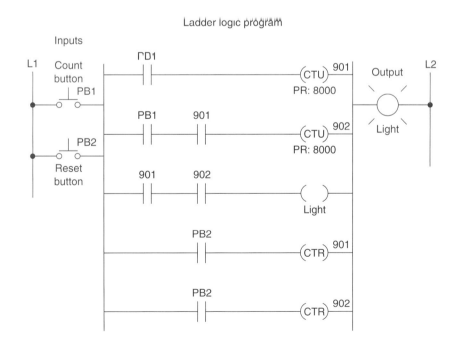

Ladder logic program

Fig. 8-21

Counting beyond the maximum count.

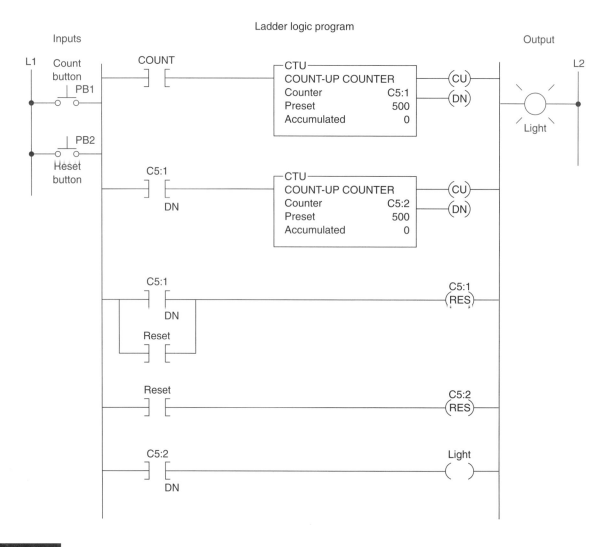

Ladder logic program

Fig. 8-22

Cascading two counters to store an extremely large number of counts.

logging of data pertaining to the operation of the process. The logic used to implement a clock as part of a PLC's program is straightforward and simple to accomplish. A single timer and counter instructions are all you need.

Figure 8-23 illustrates a timer-counter program that produces a time-of-day clock measuring time in hours and minutes. A timer instruction is programmed first with a preset value of 60 s. This timer times for a 60-s period, after which internal coil 901 is activated. The energization of coil 901 causes the counter of rung 2 to increment 1 count. On the next processor scan, the timer is reset and begins timing again. The counter of rung

2 is preset to 60 counts, and each time the timer completes its time delay, its count is incremented. When this counter reaches its preset value of 60, internal coil 902 is energized. Energization of coil 902 increments the second counter programmed in rung 3. The counter of rung 3 is preset for 24 counts. Whenever coil 902 is activated, it also resets the first counter to begin the 60-count sequence again. Whenever the second counter reaches its preset value of 24, internal coil 903 is energized to reset itself. The time of day is generated by examining the current, or accumulated, count or time for each counter or timer. The counter of rung 3 indicates the hour of day in 24-h military

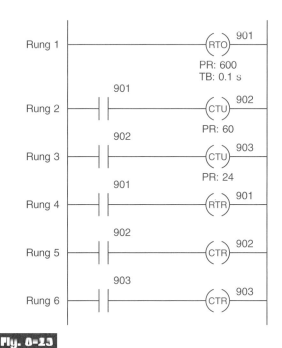

format. The current minutes are represented by the accumulated count value of the counter in rung 2. The timer of rung 1 displays the seconds of a minute as its current, or accumulated, time value.

The 24-h clock can be used to record the time of an event. Figure 8-24 illustrates the principle of this technique. In this application the time of the opening of a pressure switch is to be recorded. The circuit is set into operation by pressing the reset button and setting the clock for the time of day. This starts the 24-hour clock and switches the set indicating light on. Should the pressure switch open at any time, the clock will automatically stop and the trip indicating light will switch on. The clock can then be read to determine the time of opening of the pressure switch.

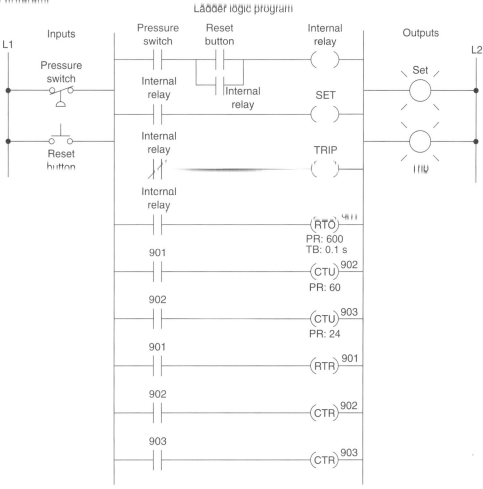

Fig. 8-24

Program for monitoring the time of an event.

8·5 INCREMENTAL ENCODER-COUNTER APPLICATIONS

The incremental encoder shown in Fig. 8-25 creates a series of square waves as its shaft is rotated. The encoder disk interrupts the light as the encoder shaft is rotated to produce the square wave output waveform.

The number of square waves obtained from the output of the encoder can be made to correspond to the mechanical movement required. For example, to divide a shaft revolution into 100 parts, an encoder could be selected to supply 100 square wave cycles per revolution. By using a counter to count those cycles, we could tell how far the shaft had rotated. Figure 8-26 illustrates an example of cutting objects to a specified size. The object is advanced for a specified distance and measured by encoder pulses to determine the correct length for cutting.

The program in part (b) of Fig. 8-27 illustrates the use of a counter for *length* measurement. This system accumulates the total

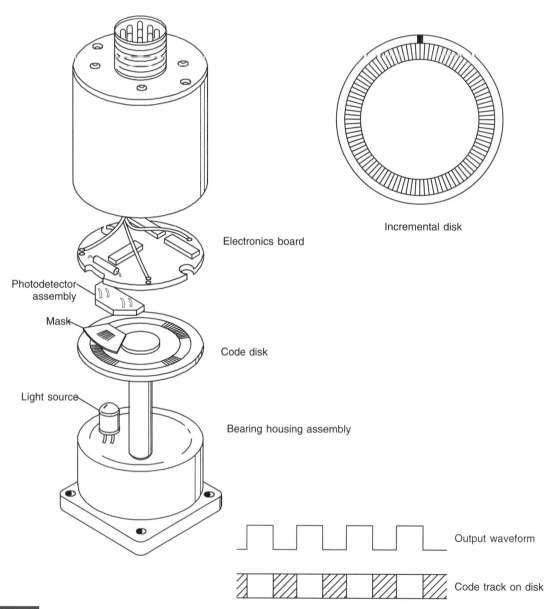

Electronics board

Photodetector assembly

Mask

Code disk

Light source

Bearing housing assembly

Incremental disk

Output waveform

Code track on disk

Fig. 8-25

Incremental encoder. *(Courtesy of BEI Motion Systems Company.)*

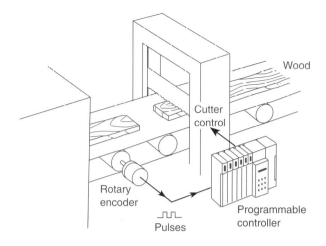

Fig. 8-26

Cutting objects to a specified size.

length of random pieces of bar stock moved on a conveyor. The operation of the program can be summarized as follows:

◆ Count input pulses are generated by the magnetic sensor, which detects passing teeth on a conveyor drive sprocket. If 10 teeth per foot of conveyor motion pass the sensor, the accumulated count of the counter would indicate feet in tenths.

◆ The photoelectric sensor monitors a reference point on the conveyor. When activated, it prevents the unit from counting, thus permitting the counter to accumulate counts only when bar stock is moving.

◆ The counter is reset by closing the reset button.

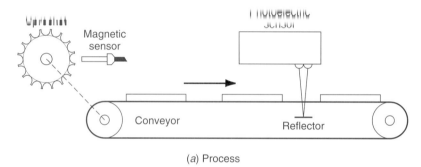

(a) Process

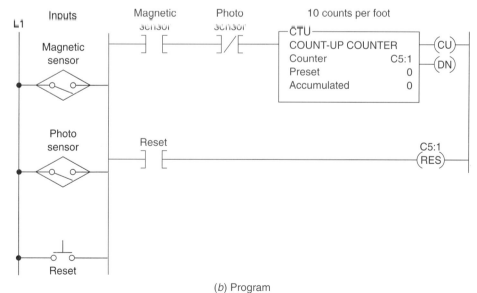

(b) Program

Fig. 8-27

Counter used for length measurement.

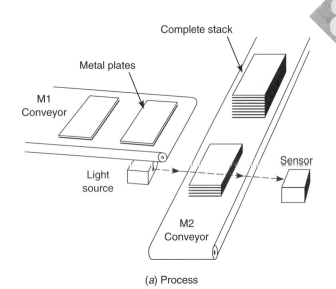

Complete stack

Metal plates

M1
Conveyor

Light
source

Sensor

M2
Conveyor

(a) Process

8•6 COMBINING COUNTER AND TIMER FUNCTIONS

There are many PLC applications in which both the counter function and the timer function are used. Figure 8-28 illustrates an automatic stacking program that requires both a timer and counter. In this process, conveyor M1 is used to stack metal plates onto conveyor M2. The photoelectric sensor provides an input pulse to the PLC counter each time a metal plate drops from conveyor

Ladder logic program

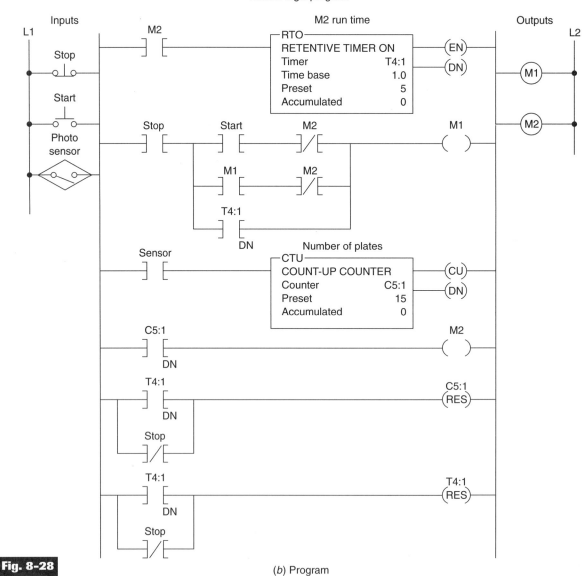

(b) Program

Fig. 8-28

Automatic stacking program.

M1 to M2. When 15 plates have been stacked, conveyor M2 is activated for 5 s by the PLC timer. The operation of the program can be summarized as follows:

◆ When the start button is pressed, conveyor M1 begins running.

◆ After 15 plates have been stacked, conveyor M1 stops and conveyor M2 begins running.

◆ After conveyor M2 has been operated for 5 s, it stops and the sequence is repeated automatically.

◆ The done bit of the timer resets the timer and counter, and provides a momentary pulse to automatically restart conveyor M1.

Figure 8-29 shows a motor lock-out program. This program is designed to prevent a

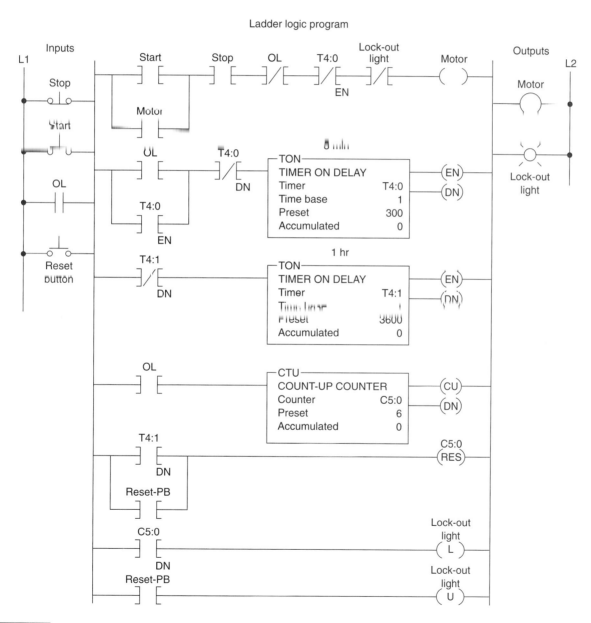

Ladder logic program

Fig. 8-29

Motor lock-out program.

machine operator from starting a motor that has tripped off more than 5 times in an hour. The operation of the program can be summarized as follows:

◆ The normally open overload (OL) relay contact momentarily closes each time an overload current is sensed.

◆ Every time the motor stops due to an overload condition, the motor start circuit is locked out for 5 min.

◆ If the motor trips off more than 5 times in an hour, the motor start circuit is permanently locked out and cannot be started until the reset button is actuated.

◆ The lock-out pilot light is switched on whenever a permanent lock-out condition exists.

Figure 8-30 shows a product part flow rate program. This program is designed to indicate how many parts per minute pass a given process point. The operation of the program can be summarized as follows:

◆ When the start switch is closed, both the timer and counter are enabled.

◆ The counter is pulsed for each part passing the parts sensor.

◆ The counting begins and the timer starts timing through its 1-min time interval.

◆ At the end of 1 min, the timer done bit causes the counter rung to go false. Sensor pulses continue but do not affect the PLC counter. The number of parts for the past minutes are represented by the accumulated value of the counter.

◆ The sequence is reset by momentarily opening and closing the start switch.

A timer is sometimes used to drive a counter when an extremely long time delay period is required. For example, if you require a timer to time to 1,000,000 s, you can achieve this by using a single timer and counter. Figure 8-31 shows how the timer and counter would be programmed for such a purpose. Timer T4:0 has a preset value of 10,000, and counter T5:0 has a preset value of 100. Each time timer T4:0 input contact closes for 10,000 s, its done bit resets timer T4:0 and increments counter C5:0 by 1. When the done bit of timer T4:0 has turned on and off 100 times, the output light becomes energized. Therefore, the output light turns on after $10,000 \times 100$, or 1,000,000, seconds after the timer input contact closes.

Ladder logic program

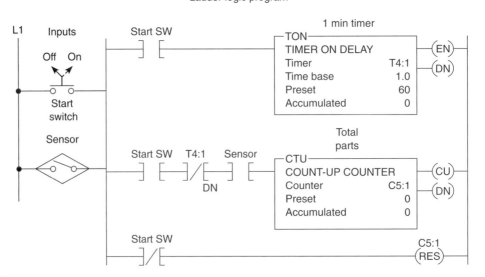

Fig. 8-30

Product flow rate program.

Ladder logic program

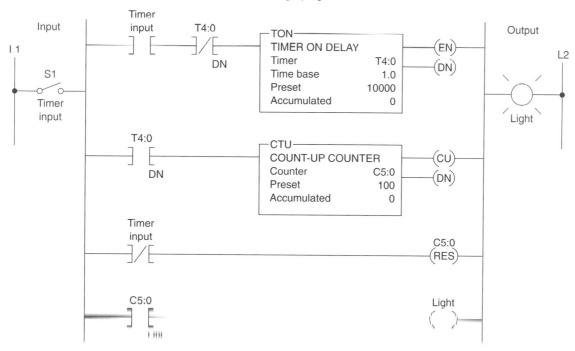

Fig. 8-31

Timer driving a counter to produce an extremely long time-delay period.

Questions

1. Name the three forms of PLC counter instructions and explain the basic operation of each.

2. State four pieces of information usually associated with a PLC counter instruction.

3. In a coil-formatted PLC counter instruction, what rule applies to the addressing of the CTU and CTR coils?

4. When is the output of a PLC counter energized?

5. When does the PLC counter instruction increment or decrement its current count?

6. The counter instructions of PLCs are normally retentive. Explain what this means.

7. **A.** Compare the operation of a standard PLC EXAMINE FOR ON contact with that of an off-to-on transitional contact.

   B. What is the normal function of a transitional contact used in conjunction with a counter?

8. Identify the type of counter you would choose for each of the following situations:

   a. Count the total number of parts made during each shift.

   b. Keep track of the current number of parts in a stage of a process as they enter and exit.

   c. There are 10 parts in a full hopper. As parts leave, keep track of the number of parts remaining in the hopper.

9. Describe the basic programming process involved in the cascading of two counters.

10. In addition to count measurement, what other type of measurement is commonly performed using a counter?

11. **A.** When is the overflow bit of an up-counter set?

    B. When is the underflow bit of a down-counter set?

Problems

1. Study the ladder logic program in Fig. 8-32 and answer the questions that follow:

   a. What type of counter has been programmed?

   b. What outputs are *real* and what outputs are *internal*?

   c. What inputs are *real* and what inputs are *internal*?

   d. When would output 009 be energized?

   e. When would output 010 be energized?

   f. Suppose your accumulated value is 024 and you lose ac line power to the controller. When power is restored to your controller, what will your accumulated value be?

   g. Rung 4 goes true and while it is true, rung 1 goes through five false-to-true transitions of rung conditions. What is the accumulated value of the counter after this sequence of events?

   h. When will the count be incremented?

   i. When will the count be reset?

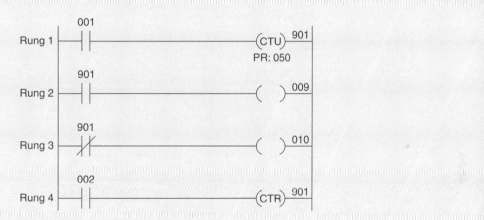

Fig. 8-32

2. Study the ladder logic program in Fig. 8-33 and answer the questions that follow:

   a. Suppose input 001 is switched from off to on and remains on. How will the status of output 009 be affected?

   b. Suppose input 001 is then switched from on to off and remains off. How will the status of output 009 be affected?

Fig. 8-33

3. Study the ladder logic ladder program in Fig. 8-34 and answer the questions that follow:

   a. What type of counter has been programmed?

   b. What input address will cause the counter to increment?

   c. What input address will cause the counter to decrement?

   d. What input address will reset the counter to a count of zero?

   e. When would output 009 be energized?

   f. Suppose the counter is first reset, and then input 001 is actuated 15 times and input 002 is actuated 5 times. What is the accumulated count value?

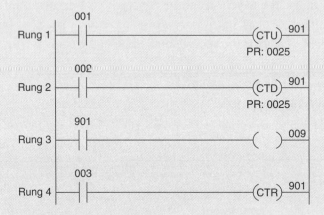

Fig. 8-34

4. Design a PLC program and prepare a typical I/O connection diagram and ladder logic program for the following counter specifications:

   ⊙ Counts the number of times a pushbutton is closed.

   ⊙ Decrements the accumulated value of the counter each time a second pushbutton is closed.

   ⊙ Turns on a light any time the accumulated value of the counter is less than 20.

   ⊙ Turns on a second light when the accumulated value of the counter is equal to or greater than 20.

   ⊙ Resets the counter to zero when a selector switch is closed.

5. Design a PLC program and prepare a typical I/O connection diagram and ladder logic program that will execute the following control circuit correctly:

- Turns on a nonretentive timer when a switch is closed (preset value of timer is 10 s).
- Timer is automatically reset by a programmed transitional contact when it times out.
- Counter counts the number of times the timer goes to 10 s.
- Counter is reset automatically by a second programmed transitional contact at a count of 5.
- Latches on a light at the count of 5.
- Resets light to off when a selector switch is closed.

6. Design a PLC program and prepare a typical I/O connection diagram and ladder logic program that will execute the industrial control process in Fig. 8-35 correctly. The sequence of operation is as follows:

- Product in position (limit switch LS1 contacts close).
- The start button is pressed and the conveyor motor starts to move the product forward toward position A (limit switch LS1 contacts open when the actuating arm returns to its normal position.)
- The conveyor moves the product forward to position A and stops (position detected by eight off-to-on output pulses from the encoder, which are counted by an up-counter).
- A time delay of 10 s occurs, after which the conveyor starts to move the product to limit switch LS2 and stops (LS2 contacts close when the actuating arm is hit by the product).
- An emergency stop button is used to stop the process at any time.
- If the sequence is interrupted by an emergency stop, counter and timer are reset automatically.

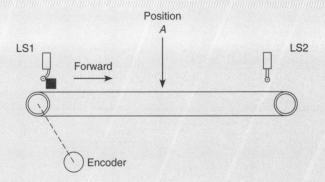

Fig. 8-35

7. Answer the following questions with reference to the up/down-counter program shown in Fig. 8-19. Assume that the following sequence of events occur:

- Input *C* is momentarily closed.
- Twenty on/off transitions of input *A* occur.
- Five on/off transitions of input *B* occur.

As a result:

    a. What is the accumulated count of counter CTU?
    b. What is the accumulated count of counter CTD?
    c. What is the state of output *A*?
    d. What is the state of output *B*?
    e. What is the state of output *C*?

8. Write a program to implement the process illustrated in Fig. 8-36. An up-counter must be programmcd as part of a batch-counting operation to sort parts automatically for quality control. The counter is installed to divert 1 part out of every 1000 for quality control or inspection purposes. The circuit operates as follows:

- A start/stop pushbutton station is used to turn the conveyor motor on and off.
- A proximity sensor counts the parts as they pass by on the conveyor.
- When 1000 is reached, the counter's output activates the gate solenoid, diverting the part to the inspection line.
- The gate solenoid is energized for 2 s, which allows enough time for the part to continue to the quality control line.
- The gate returns to its normal position when the 2-s time period ends.
- The counter resets to zero and continues to accumulate counts.
- A reset pushbutton is provided to reset the counter manually.

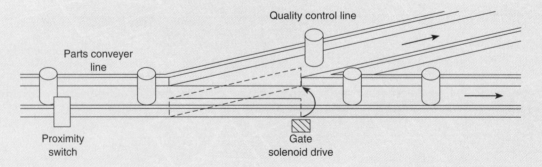

Fig. 8-36

9. Write a program that will increment a counter's accumulated value 1 count every 60 s. A second counter's accumulated value will increment 1 count every time the first counter's accumulated value reaches 60. The first counter will reset when its accumulated value reaches 60, and the second counter will reset when its accumulated value reaches 12.

10. Write a program to implement the process illustrated in Fig. 8-37. A company that makes electronic assembly kits needs a counter to count and control the number of resistors placed into each kit. The controller must stop the take-up spool at a predetermined amount of resistors (100). A worker on the floor will then cut the resistor strip and place it in the kit. The circuit operates as follows:

- A start/stop pushbutton station is used to turn the spool motor drive on and off manually.
- A through-beam sensor counts the resistors as they pass by.
- A counter preset for 100 (the amount of resistors in each kit) will automatically stop the take-up spool when the accumulated count reaches 100.
- A second counter is provided to count the grand total used.
- Manual reset buttons are provided for each counter.

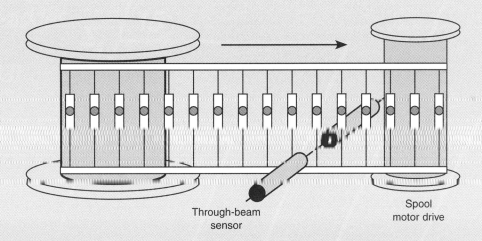

Through-beam sensor

Spool motor drive

Fig. 8-37

11. Write a program that will latch on a light 20 s after an input switch has been turned on. The timer will continue to cycle up to 20 s and reset itself until the input switch has been turned off. After the third time the timer has timed to 20 s, the light will be unlatched.

12. Write a program that will turn a light on when a count reaches 20. The light is then to go off when a count of 30 is reached.

13. Write a program to implement the box-stacking process illustrated in Fig. 8-38. This application requires the control of a conveyor belt that feeds a mechanical stacker. The stacker can stack various numbers of cartons of ceiling tile onto each pallet (depending on the pallet size and the preset value of the counter). When the required number of cartons has been stacked, the conveyor is stopped until the loaded pallet is removed and an empty pallet is placed onto the loading area. A photoelectric sensor will be used to provide count pulses to the counter after each carton passes by. In addition to a conveyor motor start/stop station, a remote reset button is provided to allow the operator to reset the system from the forklift after an empty pallet is placed onto the loading area. The operation of this system can be summarized as follows:

- The conveyor is started by pressing the start button.
- As each box passes the phototelectric sensor, a count is registered.
- When the preset value is reached (in this case, l2), the conveyor belt turns off.
- The forklift operator removes the loaded pallet.
- After the empty pallet is in position, the forklift operator presses the remote reset button, which then starts the whole cycle over again.

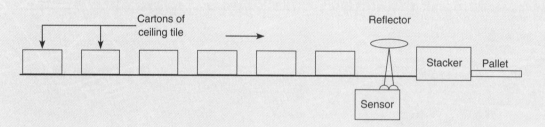

Fig. 8-38

Chapter 8

14. Write a program to operate a light according to the following sequence:

- A momentary pushbutton is pressed to start the sequence.
- The light is switched on and remains on for 2 s.
- The light is then switched off and remains off for 2 s.
- A counter is incremented by 1 after this sequence.
- The sequence then repeats for a total of 4 counts.
- After the fourth count, the sequence will stop and the counter will be reset to zero.

Program Control Instructions

After completing this chapter, you will be able to:

◆ State the purpose of program control instructions

◆ Describe the operation of the *master control reset* (MCR) instruction, and develop an elementary program illustrating its use

◆ Describe the operation of the *jump* (JMP) instruction and the *label* (LBL) instruction

◆ Explain the function of subroutines

◆ Describe the immediate input and output instructions function

◆ Describe the forcing capability of the PLC

◆ Describe safety considerations built into PLCs and programmed into a PLC installation

◆ Describe the function of the selectable time-interrupt and fault routine files

◆ Explain how the temporary end instruction can be used to troubleshoot a program

The program control instructions covered in this chapter are used to alter the program scan from its normal sequence. The use of program control instructions can shorten the time required to complete a program scan. Portions of the program not being utilized at any particular time can be jumped over, and outputs in specific zones in the program can be left in their desired states. Typical industrial program control applications are explained.

Several output-type instructions, which are often referred to as *override* instructions, provide a means of executing sections of the control logic if certain conditions are met. These program control instructions allow for greater program flexibility and greater efficiency in the program scan. Portions of the program not being utilized at any particular time can be jumped over, and outputs in specific zones in the program can be left in their desired states.

Instructions comprising the override instruction group include the *master control reset* (MCR), *zone control last* (ZCL) *state,* and *jump* (JMP) instructions. These operations are accomplished by using a series of conditional and unconditional branches and return instructions. They all operate over a user-specified range, section, or zone of processor logic. The size of the zone is specified in some manner as part of the instruction.

Hardwired master control relays are used in relay circuitry to provide input/output power shutdown of an entire circuit. Figure 9-1 shows a typical hardwired master control relay circuit. In this circuit, unless the master control relay coil is energized, there is no power flow to the load side of the MCR contacts.

The master control relay circuit shown in Fig. 9-1 could not be programmed into the PLC as it appears because it contains two vertical contacts. For this reason, most PLC manufacturers include some form of master control relay as part of their instruction set. These instructions function in a similar manner to the hardwired master control relay; that is, when the instruction is true, the circuit functions normally, and when the instruction is false, outputs are switched off. Since these instructions are not hardwired but programmed, for safety reasons, they should *not* be used as a substitute for a

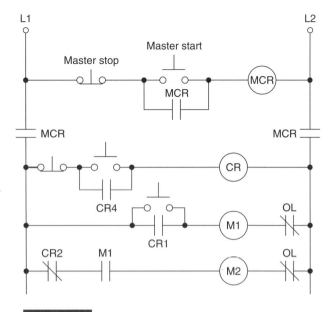

Fig. 9-1
Hardwired master control relay circuit.

hardwired master control relay, which provides *emergency* I/O power shutdown.

The master control reset (MCR) instruction can be programmed to control an entire circuit or to control only selected rungs of a circuit. In the program of Fig. 9-2, the MCR is programmed to control an entire circuit. When the MCR instruction is false, or de-energized, all *nonretentive* (nonlatched) rungs below the MCR will be *de-energized* even if the programmed logic for each rung is true. All *retentive* rungs will remain in their *last state.* The MCR instruction establishes a zone in the user program in which all nonretentive outputs can be turned off simultaneously. Therefore, retentive instructions should not normally be placed within an MCR zone because the MCR zone maintains retentive instructions in the last active state when the instruction goes false.

For Allen-Bradley PLC-5 and SLC-500 controllers, a *master control reset* (MCR) instruction sets up a zone or multiple zones in a program. The MCR instruction is used in pairs to disable or enable a zone within a ladder program and has no address. Figure 9-3 on p. 232 shows the programming of a typical MCR zone. The operation of the program can be summarized as follows:

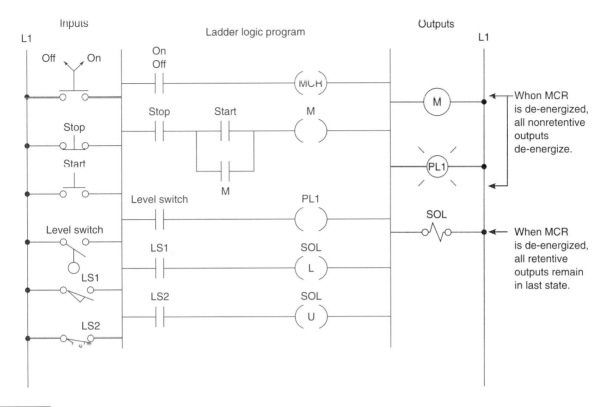

Fig. 9-2

Master control reset (MCR) instruction programmed to control an entire circuit.

◆ The MCR zone is enclosed by a *start fence,* which is a rung with a conditional MCR, and an *end fence,* which is a rung with an unconditional MCR.

◆ When the MCR in the start rung is true (input *A* is true), outputs act according to their rung logic as if the zone did not exist.

◆ When the MCR in the *start fence* is false, all rungs within the zone are treated as false. The scan ignores the inputs and de-energizes all nonretentive outputs (that is, the *output energize* instruction, the on-delay timer, and the off-delay timer). All retentive devices, such as latches, retentive timers, and counters, remain in their last state.

◆ When input *A* is false, output *A* and T4:1 will be false and output *B* will remain in its last state. The input conditions in each rung will have no effect on the output conditions.

◆ Allen-Bradley MCRs cannot be nested in a program; that is, it is *not* possible to use an MCR zone inside another MCR zone.

◆ Multiple MCR zones are permitted in a program.

A common application of an MCR zone control involves examining one or more fault bits as part of the start fence and enclosing the portion of the program you want de-energized in case of a fault in the MCR zone. In case of a detected fault condition, the outputs in that zone would be de-energized automatically.

Allen-Bradley PLC-2 controllers include a *zone control last* (ZCL) state instruction that is similar to the MCR instruction. The differences between these two instructions are described by their respective names. The MCR *(master control reset)* will, when the zone is logic false, reset all nonretentive outputs to the off state. The ZCL will, when the zone is logic false, leave all outputs in their last state.

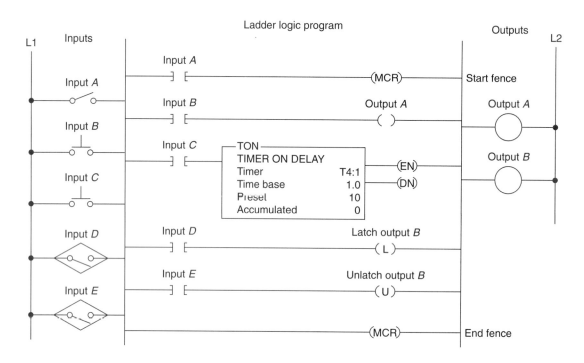

Fig. 9-3

MCR instruction programmed to control a fenced zone.

Figure 9-4 shows a programming example using a ZCL instruction. A ZCL output with conditional inputs is placed at the start of the fenced zone, and a ZCL output with no conditional inputs is placed at the end. If the Man/Auto input is true, the outputs within the zone are controlled by their respective rung input conditions. If the Man/Auto input instruction goes false, the outputs within the zone will be held in their last state.

The ZCL provides a method of determining the particular state of an output at the time of system failure and is useful in systems employing first failure annunciators. A first failure annunciator is a circuit that informs system operators which input device gave a warning signal that resulted in system shutdown. The program of Figs. 9-5a and b illustrates the operation of a first failure annunciator. Three monitoring sensor devices are used: level switch, temperature switch, and pressure switch. Activation of any one of these devices indicates an unsafe condition and triggers a shutdown of the mixing tank process. Once the system has

been shut down, it may be impossible to determine which sensor was activated. With the ZCL instruction, when the alarm is activated, the zone controlled by the ZCL instruction freezes the state of the indicators at the time of system failure. The advantage of this type of system is that the input sensing devices can change state after the alarm has latched and the data will still be recorded by the ZCL instruction.

9.2 JUMP INSTRUCTIONS AND SUBROUTINES

As in computer programming, it is sometimes desirable to be able to jump over certain program instructions if certain conditions exist. The jump instruction is an output instruction used for this purpose. The advantage to the jump instruction is the ability to reduce the processor scan time by jumping over instructions not pertinent to the machine's operation at that instant. Other useful functions of the jump instruction are the following:

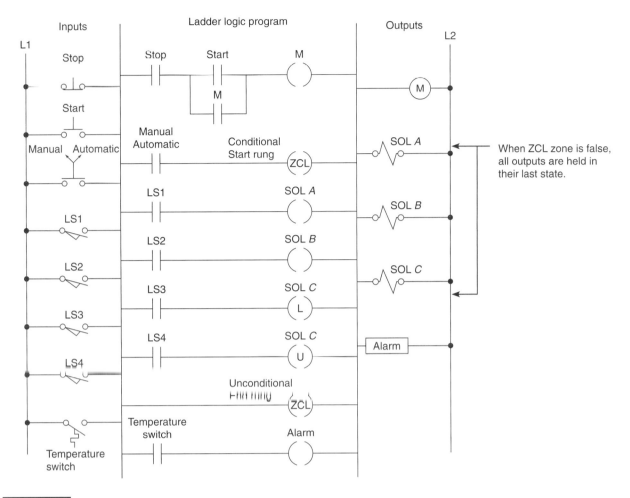

Fig. 9-4

ZCL instruction programmed to control a fenced zone.

◆ The programmable controller can hold more than one program and scan only the program appropriate to operator requirements.

◆ Sections of a program can be jumped when a production fault occurs.

Most PLC manufacturers include a JUMP instruction as part of their instruction set. Some manufacturers provide a SKIP instruction, which is essentially the same as the JUMP instruction. By using the JUMP instruction, you can branch or skip to different portions of a program (as illustrated in Fig. 9-6 on p. 234) and freeze all affected outputs in their last state. Jumps are normally allowed in both the forward and backward directions. Jumping over counters and timers will stop them from being incremented.

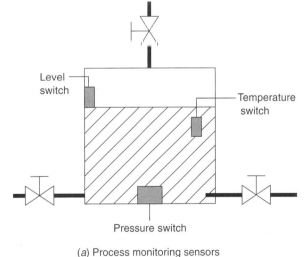

(a) Process monitoring sensors

Fig. 9-5

First failure annunciator program.

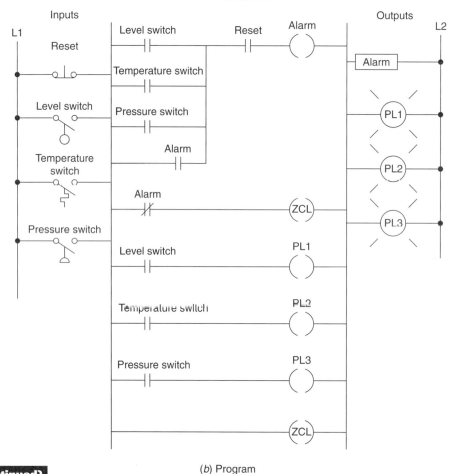

(b) Program

Fig. 9-5 (continued)

First failure annunciator program.

With Allen-Bradley programmable controllers, the *jump* (JMP) instruction and the *label* (LBL) instruction are employed together so the scan can jump over a portion of the program. The *label* is a target for the *jump*, it is the first instruction in the rung, and it is always true. A *jump* jumps to a label with the same address. The area that the processor jumps over is defined by the locations of the *jump* and *label* instructions in the program. If the *jump* coil is energized, all logic between the *jump* and *label* instructions is bypassed and the processor continues scanning after the LBL instruction.

Figure 9-7 shows a simple example of a *jump-to-label* program. The *label* (LBL) instruction is used to identify the ladder rung that is the target destination of the *jump* instruction.

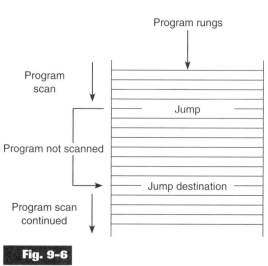

Fig. 9-6

Jump operation.

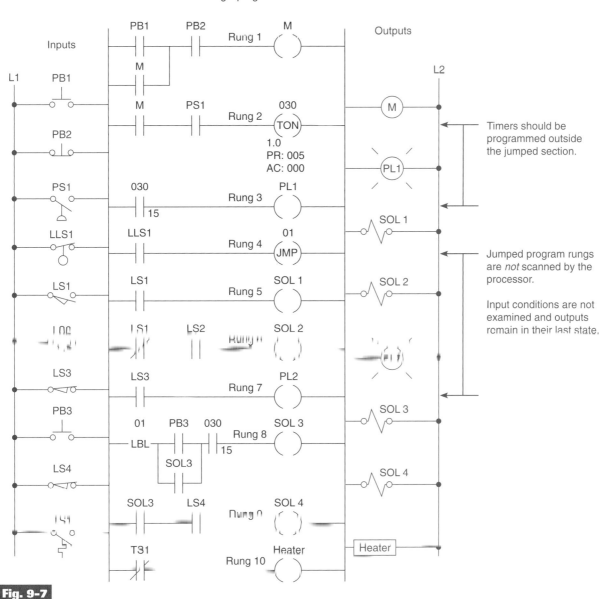

Ladder logic program

Fig. 9-7

Jump-to-label program.

The label address number must match that of the *jump* instruction with which it is used. The *label* instruction does not contribute to logic continuity, and for all practical purposes is always logically true. When rung 4 has logic continuity, the processor is instructed to jump to rung 8 and continue to execute the main program from that point. Jumped rungs 5, 6, and 7 are not scanned by the processor. Input conditions are not examined and outputs controlled by these rungs remain in their last state. Any timers or counters programmed within the jump area cease to function and will not update themselves during this period. For this reason they should be programmed outside the jumped section in the main program zone.

You can jump to the same label from multiple jump locations, as illustrated in the program of Fig. 9-8 on p. 236. In this example, there are two jump instructions numbered 20. There is a single label numbered 20. The scan can then jump from either *jump* instruction to *label 20*, depending on whether input *A* or input *D* is true.

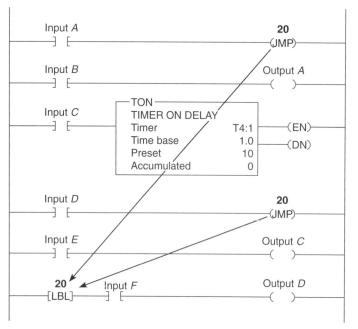

Fig. 9-8

Jump-to-label from two locations.

It is possible to jump backward in the program, but this should not be done an excessive number of times. Care must be taken that the scan does not remain in a loop too long. The processor has a watchdog timer that sets the maximum allowable time for a total program scan. If this time is exceeded, the processor will indicate a fault and shut down.

Again, as in computer programming, another valuable tool in PLC programming is to be able to escape from the main program and *go to* a program *subroutine* to perform certain functions and *then return* to the main program. Where a machine has a portion of its cycle that must be repeated several times during one machine cycle, the subroutine can save a great deal of duplicate programming. The subroutine concept is the same for all programmable controllers, but the method used to call and return from a subroutine uses different commands, depending on the PLC manufacturer. Allen-Bradley PLC-2 controllers have an *area* between the end of the main program and the message storage area set aside for subroutines. The subroutine will be acted on when the rung

containing the *jump-to-subroutine* (JSR) is true. The CPU will then look for the destination address at an LBL in the subroutine area (Fig. 9-9). To jump to a subroutine, a JSR output instruction is programmed with a two-digit octal address to indicate which subroutine is being called. The subroutine must always be completed with a return. This return rung is always unconditional. The exit from the subroutine is always returned to the rung following the JSR in the main application program. When the rung containing the JSR goes false, all outputs in the subroutine area are held in their last state, either energized or de-energized.

Figure 9-10 shows an Allen-Bradley PLC-2 program that uses the *jump-to-subroutine* (JSR), *label*, and *return* (RET) instructions. This program is similar to the one you will see in a later chapter that is used for converting Celsius temperature to Fahrenheit. In the application programmed in Fig. 9-10, we want to record the converted temperature reading every 5 s. When input *D* is true, the processor jumps to the label instruction 02 in the subroutine area and begins executing that subroutine. When the processor scan reaches the *return* instruction, the *return* instruction will send the scan back to the first instruction immediately following the *jump-to-subroutine* instruction.

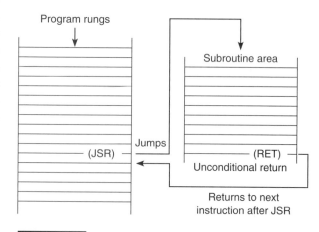

Fig. 9-9

Jump-to-subroutine operation.

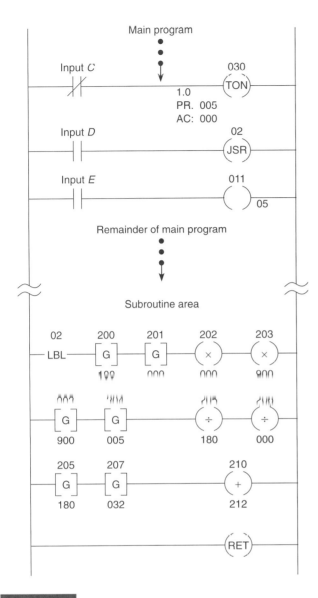

Fig. 9-10

Jump-to-subroutine and *return* program.

The processor then continues to execute the remainder of the main program.

Figure 9-11 shows a materials conveyor system with a flashing pilot light as a subroutine. If the weight on the conveyor exceeds a preset value, the solenoid is de-energized and the alarm light will begin flashing. When the weight sensor switch closes, the JSR is activated and the processor scan jumps to the subroutine area. The subroutine is continually scanned and the light flashes. When the sensor switch opens, the processor will no longer scan the subroutine area and the alarm light will return to the on state.

Allen-Bradley PLC-5 and SLC-500 controller subroutines are located in different program *files* from the main program. The main program is located in program file 2, whereas subroutines are assigned to other program file numbers. Each subroutine must be programmed in its own program file by assigning it a unique file number. Figure 9-12 on p. 238 illustrates the use of the *jump-to-subroutine* (JSR), *subroutine* (SBR), and *return* (RET) instructions. The procedure for setting up a subroutine is as follows:

◆ Note each ladder location where a subroutine should be called.

◆ Create a subroutine file for each location. Each subroutine file should begin with an SBR instruction.

◆ At each ladder location where a subroutine is called, program a JSR instruction specifying the subroutine file number.

◆ The RET instruction is optional.
  - The end of a subroutine program will cause a return to the main program.
  - If you want to end a subroutine program before it executes to the end of program file, a conditional *return* (RET) instruction may be used.

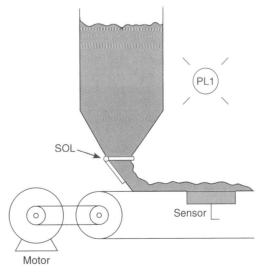

(a) Process

Fig. 9-11

Flashing pilot light subroutine.

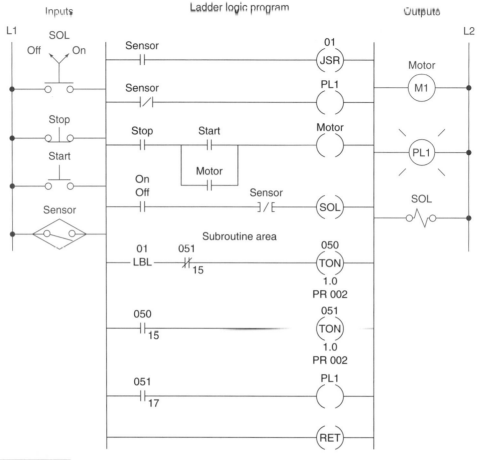

Fig. 9-11 (continued)

Flashing pilot light subroutine.

(b) Program

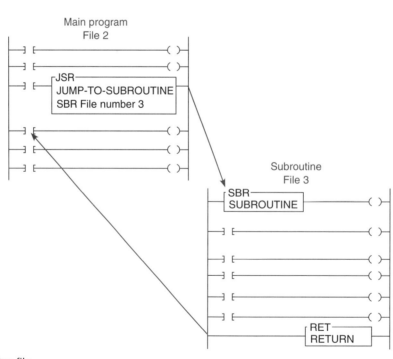

Fig. 9-12

Setting up a subroutine file.

An optional SBR instruction is the header instruction that stores incoming parameters. This feature lets you *pass* selected values to a subroutine before execution so the subroutine can perform mathematical or logical operations on the data and return the results to the main program. For example, the program shown in Fig. 9-13 will cause the scan to jump from the main program file to program file 4 when input A is true. When the scan jumps to program file 4, data will also be passed from N7:30 to N7:40. When the scan returns to the main program from program file 4, data will be passed from N7:50 to N7:60.

Nesting subroutines allow you to direct program flow from the main program to a subroutine and then to another subroutine, as illustrated in Fig. 9-14 on p. 240. Nested subroutines make complex programming easier and program operation faster because the pro-grammer does not have to continually return from one subroutine to enter another. Programming nested subroutines may cause scan time problems because while the subroutine is being scanned, the main program is not. Excessive delays in scanning the main program may cause the outputs to operate later than required. This situation may be avoided by updating critical I/O using immediate input and/or output instructions.

9.3 IMMEDIATE INPUT AND IMMEDIATE OUTPUT INSTRUCTIONS

The immediate input and immediate output instructions interrupt the normal program scan to update the input image table file with current input data or to update an output module group with the current output

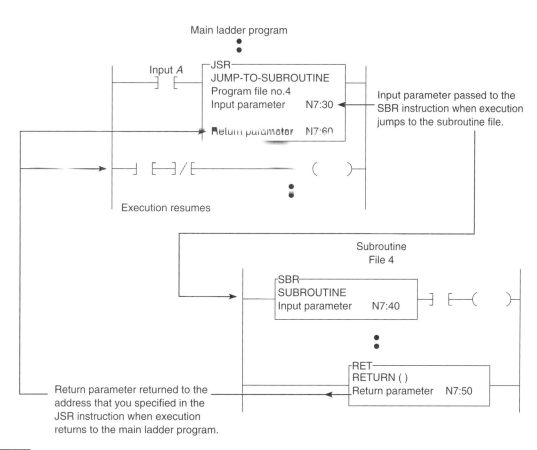

Fig. 9-13

Passing subroutine parameters.

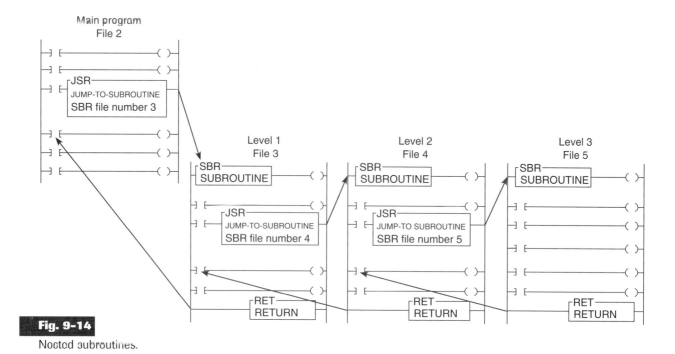

Fig. 9-14

Nested subroutines.

image table file data. These instructions are intended to be used only in areas where time or timing is critical.

The *immediate input* (I or IIN) instruction is a special version of the EXAMINE IF CLOSED instruction used to read an input condition before the I/O update is performed. This operation interrupts the program scan when it is executed. After the immediate input instruction is executed, normal program scan resumes. This instruction is used with critical input devices that require updating in advance of the I/O scan.

The operation of the immediate input instruction is illustrated in Fig. 9-15. When the program scan reaches the immediate input instruction, the scan is interrupted and the bits of the addressed word are updated. The immediate input is most useful if the instruction associated with the critical input device is at the middle or toward the end of the program. The immediate input is not needed near the beginning of the program since the I/O scan has just occurred at that time. Although the *immediate input*

instruction speeds the updating of bits, its scan-time interruption increases the total scan time of the program.

The *immediate output* (IOT) instruction is a special version of the OUTPUT ENERGIZE instruction used to update the status of an output device before the I/O update is performed. The *immediate output* is used with critical output devices that require updating in advance of the I/O scan. The operation of the *immediate output* instruction is illustrated in Fig. 9-16 on p. 242. When the program scan reaches the *immediate output* instruction, the scan is interrupted and the bits of the addressed word are updated.

Processor communication with the local chassis is many times faster than communication with the remote chassis. This is due to the fact that local I/O scan is synchronous with the program scan and communication is in *parallel* with the processor, whereas the remote I/O scan is asynchronous with the program scan and communication with remote I/O is *serial*. For this reason, fast-acting devices should be wired into the local chassis.

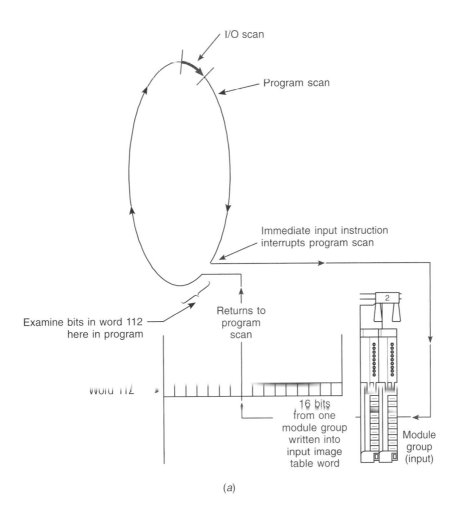

I/O scan

Program scan

Immediate input instruction interrupts program scan

Examine bits in word 112 here in program

Returns to program scan

Word 112

16 bits from one module group written into input image table word

Module group (input)

(a)

12
(IIN)

(b) When the program scan reaches a true IIN instruction, the scan is interrupted and the processor updates sixteen bits in the input image table at the location indicated on the IIN instruction. The two-digit address on the IIN instruction is comprised of the rack number (first digit) and the I/O group number containing the input or inputs that need immediate updating (second digit).

Fig. 9-15

Immediate input instruction.

9·4 FORCING EXTERNAL I/O ADDRESSES

The forcing capability of a PLC allows the user to turn an external input or output on or off from the keyboard of the programmer.

This step is accomplished regardless of the actual state of the field device (limit switch, etc.) for an input or logic rung for an output. This capability allows a machine or process to continue operation until a faulty field device can be repaired. It is also valuable during start-up and troubleshooting of a machine or process to simulate the action of portions of the program that have not yet been implemented.

Forcing inputs manipulates the input image table file bits and thus affects *all* areas of the program that use those bits. The forcing of inputs is done just after the input scan. When we force an input address, we are forcing the status bit of the instruction at the I/O address to an on or off state. Figure 9-17 on p. 243 illustrates how an input is forced on. The processor ignores the actual state of the limit switch (which is off) and considers input 003 as being in the on state.

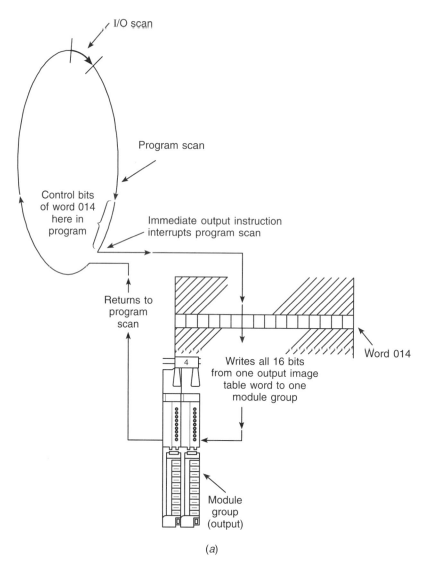

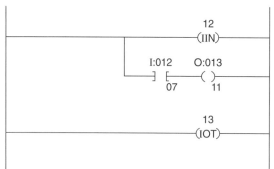

(a)

(b) When the program scan reaches a true IOT instruction, the scan is interrupted and the data in the ouput image table at the word address on the instruction are transferred to the real-world outputs. In this example, the IOT instruction follows the *output energize* instruction. Thus, the output image table word is updated first, and then the data are transferred to the real-world outputs.

Fig. 9-16

Immediate output instruction.

The program scan records this, and the program is executed with this forced status. In other words, the program is executed as if the limit switch were actually closed.

Forcing outputs affects *only* the addressed output terminal. Therefore, since the output image table file bits are unaffected, your program will be unaffected. The forcing of outputs is done just before the output image table file is updated. When we force an output address, we are forcing only the output terminal to an on or off state. The status bit of the output instruction at the address is usually not affected. Figure 9-18 on p. 244 illustrates how an output is forced on. The programming terminal acts in conjunction with the processor to turn output 005 on even though the output image table file indicates that the user logic is

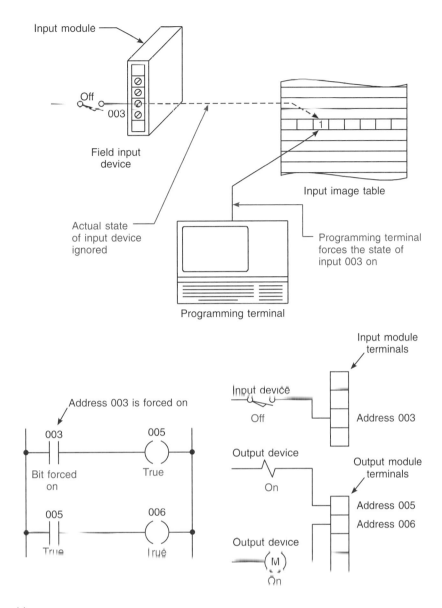

Forcing an input address on.

setting the point to off. Output 006 remains off because the status bit of output 005 is not affected.

Care should be exercised when using forcing functions. Forcing functions should be used only by personnel who completely understand the circuit and the process machinery or driven equipment. An understanding of the potential effect that forcing given inputs or outputs will have on machine operation is essential to avoid possible personal injury and equipment damage. Before using a force function, the user should check whether the "force" acts on the I/O point only, or whether it acts on the user logic as well as on the I/O point. Most programming terminals provide some visible means of alerting the user that a force is in effect.

In cases where rotating equipment is involved, the force instruction can be extremely dangerous. For example, if maintenance personnel are performing routine maintenance on a de-energized motor, the machine may suddenly become energized by someone forcing the motor to turn on. This is why a hardwired master control circuit is required for the I/O rack. The hardwired circuit will provide a method of physically removing

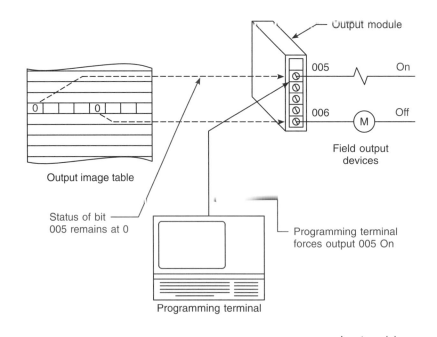

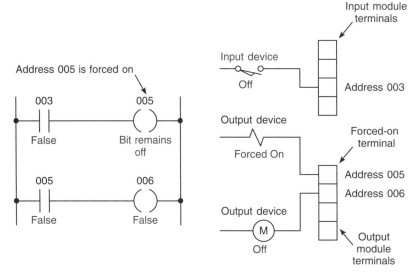

Fig. 9-18

Forcing an output address on.

power to the I/O system, thereby ensuring that it is impossible to energize any inputs or outputs when the master control is off.

9.5 SAFETY CIRCUITRY

Sufficient emergency circuits must be provided to stop either partially or totally the operation of the controller or the controlled machine or process. These circuits should be hardwired outside the controller so that, in the event of total controller failure, independent and rapid shutdown is available.

Figure 9-19 shows a typical safety wiring diagram for a PLC installation. A main disconnect switch is installed on the incoming power lines as a means of removing power from the entire programmable controller system.

The main power disconnect switch should be located where operators and maintenance personnel have quick and easy access to it.

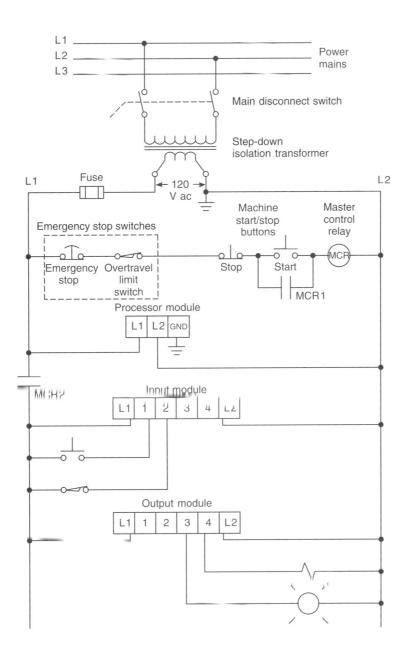

Fig. 9-19

Typical PLC safety wiring diagram.

Ideally, the disconnect switch is mounted on the outside of the PLC enclosure so that it can be accessed without opening the enclosure. In addition to disconnecting electrical power, all other sources of power (pneumatic and hydraulic) should be de-energized, locked out, and tagged before working on a machine or process controlled by the controller. An isolation transformer is used to isolate the controller from the main power distribution system and step the voltage down to 120 V ac.

A hardwired master control relay is included to provide a convenient means for emergency controller shutdown. Since the master control relay allows the placement of several emergency-stop switches in different locations, its installation is important from a safety standpoint. Overtravel limit switches or mushroom head pushbuttons are wired in series so that when one of them opens, the master control is de-energized. This removes power to input and output device circuits. Power continues to be supplied to the controller power supply

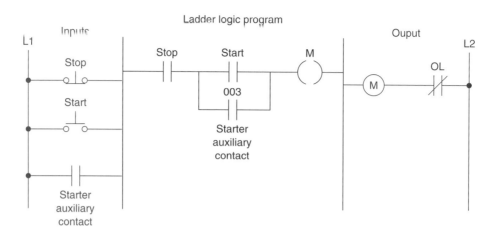

Ladder logic program

(a)

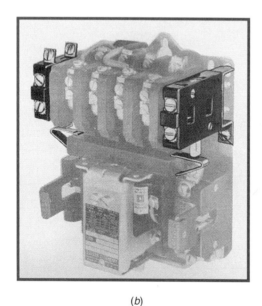

(b)

Fig. 9-20

Motor starter program using the auxiliary contact.

so that any diagnostic indicators on the processor module can still be observed. Note that the master control relay is *not* a substitute for a disconnect switch. When replacing any module, replacing output fuses, or working on equipment, the main disconnect switch should be pulled and locked out.

The master control relay must be able to inhibit all machine motion by removing power to the machine I/O devices when the relay is de-energized. Any part can fail, including the switches in a master control relay circuit. The failure of one of these switches would most likely cause an open circuit, which would be a safe power-off failure.

However, if one of these switches shorts out, it no longer provides any safety protection. These switches should be tested periodically to ensure that they will stop machine motion when needed. Never alter these circuits to defeat their function. Serious injury or machine damage could result.

Certain safety considerations should be developed as part of the PLC program. A PLC program for any application will be only as safe as the time and thought spent on both personal and hardware considerations make it. One such consideration involves the use of a motor starter *seal-in* contact (Fig. 9-20) in place of the programmed contact referenced to the output coil instruction. The use of the field-generated starter auxiliary contact status in the program is more costly in terms of field wiring and hardware, but it is *safer* because it provides positive feedback to the processor about the exact status of the motor. Assume, for example, that the OL contact of the starter opens under an overload condition. The motor, of course, would stop operating since power would be lost to the starter coil. If the program was written using an NO contact instruction referenced to the output coil instruction as the seal-in for the circuit, the processor would never know power has been lost to the motor. When the OL was reset, the motor would restart instantly, creating a potentially unsafe operating condition.

Another safety consideration concerns the wiring of stop buttons. A stop button is

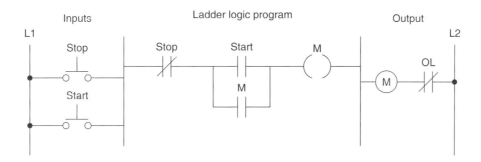

Fig. 9-21
Normally open pushbutton stop configuration.

generally considered a safety function as well as an operating function. As such *it should be wired using an NC contact and programmed to examine for an on condition.* Using an NO contact programmed to examine for an off condition (Fig. 9-21) will produce the same logic but is not considered to be as safe. Assume the latter configuration is used. If, by some chain of events, the circuit between the button and the input point were to be broken, the stop button could be depressed forever, but the PLC logic could never react to the stop command since the input would never be true. The same holds true if power were lost to the stop button control circuit. If the NC wiring configuration is used, the input point receives power continuously unless the stop function is desired. Any faults occurring with the stop circuit wiring, or a loss of circuit power, would effectively be equivalent to an intentional stop.

9•6 SELECTABLE TIMED INTERRUPT

The *selectable timed interrupt* (STI) function allows you to interrupt the scan of the main program file automatically, on a time basis, to scan a specified subroutine file. For Allen-Bradley SLC-500 controllers, the time base at which the program file is executed and the program file assigned as the selectable timed interrupt file are determined by the values stored in words 30 and 31 in the status section of the data files. The value in word 30 stores the time base, which may be from 1 through 32,767 ms, at 1-ms intervals. Word 31 stores the program file assigned as the selectable interrupt file, which may be any program file from 3 through 999. Entering a 0 in the time-base word disables the selectable timed interrupt.

Programming the selectable timed interrupt is done when a section of program needs to be executed on a time basis rather than on an event basis. For example, a program may require certain calculations to be executed at a repeatable time interval for accuracy. These calculations can be accomplished by placing this programming in the selectable timed-interrupt file. This instruction can also be used for process applications that require periodic lubrication.

The *immediate input* and *immediate output* instructions are often located in a selectable timed-interrupt file, so that section of program is updated on a time basis. This process could be done on a high-speed line, when items on the line are being examined and the rate at which they pass the sensor is faster than the scan time of the program. In this way, the item can be scanned multiple times during the program scan, and the appropriate action may be taken before the end of the scan.

The *selectable timed disable* (STD) instruction is generally paired with the *selectable timed enable* (STE) instruction to create zones in which STI interrupts *cannot* occur. Figure 9-22 on p. 248 illustrates the use of the STD and STE instructions. In this program, the STI instruction is assumed to be in effect. The STD

Program Control Instructions

247

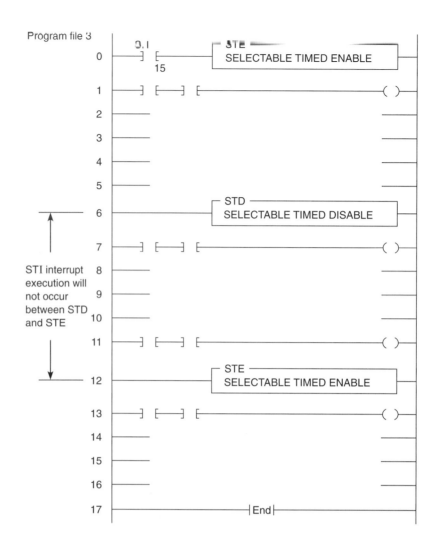

Program file 3

Fig. 9-22

Selectable timed disable (STD) and *selectable timed enable* (STE) instructions.

and STE instructions in rungs 6 and 12 are included in the ladder program to avoid having STI subroutine execution at any point in rungs 7 through 11. The STD instruction (rung 6) resets the STI enable bit and the STE instruction (rung 12) sets the enable bit again. The first pass bit S:1/15 and the STE instruction in rung 0 are included to ensure that the STI function is initialized following a power cycle.

9.7 FAULT ROUTINE

Allen-Bradley PLC-5 and SLC-500 controllers allow you to designate a subroutine file as a fault routine. If used, it determines how the processor responds to a programming error. The program file assigned as the fault routine is determined by the value stored in word 29 in the status file. Entering a 0 in word 29 disables the fault routine.

There are two kinds of major faults that result in a processor fault: recoverable and nonrecoverable faults. For the PLC-5, bits 00 through 07 in the major-fault word, word 11 in the status file, indicate recoverable faults; bits 08 through 15 indicate nonrecoverable faults. When the processor detects a major fault, it looks for a fault routine. If a fault routine exists, it is executed; if one does not exist, the processor shuts down. When there is a fault routine, and the fault is *recoverable,* the fault routine is executed. If the fault is *nonrecoverable,* the fault routine is scanned once and shuts down. Either way, the fault routine allows for an orderly shutdown.

9•8 TEMPORARY END (TND) INSTRUCTION

The *temporary end* (TND) instruction is an output instruction used to progressively debug a program or conditionally omit the balance of your current program file or subroutines. When rung conditions are true, this instruction stops the program scan, updates the I/O, and resumes scanning at rung 0 of the main program file.

Figure 9-23 illustrates the use of the TND instruction in troubleshooting a program. The TND instruction lets your program run only up to this instruction. You can move it progressively through your program as you debug each new section. You can program the TND instruction unconditionally, or you can condition its rung according to your debugging needs.

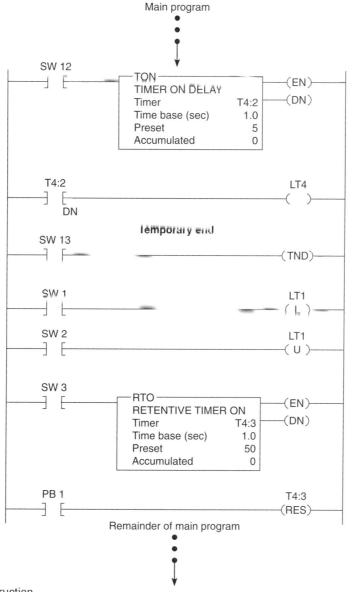

Fig. 9-23

Temporary end (TND) instruction.

Questions

1. **A.** Two MCR output instructions are to be programmed to control a section of a program. Explain the programming procedure to be followed.

   B. State how the status of the output devices within the fenced zone will be affected when the MCR instruction makes a false-to-true transition.

   C. State how the status of the output devices within the fenced zone will be affected when the MCR instruction makes a true-to-false transition.

2. Explain the difference between the MCR and ZCL program control instructions.

3. What is the main advantage of the *jump* instruction?

4. What types of instructions are not normally included inside the jumped section of a program? Why?

5. **A.** What is the purpose of the *label* instruction in the *jump-to-label* instruction pair?

   B. When the *jump-to-label* instruction is executed, in what way are the jumped rungs affected?

6. **A.** Explain what the *jump-to-subroutine* instruction allows the program to do.

   B. In what type of machine operation can this instruction save a great deal of duplicate programming?

7. What advantage is there to the nesting of subroutines?

8. **A.** When are the *immediate input* and *immediate output* instructions used?

   B. Why is it of little benefit to program an *immediate input* or *immediate output* instruction near the beginning of a program?

9. **A.** What does the forcing capability of a PLC allow the user to do?

   B. Outline two practical uses for forcing functions.

   C. Why should extreme care be exercised when using forcing functions?

10. Why should emergency stop circuits be hardwired instead of programmed?

11. State the function of each of the following in the basic safety wiring for a PLC installation:

    a. Main disconnect switch   **c.** Emergency stops
    b. Isolation transformer   **d.** Master control relay

12. When programming a motor starter circuit, why is it safer to use the starter seal-in auxiliary contact in place of a programmed contact referenced to the output coil instruction?

13. When programming stop buttons, why is it safer to use an NC button programmed to examine for an on condition than an NO button programmed to examine for an off condition?

14. Explain the *selectable timed interrupt* function.

15. Explain the function of the *fault routine* file.

16. How is the *temporary end* instruction used to troubleshoot a program?

Problems

1. Answer the questions, in sequence, for the MCR program in Fig. 9-24, assuming the program has just been entered and the PLC placed in the RUN mode with all switches turned off.

   a. Switches S2 and S3 are turned on. Will outputs PL1 and PL2 come on? Why?

   b. With switches S2 and S3 still on, switch S1 is turned on. Will output PL1 or PL2 or both come on? Why?

   c. With switches S2 and S3 still on, switch S1 is turned off. Will both outputs PL1 and PL2 de-energize? Why?

   d. With all other switches off, switch S6 is turned on. Will the timer time? Why?

   e. With switch S6 still on, switch S5 is turned on. Will the timer time? Why?

   f. With switch S6 still on, switch S5 is turned off. What happens to the timer? If the timer was an RTO type instead of a TON, what would happen to the accumulated value?

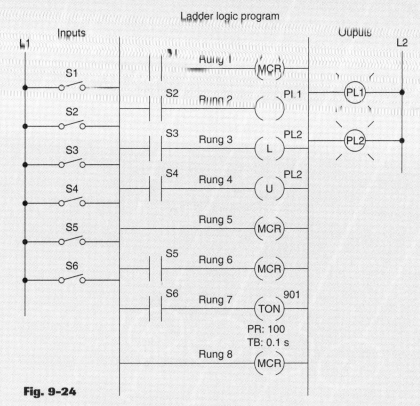

Ladder logic program

Fig. 9-24

2. Answer the questions, in sequence, for the ZCL program in Fig. 9-25, assuming that the program has just been entered and the PLC placed in the run mode with all switches turned off.

a. Switches S2 and S5 are turned on. Will any outputs come on? Why?

b. With switches S2 and S5 still on, switch S1 is turned on. What will happen to outputs PL1 and PL2? Why?

c. With switches S2 and S5 still on, switch S1 is turned off. What will happen to the outputs?

d. Switch S2 is now turned off. Will output PL1 go off? Why?

e. With all other switches off, switch S4 is turned on. Will timer 901 start timing? Why?

f. With switch S4 still on, switch S3 is turned on. Will the timer function correctly?

g. With the timer timing, switch S3 is turned off. What will happen to the timer's accumulated value?

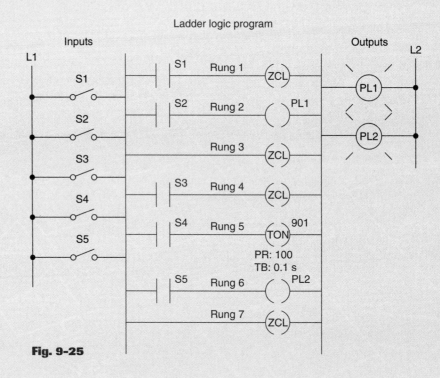

Fig. 9-25

3. Answer the questions, in sequence, for the *jump-to-label* program in Fig. 9-26. Assume all switches are turned *off after each operation.*

    a. Switch S3 is turned on. Will output PL1 be energized? Why?

    b. Switch S2 is turned on *first,* then switch S5 is turned on. Will output PL4 be energized? Why?

    c. Switch S3 is turned on and output PL1 is energized. Next, switch S2 is turned on. Will output PL1 be energized or de-energized after turning on switch S2? Why?

    d. All switches are turned on in order according to the following sequence: S1, S2, S3, S5, S4. Which pilot lights will turn on?

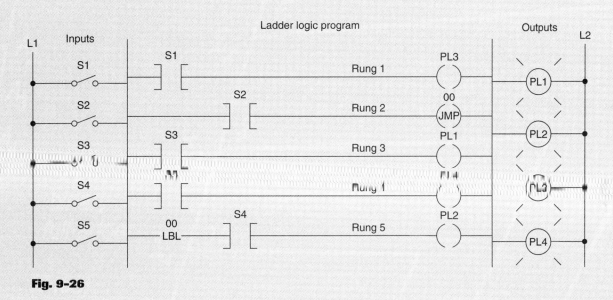

Fig. 9-26

4. Answer the questions, in sequence, for the jump-to-subroutine and return program in Fig. 9-27. Assume all switches are turned *off after each operation.*

    a. Switches S1, S3, S4, and S5 are all turned on. Which pilot light will *not* be turned on? Why?

    b. Switch S2 is turned on and then switch S4 is turned on. Will output PL3 be energized? Why?

    c. To what rung does the RET instruction return the program scan to?

    d. Is rung 6 part of the subroutine area? Why?

    e. Assume all switches are turned on. In what order will the rungs be scanned?

    f. Assume all switches are turned off. In what order will the rungs be scanned?

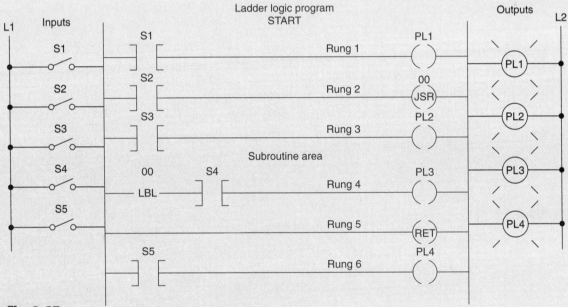

Fig. 9-27

5. Answer the questions, in sequence, for Fig. 9-28. Assume all switches are turned *off after each operation.*

    a. Switches S2, S12, and S5 are turned on in order. Will output PL5 be energized. Why?

    b. All switches, except S7, are turned off. Will RTO/31 start timing? Why?

    c. Switches S3 and S8 are turned on in order. Will pilot light PL2 come on? Why?

    d. When will timer TON/030 function?

    e. Assume all switches are turned on. In what order will the rungs be scanned?

    f. Assume all switches are turned off. In what order will the rungs be scanned?

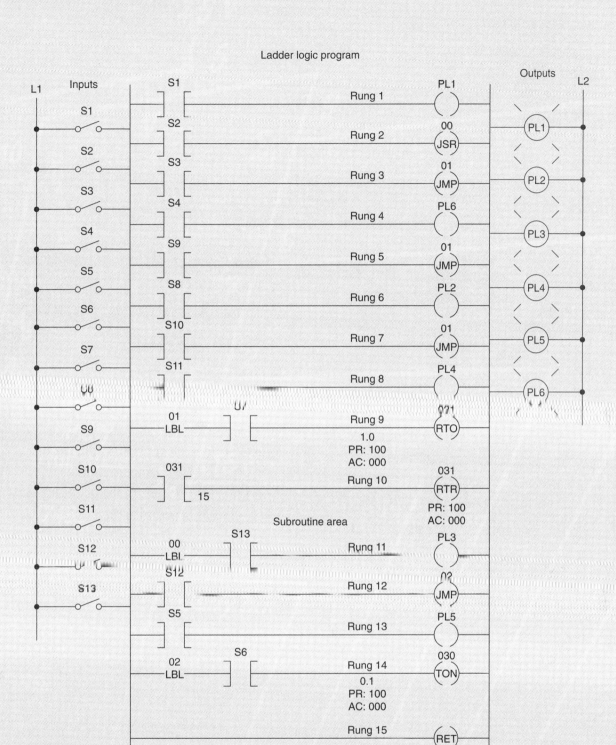

Ladder logic program

Fig. 9-28

Data Manipulation Instructions

After completing this chapter, you will be able to:

◆ Define data manipulation and apply it by writing a PLC program

◆ Describe the operation of the word-level instructions used to copy data from one memory location to another

◆ Interpret data transfer and data compare instructions as they apply to a PLC program

◆ Compare the operation of discrete I/Os with that of multibit and analog types

◆ Describe the basic operation of a closed-loop control system

Data manipulation involves the following: (1) transferring data and (2) operating on data with math functions, data conversions, data comparison, and logical operations. This chapter covers both data manipulation instructions that operate on word data and those that operate on file data, which involve multiple words. Data manipulations are performed internally in a manner similar to that used in microcomputers. Examples of processes that need these operations on a fast and continuous basis are studied.

10·1 DATA MANIPULATION

Data manipulation instructions enable the programmable controller to take on some of the qualities of a computer system. Most PLCs now have this ability to manipulate data stored in memory. This extra computer characteristic gives the PLC capabilities that go far beyond the conventional relay equivalent instructions.

Data manipulation involves transfer of data and operation on data with math functions, data conversion, data comparison, and logical operations. There are two basic classes of instructions to accomplish this: instructions that operate on *word* data, and those that operate on *file,* or *block,* data, which involve multiple words.

Each data manipulation instruction requires two or more words of data memory for operation. The words of data memory in singular form may be referred to either as *registers* or as *words,* depending on the manufacturer. The terms *table* or *file* are generally used when a *consecutive* group of related data

memory words is referenced. Figure 10-1 illustrates the difference between a word and file. The data contained in files and words will be in the form of binary *bits* represented as series of 1's and 0's.

The data manipulation instructions allow the movement, manipulation, or storage of data in either single- or multiple-word groups from one data memory area of the PLC to another. Use of these PLC instructions in applications that require the generation and manipulation of large quantities of data greatly reduces the complexity and quantity of the programming required. Data manipulation can be placed in two broad categories: *data transfer* and *data comparison.*

The manipulation of entire words is an important feature of a programmable controller. This feature enables PLCs to handle inputs and outputs containing multiple bit configurations such as analog inputs and outputs. Arithmetic functions also require data within the programmable controller to be handled in word or register format. To simplify the explanation of the various data manipulation instructions available, the instruction

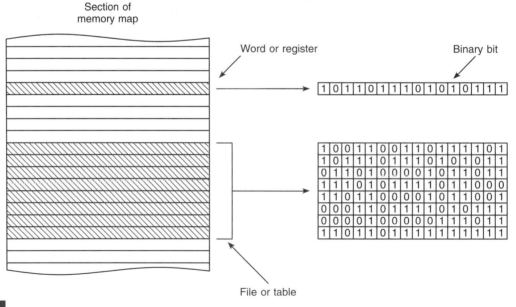

Fig. 10-1

Data files, words, and bits.

protocol for the Allen-Bradley PLC-2, PLC-5, and SLC-500 families of PLCs will be used. Again, even though the format and instructions vary with each manufacturer, the concepts of data manipulation remain the same.

•2 DATA TRANSFER OPERATIONS

Data transfer instructions simply involve the transfer of the contents from one word or register to another. Figure 10-2a and b illustrate the concept of moving numerical binary data from one memory location to another. Figure 10-2a shows that numerical data are stored in word 130 and that no information is currently stored in word 040. Figure 10-2b shows that after the data transfer has occurred, word 040 now holds a duplicate of the information that is in word 130. If word 040 had other information already stored (rather than all 0's), this information would have been replaced. When new data replace existing data in this manner, the process is referred to as *writing over the existing data.*

Data transfer instructions can address almost any location in the memory. Prestored values can be automatically retrieved and placed in any new location. That location may be the preset register for a timer or counter or even an output register that controls a seven-segment display.

The Allen-Bradley PLC-2 controller uses coil formatted data transfer instructions: GET and PUT. GET instructions tell the processor to go *get* a value stored in some word. The GET instruction is programmed in the *condition* portion of a logic rung. It is always a logic true instruction that will get an entire 16-bit word from a specific location in the data table. Figure 10-3 shows a typical GET instruction, which tells the processor to *get* the value 005 stored in word 020.

```
        020
      ┌  ┐  Word address
──────┤G ├
      └  ┘  Numeric value
        005
```

Fig. 10-3

Allen-Bradley PLC-2 GET instruction.

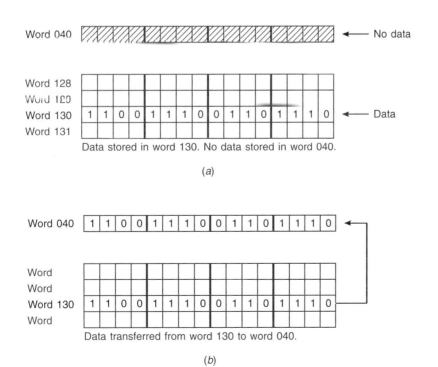

Word 040 [////////////////////////] ← No data

Word 128
Word 129
Word 130 [1 1 0 0 1 1 1 0 0 1 1 0 1 1 1 0] ← Data
Word 131

Data stored in word 130. No data stored in word 040.

(a)

Word 040 [1 1 0 0 1 1 1 0 0 1 1 0 1 1 1 0]

Word
Word
Word 130 [1 1 0 0 1 1 1 0 0 1 1 0 1 1 1 0]
Word

Data transferred from word 130 to word 040.

Fig. 10-2

Data transfer concept.

(b)

PUT instructions tell the processor where to *put* the information it obtained from the GET instruction. The PUT instruction is programmed in the *output* portion of the logic rung. A PUT instruction receives all 16 bits of data from the immediately preceding GET instruction. It is used to store the result of other operations in the memory location (word or register) specified by the PUT coil. Figure 10-4 shows a typical PUT instruction, which tells the processor to put the information obtained from the GET instruction into word 130.

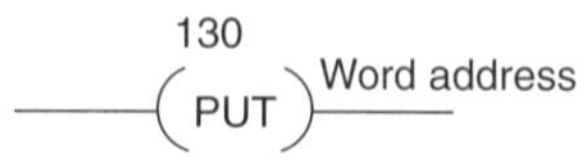

The PUT instruction is used with the GET instruction to form a data transfer rung. Figure 10-5 shows an example of a data transfer rung. When input 110/10 is true, the GET/PUT instructions tell the processor to *get* the numeric value 005 stored in word 020 and *put* it into word 130. In every case, the PUT instruction must be preceded by a GET instruction.

Allen-Bradley PLC 5 and SLC-500 controllers use a block formatted MOVE instruction to accomplish data moves. The MOVE instruction is used to copy the value in one *word* to another *word*. This instruction copies data from a *source* word to a *destination* word. Figure 10-6 shows an example of the MOVE instruction. In this example, when the rung is true, the value stored at the

130
—(PUT)— Word address

Fig. 10-4

Allen-Bradley PLC-2 PUT instruction.

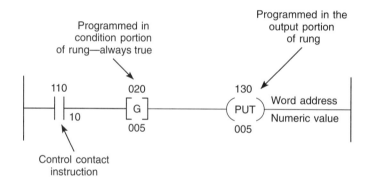

Fig. 10-5

GET/PUT data transfer rung.

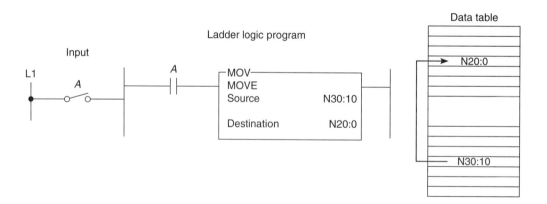

This output instruction moves the source value to the destination location.
The source value remains unchanged and no data conversion occurs.

Fig. 10-6

PLC-5 and SCL-500 block-formatted MOVE INSTRUCTION.

source address, N30:10, is copied into the destination address, N20:0. When the rung goes false, the destination address will retain the value, unless it is changed elsewhere in the program. The instruction may be programmed with input conditions preceding it, or it may be programmed unconditionally.

The *move with mask* (MVM) instruction differs slightly from the MOVE instruction because a *mask* word is involved in the move. The data being moved must pass through the mask to get to their destination address. The MVM instruction is used to copy the desired part of a 16-bit word by masking the rest of the value.

Figure 10-7 shows an example of a *mask move* (MVM) instruction. This instruction transfers data through the mask from the source address, B3:0 to the destination address B3:4. The mask may be entered as an address or in hexadecimal format, and its value will be displayed in hexadecimal. Where there is a 1 in the mask, data will pass from the source to the destination. Where there is a 0 in the mask, data in the destination will remain in their last state. The mask must be the same word size as the source and destination.

The *bit distribute* (BTD) instruction is used to move *bits* within a word or between words, as illustrated in Fig. 10-8 on p. 262. On each scan, when the rung that contains the BTD instruction is true, the processor moves the bit field from the source word to the destination word. To move data within a word, enter the same address for the source and destination. The source remains unchanged. The instruction writes over the destination with the specified bits.

Figure 10-9 on p. 262 shows an example of how the GET/PUT data transfer instructions can be used to change the preset time of an on-delay timer. In this example, delay times of 5 or 10-s are selected by means of selector switch SS1. When the selector switch is in the 10-s position, rung 2 has logic continuity and rung 3 does not. As a result, the processor is told to get the value of 010 stored in word 021 and put it into word 130. Address 130 is where the preset value of the on-delay time 030 is stored. Therefore, the preset value of the timer 030 will change from 000 to 010. When pushbutton PB1 is closed, there will be a 10-s delay period before the pilot light is energized. The timer contact address 030/15 is the *done* bit of the

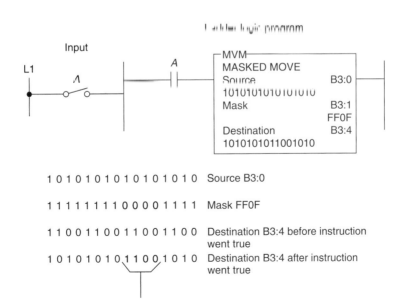

Fig. 10-7

Masked move (MVM) instruction.

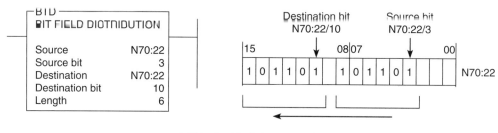

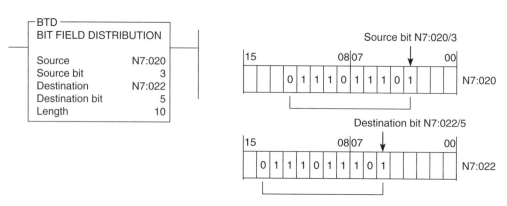

(a) Moving bits between words

(b) Moving bits between words. Bits are lost if they extend beyond the end of the destination word; the bits are not wrapped to the next higher word.

Fig. 10-8

Bit distribute (BTD) instruction.

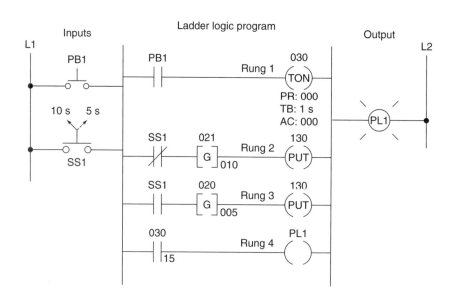

Fig. 10-9

Changing the preset value of a timer with GET/PUT instructions.

timer. Its NO contact closes when the preset value is equal to the accumulated value, thus indicating that the timer instruction has completed its function. To change the preset value of the timer to 5 s, the selector switch is turned to the 5-s position. This makes rung 3 true and rung 2 false. As a result, the preset value of the timer 030 will change from 010 to 005. Pressing pushbutton PB1 closed will now result in a 5-s time-delay period before the pilot light is energized.

Figure 10-10 shows how the *move* (MOV) data transfer instruction can be used to change the preset time of an on-delay timer. When the selector switch is in the 10-s position, rung 2 has logic continuity and rung 3 does not. As a result, the value 10 stored at the source address, N7:1, is copied into the destination address, T4:1 PRE. Therefore, the preset value of timer T4:1 will change from 0 to 10. When pushbutton

PB1 is closed, there will be a 10-s delay period before the pilot light is energized. To change the preset value of the timer to 5 s, the selector switch is turned to the 5-s position. This makes rung 3 true and rung 2 false. As a result, the preset value of timer T4:1 will change from 10 to 5. Pressing pushbutton PB1 closed will now result in a 5-s time-delay period before the pilot light is energized.

Figure 10-11 on pp. 264-265 shows an example of how the GET/PUT and *move* (MOV) data transfer instructions can be used to change the preset count of an up-counter. In this example, a limit switch programmed to operate a counter counts the products coming off a conveyor line onto a storage rack. Three different types of products are run on this line. The storage rack has room for only 300 boxes of product A or 175 boxes of product B or 50 boxes of product C. Three switches are provided to select the desired preset

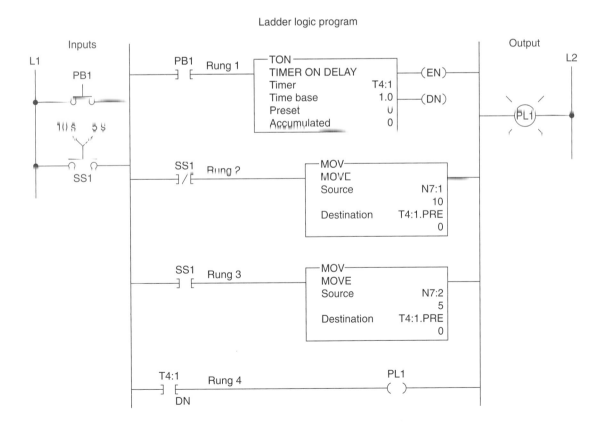

Ladder logic program

Fig. 10-10

Changing the preset value of a timer using the *move* (MOV) instruction.

counter value depending on the product line—A, B, or C—being manufactured. A reset button is provided to reset the accumulated count to zero. A pilot lamp is switched on to indicate when the storage rack is full. If more than one of the preset counter switches is closed, the *last* value is selected.

A *file* is a group of related consecutive words in the data table that have a defined start and end and are used to store information. For example, a batch process program may contain several separate recipes in different files that can be selected by an operator.

In some instances it may be necessary to shift complete files from one location to another within the programmable controller memory. Such data shifts are termed *file-to-file shifts*. File-to-file shifts are used when the data in one file represents a set of conditions that must interact with the programmable controller program several times and,

therefore, must remain *intact* after each operation. Because the data within this file must also be changed by the program action, a second file is used to handle the data changes, and the information within that file is allowed to be altered by the program. The data in the first file, however, remain constant and therefore can be used many times. Other types of data manipulation used with file instructions include word-to-file and file-to-word moves (Fig. 10-12).

Files allow large amounts of data to be scanned quickly and are useful in programs requiring the transfer, comparison, or conversion of data. Most PLC manufacturers display file instructions in block format on the programming terminal screen. With Allen-Bradley PLC-5 and SLC 500 controllers, the address that defines the beginning of a file starts with the index prefix # (Fig. 10-13 on p. 266). The # prefix is omitted in a word or element address.

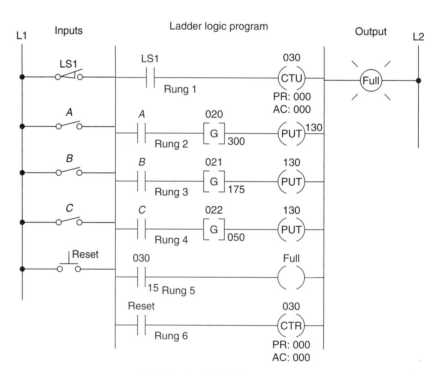

(a) Using the GET/PUT instruction

Fig. 10-11

Changing the preset value of a counter.

Ladder logic program

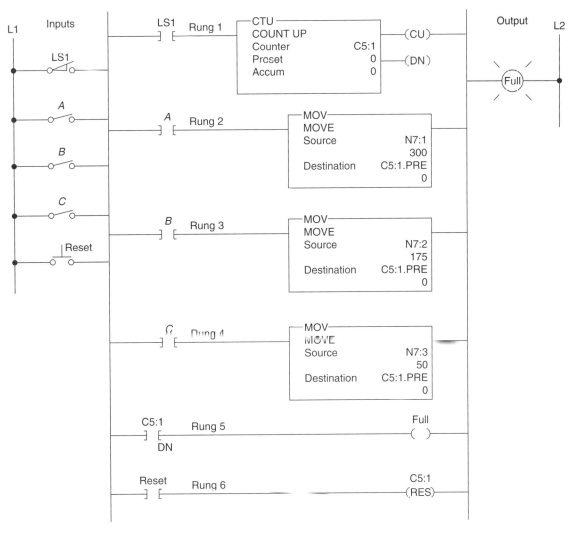

(b) Using the MOVE instruction

Fig. 10-11 (continued)

Changing the preset value of a counter.

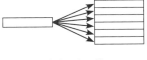

(a) Word to file

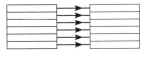

(b) File to file

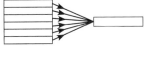

(c) File to word

Fig. 10-12

Moving data with file instructions.

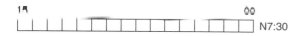

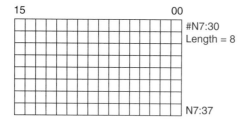

(a) Address N7:30 is a word address that represents a single word: word number 30 in integer file 7.

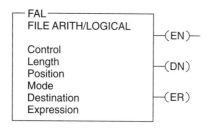

(b) Address #N7:30 represents the starting address of a group of consecutive words in integer file 7. The length shown is eight words, which is determined by the instruction where the file address is used.

Fig. 10-13

Allen-Bradley PLC-5 and SLC-500 word and file addresses.

The *file arithmetic and logic (FAL)* instruction is used to copy data from one file to another and to do file math and file logic. An example of the FAL instruction is shown in Fig. 10-14. The basic operation of the instruction is similar in all functions and requires the following parameters and addresses to be entered in the instruction:

```
┌ FAL ─────────────────┐
│  FILE ARITH/LOGICAL      ──(EN)─
│                        │
│  Control               │
│  Length                ──(DN)─
│  Position              │
│  Mode                  │
│  Destination           ──(ER)─
│  Expression            │
└────────────────────────┘
```

Fig. 10-14

FAL instruction.

◆ **Control** is the first entry and the address of the control structure in the control area (R) of processor memory. The processor uses this information to run the instruction. The default file for the control file is data file 6. Other control files may be assigned from data files 9 through 999. The control element for the

FAL instruction must be unique for that instruction and may not be used to control any other instruction. The control element is made up of three words. The control word in the FAL instruction uses four control bits: bit 15 (enable bit), bit 13 (done bit), bit 11 (error bit), and bit 10.

	Bit:	15	13	11	10
Control word:	R6:0	EN	DN	ER	UL
Length word:	R6:0.LEN	Stores file length, 1–1000			
Position word:	R6:0.POS	Position in the file			

◆ **Length** (the second word of the control element) is the second entry and represents the file length. This entry will be in words, except for the floating-point file, for which the length is in elements. (A floating-point element consists of two words.) The maximum length possible is 1000 elements. Enter any decimal number 1–1000.

◆ **Position** (the fourth word of the control element) is the third entry and represents the current location in the data block that the processor is accessing. It points to the word being operated on. The position starts with 0 and indexes to 1 less than the file length. You generally enter a 0 to start at the beginning of a file. You may also enter another position at which you want the FAL to start its operation. When the instruction resets, however, it will reset the position to 0. You can manipulate the position from the program.

◆ **Mode** is the fourth entry and represents the number of file elements operated on per program scan. There are three choices:
- The *all mode,* for which you enter an A. In the *all mode,* the instruction will transfer the complete file of data in *one* scan. The *enable* (EN) bit will go true when the instruction goes true and will follow the rung condition. When all of the data have been transferred, the *done*

(DN) bit will go true. This will be on the same scan as when the instruction goes true. If the instruction does not go to completion due to an error in the transfer of data (such as trying to store too large or too small a number for the data-table type), the instruction will stop at that point and set the *error* (ER) bit. The scan will continue, but the instruction will not continue until the error bit is reset. If the instruction goes to completion, the *enable bit* and the *done bit* will remain set until the instruction goes false, at which point the position, the *enable bit*, and the *done bit* will all be reset to 0.

- The *numeric mode,* for which you enter a decimal number (1–1000). In the numeric mode, the file operation is distributed over a *number* of program scans. The value you enter sets the number of elements to be transferred per scan. The numeric mode can decrease the time it takes to complete a program scan. Instead of waiting for the total file length to be transferred in one scan, the numeric mode breaks up the transfer of the file data into multiple scans, thereby cutting down on the instruction execution time per scan.

- The *incremental mode,* for which you enter an I. In the incremental mode, one element of data is operated on for every false-to-true transition of the instruction. The first time the instruction sees a false-to-true transition and the position is at 0, the data in the first element of the file is operated on. The position will remain at 0 and the UL bit will be set. The EN bit will follow the instruction's condition. On the second false-to-true transition, the position will index to 1, and data in the second word of the file will be operated on. The UL bit controls whether the instruction will operate just on data in the current position, or whether it will index the position and then transfer data. If the UL bit is reset, the instruction—on a false-to-true transition of the instruction—will operate on the data in the

current position and set the UL bit. If the UL bit is set, the instruction—on a false-to-true transition of the instruction—will index the position by 1 and operate on the data in its new position.

◆ **Destination** is the fifth entry and is the address where the processor stores the result of the operation. The instruction converts to the data type specified by the destination address. It may be either a file address or an element address.

◆ **Expression** is the last entry and contains addresses, program constants, and operators that specify the source of data and the operations to be performed. The expression entered determines the function of the FAL instruction. The expression may consist of file addresses, element addresses, or a constant, and may contain only one function because the FAL instruction may perform only one function.

Figure 10-15 on p. 268 shows an example of a file-to-file copy function using the FAL instruction. When input *A* goes true, data from the expression file #N7:20 will be copied into the destination file #N7:50. The length of the two files is set by the value entered in the control element word R6:1.LEN. In this instruction, we have also used the ALL mode, which means all of the data will be transferred in the first scan in which the FAL instruction does a false-to-true transition. The DN bit will also come on in that scan unless an error occurs in the transfer of data, in which case the ER bit will be set, the instruction will stop operation at that position, and then the scan will continue at the next instruction.

Figure 10-16 on p. 268 shows an example of a file-to-word copy function using the FAL instruction. With each false-to-true rung transition of input *A,* the processor reads one element of integer file N29, starting at element 0, and writes the image into element 5 of integer file N29. The instruction writes over any data in the destination.

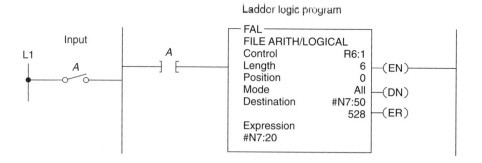

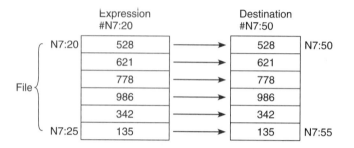

Fig. 10-15

File-to-file copy function using the FAL instruction.

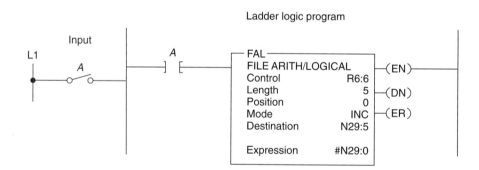

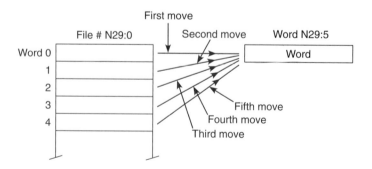

Fig. 10-16

File-to-word copy function using the FAL instruction.

Figure 10-17 shows an example of a word-to-file copy function using the FAL instruction. It is similar to the file-to-word copy function except that the instruction copies data from a word address into a file. Note that the expression (N7:100) is a word address and the destination is a file address (#N7:101). If we start with position 0, the data from N7:100 will be copied into N7:101 on the first false-to-true transition of input A. The second false-to-true transition of input A will copy the data from N7:100 into N7:102. On successive false-to-true transitions of the instruction, the data will be copied into the next position in the file until the end of the file, N7:106, is reached.

The exception to the rule that file addresses must take consecutive words in the data table are in the timer, counter, and control data files for the FAL instruction. In these three data files, if you designate a file address, the FAL instruction will take every third word in that file and make a file of preset, accumulated, length, or position data within the corresponding file type. This might be done, for example, so that recipes storing values for timer presets can be moved into the timer presets, as illustrated in Fig. 10-18.

The *file copy* (COP) instruction and the *fill file* (FLL) instruction are high-speed instructions that operate more quickly than the same operation with the FAL instruction. Unlike the FAL instruction, there is no control element to monitor or manipulate. Data conversion does not take place, so the source and destination should be the same file types. An

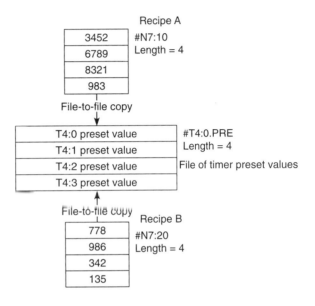

Fig. 10-18

Copying recipe and storing values for timer presets.

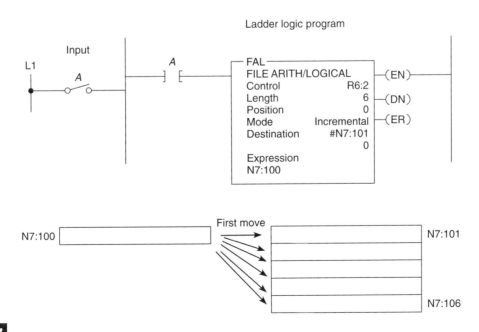

Fig. 10-17

Word-to-file copy function using the FAL instruction.

example of the file COP instruction is shown in Fig. 10-19. Both the source and destination are file addresses. When input *A* goes true, the values in file N40 are copied to file N20. The instruction copies the file length for each scan during which the instruction is true.

An example of the *fill file* (FLL) instruction is shown in Fig. 10-20. It operates in a manner similar to the FAL instruction that performs the word-to-file copy in the ALL mode. When input *A* goes true, the value in

N15:5 is copied into N20:1 through N20:6. Since the instruction transfers to the end of the file, the file will be filled with the same data value in each word.

The FLL instruction is frequently used to zero all of the data in a file, as illustrated in the program of Fig. 10-21. Momentarily pressing PB1 copies the contents of file #N10:0 into file #N12:0. Momentarily pressing PB2 then clears file #N12:0. Note that zero is entered for the source value.

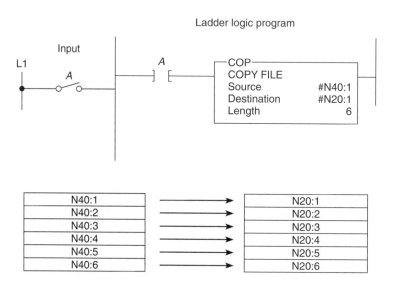

Fig. 10-19

File copy (COP) instruction.

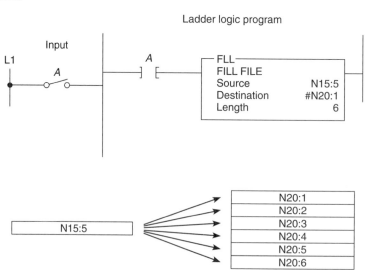

Fig. 10-20

Fill fill (FLL) instruction.

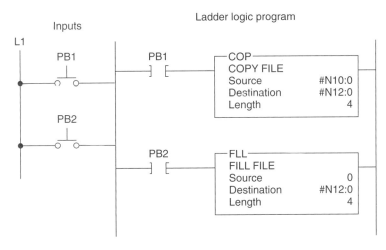

Inputs

Ladder logic program

L1

PB1

PB1

```
┌COP──────────────────┐
│ COPY FILE            │
│ Source        #N10:0 │
│ Destination   #N12:0 │
│ Length             4 │
└──────────────────────┘
```

PB2

PB2

```
┌FLL──────────────────┐
│ FILL FILE            │
│ Source             0 │
│ Destination   #N12:0 │
│ Length             4 │
└──────────────────────┘
```

Fig. 10-21

Using the FLL instruction to change all the data to 0 in a file.

10·3 DATA COMPARE INSTRUCTIONS

Data transfer operations are all output instructions, whereas data compare instructions are input instructions. Word comparison instructions have many uses in industry. The word compare instructions can be used when data needs to be compared before the next part of a process can take place. Applications where the comparison instructions are used include the following:

◆ To start an action or process when the counter is at a specific value.

◆ To verify that an input device's data (thumbwheel switch) is within range before sending the thumbwheel value to the presets of timers or counters.

◆ To verify that the parts are within tolerance.

Data compare instructions compare the data stored in two or more words (or registers) and make decisions based on the program instructions. With Allen-Bradley PLC-2 controllers, the compare instructions operate in conjunction with GET instructions. Like GET instructions, compare instructions are programmed in the condition portion of a logic rung. Numeric values in two words of memory can

be compared for each of the following conditions, depending on the PLC:

Name	Symbol
LESS THAN	(<)
EQUAL TO	(=)
GREATER THAN	(>)
LESS THAN OR EQUAL TO	(≤)
GREATER THAN OR EQUAL TO	(≥)

Data comparison concepts have already been used with the timer and counter instructions of previous units. In both of these instructions, an output was turned on or off when the accumulated value of the timer or counter equaled its preset value (AC = PR). What actually occurred was that the accumulated numeric data in one memory word was *compared* to the reset value of another memory word on each scan of the processor. When the processor saw that the accumulated value was equal to (=) the preset value, it switched the output on or off.

Figure 10-22 on p. 272 shows a logic rung that uses an EQUAL TO (=) instruction. A GET instruction is always programmed preceding any data compare instruction. The EQUAL TO instruction will have a logic true condition only when the value stored in the EQUAL

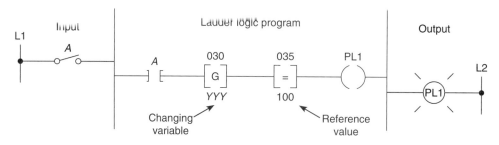

Output will be energized when input *A* is true and the value in word 030 is *equal* to the reference value *100* in word 035.

Fig. 10-22

GET/EQUAL TO logic rung.

TO instruction is the *same as* the value stored in the GET instruction. In this example, the GET value is the changing variable and is compared to the reference value of the EQUAL TO instruction for an equal to condition. Assume that input *A* is true. When the GET value *YYY* equals 100, the comparison is true and logic rung continuity is established.

Figure 10-23 shows a logic rung that uses a LESS THAN (<) instruction. The LESS THAN instruction will have a logic true condition only when the value stored in the GET instruction is *less than* the value stored in the LESS THAN instruction. In this example, if input *A* is true, rung continuity will be established any time the GET value is *less than* 100.

A GREATER THAN (>) comparison is also made with the GET/LESS THAN pair of instructions. This time the GET instruction value is the *reference* and the LESS THAN instruction value is the changing variable.

Figure 10-24 shows the program required to implement a greater than (>) comparison. In this example, if input *A* is true, rung continuity will be established any time the get value is *greater than* 100.

For a LESS THAN OR EQUAL TO (≤) comparison, the rung is programmed with one GET instruction, followed by LESS THAN (<) and EQUAL TO (=) instructions in parallel. The GET value is the changing value. The LESS THAN and EQUAL TO instructions are assigned a reference value. Figure 10-25 shows the program required to implement a LESS THAN OR EQUAL TO (≤) comparison. In this example, if input *A* is true, rung continuity will be established any time the GET value is either *less than* or *equal to* 100. Note that both the LESS THAN (<) and EQUAL TO (=) instructions use the same word address and have the same reference value.

The GREATER THAN OR EQUAL TO (≥) comparison rung is also programmed with

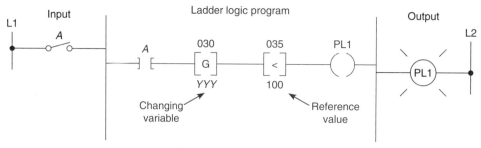

Output will be energized when input *A* is true and the value in word 030 is *less than* the reference value *100* in word 035.

Fig. 10-23

GET/LESS THAN logic rung.

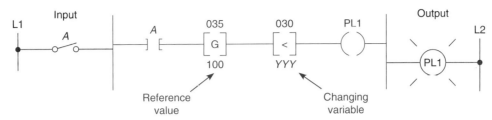

Fig. 10-24

GET/GREATER THAN logic rung.

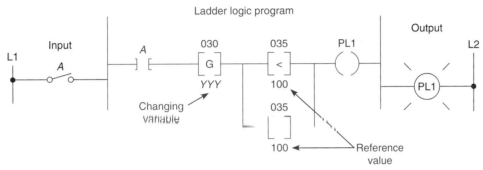

Fig. 10-25

GET/LESS THAN OR EQUAL TO logic rung.

one GET instruction, followed by LESS THAN and EQUAL TO instructions connected in parallel. In this instance however, the GET instruction value is the *reference value*. The LESS THAN and EQUAL TO values change, and are compared to the GET value.

Figure 10-26 shows the program required to implement a GREATER THAN OR EQUAL TO (≥) comparison. In this example, if input A is true, rung continuity will be established any time the LESS THAN OR EQUAL TO values are *greater than* or *equal to* 100.

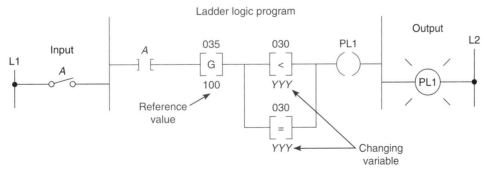

Fig. 10-26

GET/GREATER THAN OR EQUAL TO logic rung.

Allen-Bradley PLC-5 and SLC-500 controllers use block formatted data-comparison instructions. Their instruction set provides for an extensive comparison of data values. Figure 10-27 shows some of the compare instructions available for use with these controllers.

Figure 10-28 shows an example of the *equal* (EQU) input-comparison instruction. In this application, when the accumulated value of counter T4:0 stored in source A's address equals the value stored in source B's address, N7:40, the instruction is true and the output is energized. With the *equal* instruction, the floating-point data type is not recommended because of the exactness required. One of the other comparison instructions, such as the limit test, is preferable.

Figure 10-29 shows an example of the *not equal* (NEQ) input-comparison instruction. When the value stored at source A's address, N7:5, is not equal to 25, the output will be

If you want to...	Then use this instruction	Mnemonic
Test if two values are equal	EQUAL	EQU
Test if one value is not equal to a second value	NOT EQUAL	NEQ
Test if one value is greater than a second value	GREATER THAN	GRT
Test if one value is less than a second value	LESS THAN	LES
Test if one value is greater than or equal to a second value	GREATER THAN OR EQUAL	GEQ
Test if one value is less than or equal to a second value	LESS THAN OR EQUAL	LEQ
Test if one value is inside or outside the limit range of two other values	LIMIT TEST	LIM
Pass one bit pattern through a mask and test if it is equal to another bit pattern	MASKED COMPARISON FOR EQUAL	MEQ

Fig. 10-27

Available comparison instructions for Allen-Bradley PLC-5 and SLC-500 controllers.

Ladder logic program

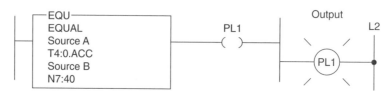

Fig. 10-28

EQUAL logic rung.

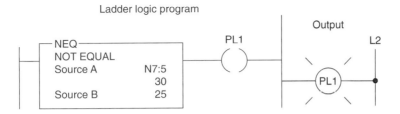

Fig. 10-29

NOT EQUAL (NEQ) logic rung.

true; otherwise, the output will be false. In all input-comparison instructions, source A and source B can either be values or addresses that contain values.

Figure 10-30 shows an example of the *greater than* (GRT) input-comparison instruction. The instruction is either true or false, depending on the values being compared. In this application, when the accumulated value of timer T4:10, stored at the address of source A, is greater than the constant 200 of source B, the output will be on; otherwise, it will be off.

Figure 10-31 shows an example of the *less than* (LES) input-comparison instruction. In this application, when the accumulated value of counter C5:10, stored at the address of source A, is less than the constant 350 of source B, the output will be on; otherwise, it will be off.

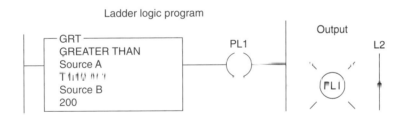

Fig. 10-30

GREATER THAN (GRT) logic rung.

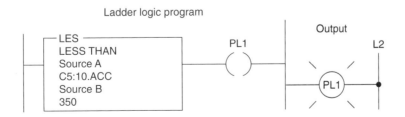

Fig. 10-31

LESS THAN (LES) logic rung.

Figure 10-32 shows an example of the *greater than or equal* (GEQ) input-comparison instruction. If the value stored at the address of source A, N7:55, is greater than or equal to the value stored at the address of source B, N7:12, the output will be true; otherwise, it will be false.

Figure 10-33 shows an example of the *less than or equal* (LEQ) input-comparison instruction. In this application, if the accumulated count of counter C5:1 is less than or equal to 457, the pilot light will turn on.

The *limit test instruction* (LIM) shown in Fig. 10-34 compares a test value to values in the low limit and the high limit. The limit test instruction is said to be *circular* because it can function in either of two ways:

◆ If the *high* limit has a *greater* value than the *low* limit, then the instruction is true if the value of the test is between or equal to the values of the high limit and the low limit.

◆ If the value of the *low* limit is *greater* than the value of the *high* limit, the

instruction is true if the value of the test is equal to or less than the low limit or equal to or greater than the high limit.

In Fig. 10-34a, the high limit has a value of 50, and the low limit has a value of 25. The instruction is true, then, for values of the test from 25 through 50. The instruction as shown is true because the value of the test is 48. In Fig. 10-34b, the high limit has a value of 50, and the low limit has a value of 100. The instruction is true, then, for test values of 50 and less than 50, and for test values of 100 and greater than 100. The instruction as shown is true because the test value is 125. Applications where the limit test instruction is used include the following:

◆ To allow the mixer to start as long as the temperature is within range.

◆ To allow a process to happen as long as the temperature is outside the range.

Figure 10-35 shows an example of the *masked comparison for equal* (MEQ) input-comparison instruction. This instruction compares a value from a source address with

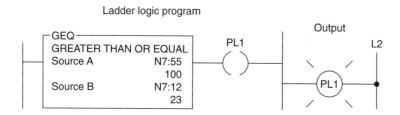

Fig. 10-32

GREATER THAN OR EQUAL (GEQ) logic rung.

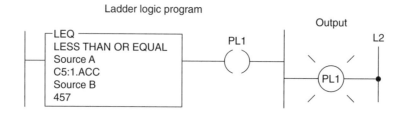

Fig. 10-33

LESS THAN OR EQUAL (LEQ) logic rung.

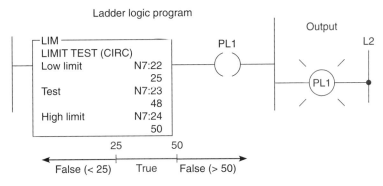

(a) High limit has a greater value than the low limit. Instruction is true for test values from 25 through 50. Instruction is false for test values less than 25 or greater than 50.

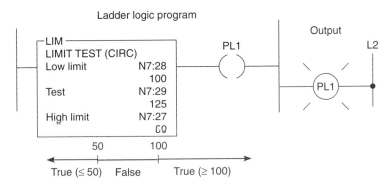

(b) Low limit has a greater value than the high limit. Instruction is false for test values greater than 50 and less than 100. Instruction is true for test values equal to or less than 50 or equal to or greater than 100.

Fig. 10-34

Limit test (LIM) instruction logic rung.

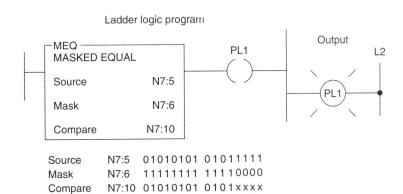

Source   N7:5   01010101 01011111
Mask   N7:6   11111111 11110000
Compare   N7:10   01010101 0101xxxx

Result: The instruction is true because reference bits xxxx are not compared.

Fig. 10-35

Masked comparison for equal (MEQ) logic rung.

data at a compare address and allows portions of the data to be masked. If the data at the source address match the data at the compare address bit-by-bit (less masked bits), the instruction is true. The instruction goes false as soon as it detects a mismatch. A mask passes data when the mask bits are set (1); a mask blocks data when the mask bits are reset (0). The mask must be the same element size (16 bits) as the source and compare address. You must set mask bits to 1 to compare data. Bits in the compare address that correspond to 0's in the mask are not compared. If you want the ladder program to change mask value, store the mask at a data address. Otherwise, enter a hexadecimal value for a constant mask value.

Applications where the masked compare for equal instruction is used include the following:

◆ To compare the correct position of up to 16 limit switches when the source contains the limit switch address and the compare stores their desired states. The mask can block out the switches you don't want to compare. (Position switches must be in for start-up.)

◆ When a two-digit thumbwheel is connected to an input module and you want to compare the 8 bits to a desired state, you can block out the other inputs connected to the module and make the comparison to just the two digits.

10·4 DATA MANIPULATION PROGRAMS

As mentioned, data manipulation instructions give new dimension and flexibility to the programming of control circuits. For example, consider the original relay-operated, time-delay circuit for Fig. 10-36. This circuit uses three pneumatic time-delay relays to control four solenoid valves. When the start pushbutton is pressed, solenoid A is

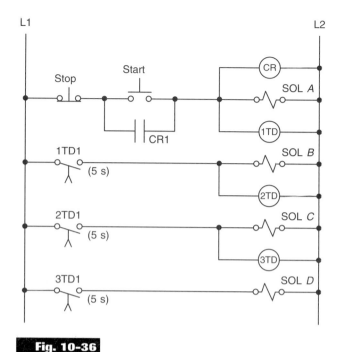

Fig. 10-36

Original relay time-delay circuits.

energized immediately, solenoid B is energized 5 s later, solenoid C is energized 10 s later, and solenoid D is energized 15 s later.

This circuit could be implemented using a conventional PLC program and three internal timers. However, the same circuit can be programmed using only *one* internal timer along with *data compare* statements, and this will result in a savings of memory words. Figure 10-37 shows the program required to implement the circuit using only one internal timer. Assume that the stop button is closed. When the start button is pushed, SOL A output will energize. As a result, solenoid A will switch on; SOL A contact will close to seal in output SOL A; contact SOL A will close to start on-delay timer TON 030. The timer has been preset to 15 s, and the accumulated time will be stored in word 030. Output SOL D will energize (through the *done* bit 030/15) after a total time delay of 15 s to energize solenoid D. Output SOL B will energize after a total time delay of 5 s when the accumulated time becomes equal to and then greater than the GET reference time (020/005) of 5 s.

Ladder logic program

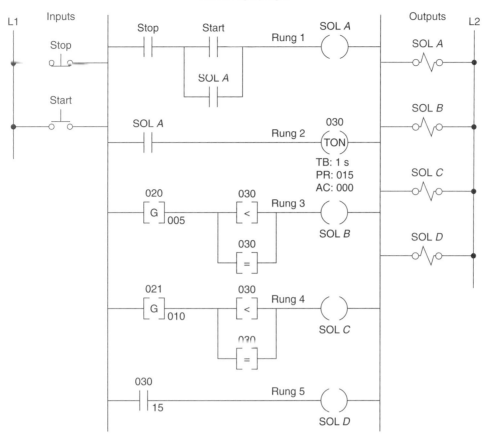

Fig. 10-37

Multiple timers using GET/GREATER THAN OR EQUAL TO data compare statements.

This in turn, will energize solenoid *D*. Output SOL *C* will energize after a total time delay of 10 s when the accumulated time becomes equal to and then greater than the get reference time (021/010) of 10 s. This, in turn, will energize solenoid *C*. Figure 10-38 on p. 280 shows the same operation programmed using the GREATER THAN OR EQUAL TO instruction.

Figure 10-39 on p. 280 shows an on-delay timer program that uses the EQUAL TO instruction. When the switch (S1) is closed, timer 031 will begin timing. Both GET instructions are addressed to get the accumulated value from the timer while it is running. The EQUAL TO instruction in rung 2 has the BCD value of 005 stored in address 022.

When the accumulated value of the timer is equal to 005, the EQUAL TO instruction in rung 2 will become logic true for 1 s. As a result, the *latch* output will energize to switch the light on. Then, when the accumulated value of the timer reaches 015, the EQUAL TO instruction (023/015) in rung 3 will be true for 1 s. As a result, *unlatch* output will energize to switch the light off. Therefore, when the switch is closed, the light will come on after 5 s, stay on for 10 s, and then turn off. Figure 10-40 on p. 281 shows the same operation programmed using the EQUAL instruction.

Figure 10-41 on p. 281 shows an up-counter program that uses the LESS THAN instruction. Up-counter 031 of rung 1 will increment

Ladder logic program

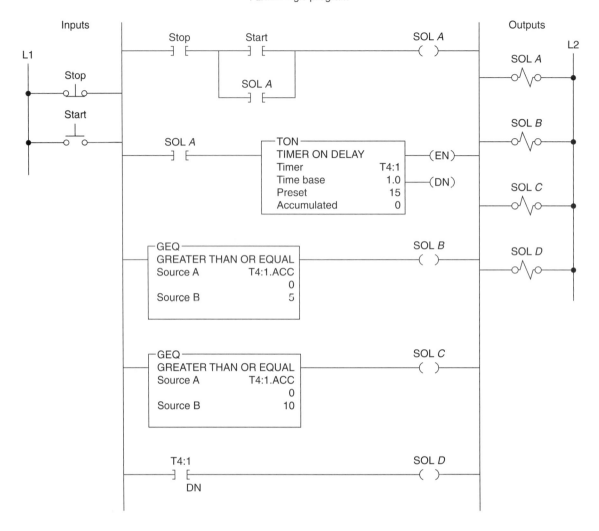

Fig. 10-38

Multiple timers using the GREATER THAN OR EQUAL TO instruction.

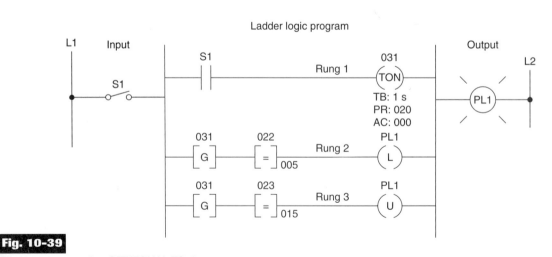

Fig. 10-39

Timer program using GET/EQUAL TO data compare statements.

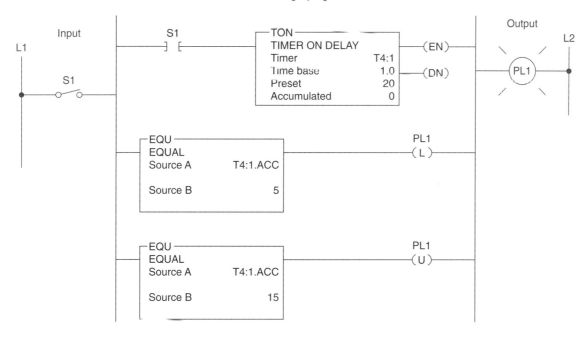

Fig. 10-40

Timer program using the EQUAL instruction.

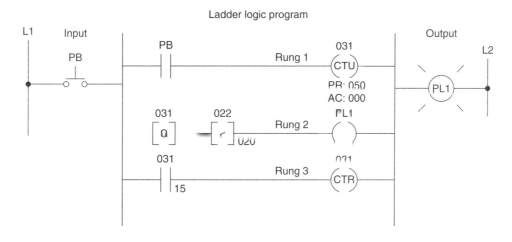

Fig. 10-41

Counter program using GET/LESS THAN data compare statements.

by 1 for every false-to-true transition of push-button input. Rung 2 contains a GET instruction that is addressed to the accumulated value of the counter (031) and a LESS THAN instruction that has the BCD value 020 stored in address 022. The LESS THAN instruction will be true as long as the value contained in the GET instruction is less than the value stored in the LESS THAN instruction.

Therefore, the output and the light will be on when the accumulated value of the counter is between 000 and 019. As soon as the counter's accumulated value is 020, the LESS THAN instruction will go false, turning off the output and the light. When the counter's accumulated value reaches its preset value of 050, the counter reset will be energized through the done bit 15 of word 031 to reset

Data Manipulation Instructions

the accumulated count to zero. Figure 10-42 shows the same operation programmed using the LESS THAN instruction.

The counter program of Fig. 10-41 on p. 281 can be modified as shown in Fig. 10-43 to achieve a GREATER THAN instruction. In this instance, the GET instruction has a fixed reference value of 020 stored in address 022, and the LESS THAN instruction is addressed to the counter's accumulated value. The output and the light will now be on when the counter's accumulated value is from 021 to 050, at which point rung 3 will reset the counter.

The use of comparison instructions is generally straightforward. However, one common programming error involves the use of these instructions in a PLC program to control the

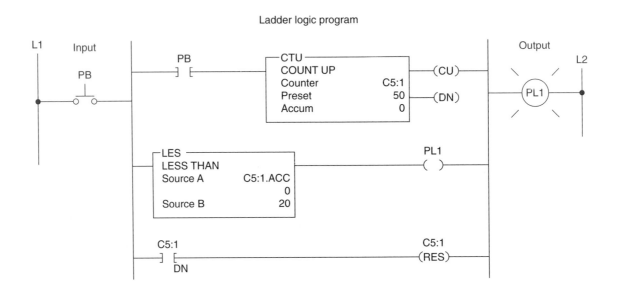

Fig. 10-42

Counter program using the LESS THAN instruction.

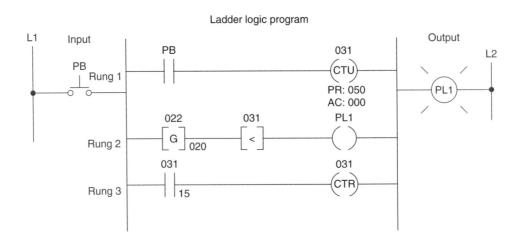

Fig. 10-43

Counter program using the GET/GREATER THAN data compare statement.

flow of a raw material into a vessel. The receiving vessel has its weight monitored continuously by the PLC program as it fills. When the weight reaches a preset value, the flow is cut off. While the vessel fills, the PLC performs a comparison between the vessel's current weight and a predetermined constant programmed in the processor. If the programmer uses only the EQUAL TO instruction, problems may result. As the vessel fills, the comparison for equality will be false. At the instant the vessel weight reaches the desired preset value of the EQUAL TO instruction, the instruction becomes true and the flow is stopped. However, should the supply system leak additional material into the vessel, the total weight of the material could rise *above* the preset value, causing the instruction to go false and the vessel to overfill. The simplest solution to this problem is to program the comparison instruction as GREATER THAN OR EQUAL TO. This way any excess material entering the vessel will not affect the filling operation. It may be necessary, however, to include additional programming to indicate a serious overfill condition.

10.5 NUMERICAL DATA I/O INTERFACES

The expanding data manipulation processing capabilities of PLCs led to a new class of I/O interfaces known as numerical data I/O interfaces. In general, numerical data I/O interfaces can be divided into two groups: those that provide interface to *multibit* digital devices and those that provide interface to *analog* devices.

The multibit digital devices are like the discrete I/O because processed signals are discrete (on/off). The difference is that, with the discrete I/O, only a *single* bit is required to read an input or control an output. Multibit interfaces allow a *group* of bits to be input or output *as a unit.* They are used to accommodate devices that require BCD inputs or outputs.

Figure 10-44 on p. 284 shows a BCD input interface module connected to *thumbwheel switches* (TWS). The BCD-input module allows the processor to accept 4-bit digital codes. This interface inputs data into specific register or word locations in memory to be used by the control program. Register input modules generally accept voltages in the range of 5 V dc (TTL) to 24 V dc. They are grouped in a module containing 16 or 32 inputs, corresponding to one or two I/O registers, respectively. Data manipulation instructions are used to access the data from the module that provides data for registers used in the control program. This allows a person to change set points or presets *externally* without modifying the control program.

Figure 10-45 on p. 285 shows a BCD output interface module connected to a seven-segment LED display board. This interface is used to output data from a specific register or word location in memory. Register output modules generally provide voltages that range from 5 V dc (TTL) to 30 V dc and have 16 or 32 output lines corresponding to one or two I/O registers, respectively. This type of module enables a PC to operate devices that require BCD-coded signals. The BCD output module can also be used to drive small dc loads that have current requirements in the 0.5-A range.

Figure 10-46 on p. 286 shows a PLC program that uses a BCD input interface module connected to thumbwheel switches, and a BCD output interface module connected to an LED display board. The decimal setting of the thumbwheel switches is monitored by the LED display board. In this program, the setting of the thumbwheel switches is compared to that of the reference number (100) stored in word 035. Output 011/12 will be energized when input 110/02 is true and the value of the thumbwheel switches is equal to the reference value 100 in word 035. Figure 10-47 on p. 286 shows the same operation programmed using the MOVE and EQUAL instructions.

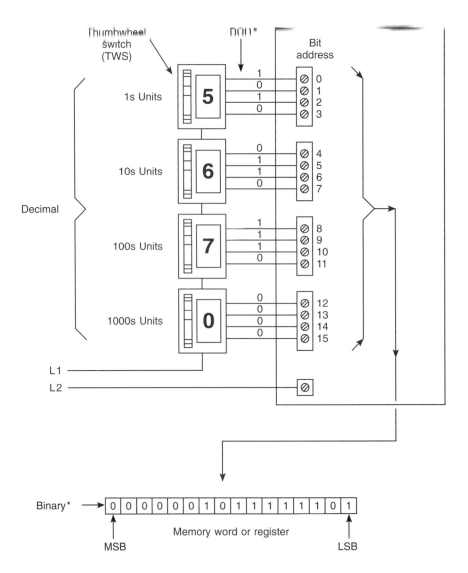

Thumbwheel switch (TWS)

Bit address

1s Units — 5
1 — 0
0 — 1
1 — 2
0 — 3

10s Units — 6
0 — 4
1 — 5
1 — 6
0 — 7

Decimal

100s Units — 7
1 — 8
1 — 9
1 — 10
0 — 11

1000s Units — 0
0 — 12
0 — 13
0 — 14
0 — 15

L1

L2

Binary* → | 0 | 0 | 0 | 0 | 0 | 0 | 1 | 0 | 1 | 1 | 1 | 1 | 1 | 1 | 0 | 1 |

Memory word or register

MSB   LSB

*In this illustration, it is assumed that the PLC processor is programmed to convert the BCD value into an equivalent binary value, and then to load that binary value into the register.

(a)

(b)

Fig. 10-44

BCD input interface module. (a) Module. (b) Four BCD thumbwheel switches. *(Courtesy of Cincinnati Milacron)*

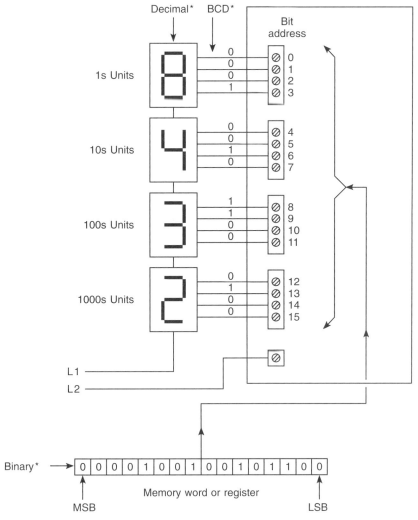

Decimal*   BCD*

Bit address

1s Units

0   ⊘ 0
0   ⊘ 1
0   ⊘ 2
1   ⊘ 3

10s Units

0   ⊘ 4
0   ⊘ 5
1   ⊘ 6
0   ⊘ 7

100s Units

1   ⊘ 8
1   ⊘ 9
0   ⊘ 10
0   ⊘ 11

1000s Units

0   ⊘ 12
1   ⊘ 13
0   ⊘ 14
0   ⊘ 15

⊘

L1

L2

Binary* → | 0 | 0 | 0 | 0 | 1 | 0 | 0 | 1 | 0 | 0 | 1 | 0 | 1 | 1 | 0 | 0 |

Memory word or register

MSB   LSB

*In this illustration, it is assumed that the PLC
processor is programmed to convert the binary value
in the register into an equivalent BCD value, and
that the LED display board is responsible for
encoding the programmable controller BCD output
to produce the correct decimal digit on each display.

(a)

(b)

Fig. 10-45

BCD output interface module.
(a) Module. (b) BCD to seven-
segment display. *(Courtesy of
Cincinnati Milacron)*

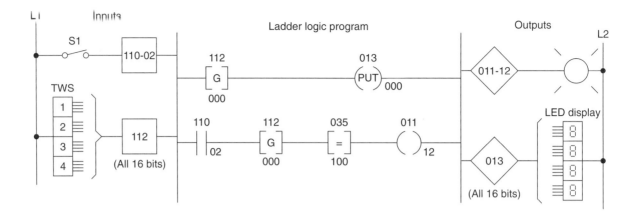

Fig. 10-46

BCD I/O program using GET/PUT and GET/EQUAL TO data compare statements.

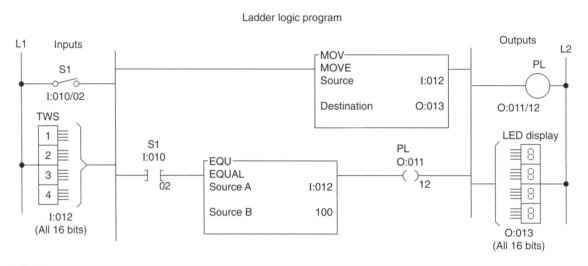

Fig. 10-47

BCD I/O program using MOVE and EQUAL Instructions.

Input and output modules can be addressed either at the bit level or at the word level, whichever is more convenient in the ladder logic program. Analog modules convert analog signals to 16-bit digital signals (input) or 16-bit digital signals to analog values (output). For example, they are used to interface with thermocouples and pressure transducers.

An analog I/O will allow monitoring and control of analog voltages and currents. Figure 10-48 shows how an analog input interface operates. The analog input module contains the circuitry necessary to accept analog voltage or current signals from field devices. These voltage or current inputs are converted from an analog to a digital value by an A/D converter circuit. The conversion value, which is proportional to the analog signal, is passed through the controller's data bus and stored in a specific register or word location in memory for later use by the control program.

An analog output interface module receives numerical data from the processor; this data is then translated into a proportional voltage or current to control an analog field device.

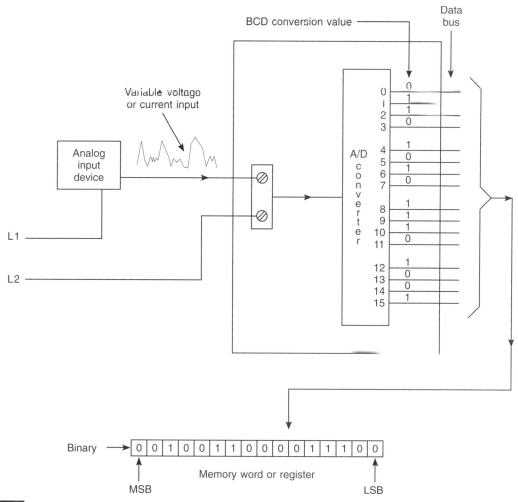

BCD conversion value

Data bus

Variable voltage or current input

Analog input device

L1

L2

A/D converter

0 — 0
I — 1
2 — 1
3 — 0
4 — 1
5 — 0
6 — 1
7 — 0
8 — 1
9 — 1
10 — 1
11 — 0
12 — 1
13 — 0
14 — 0
15 — 1

Binary → | 0 | 0 | 1 | 0 | 0 | 1 | 1 | 0 | 0 | 0 | 0 | 1 | 1 | 1 | 0 | 0 |

MSB

Memory word or register

LSB

Fig. 10-48

Analog input interface module.

Figure 10-49 on p. 288 shows how an analog output interface operates. Data from a specific register or word location in memory is passed through the controller's data bus to a D/A converter. The analog output from the converter is then used to control the analog output device. These output interfaces normally require an external power supply that meets certain current and voltage requirements.

10.6 SET-POINT CONTROL

Set-point control in its simplest form compares an input value, such as an analog or thumbwheel inputs, to a set-point value. A discrete output signal is provided if the input value is less than, equal to, or greater than the set-point value.

The temperature control program of Fig. 10-50 on p. 289 (Allen-Bradley PLC-2 protocol) is one example of set-point control. In this application, a PLC is to provide for simple off/on control of the electric heating elements of an oven. The oven is to maintain an average set-point temperature of 600°F, with a variation of about 1% between the off and on cycles. Therefore, the electric heaters will be turned on when the temperature of the oven is 597°F or less and stay on until the temperature rises to 603°F or more. The electric heaters stay off until the temperature drops down to 597°F, at which time the cycle repeats itself. Rung 2

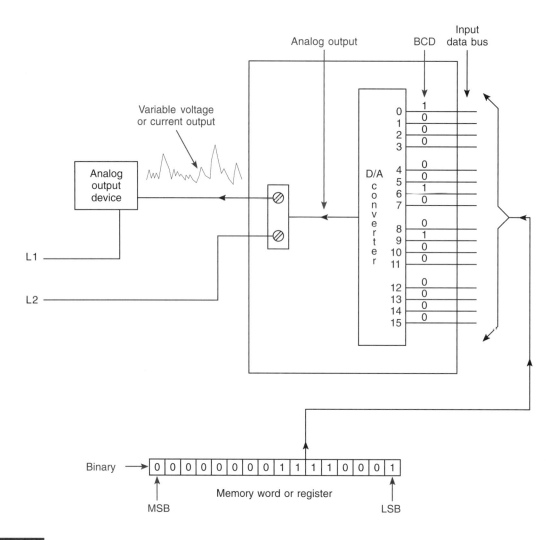

Analog output

Variable voltage
or current output

BCD

Input
data bus

Analog
output
device

L1

L2

D/A
converter

0	1
1	0
2	0
3	0
4	0
5	0
6	1
7	0
8	0
9	1
10	0
11	0
12	0
13	0
14	0
15	0

Binary →

0	0	0	0	0	0	0	0	1	1	1	1	0	0	0	1

Memory word or register

MSB

LSB

Fig. 10-49

Analog output interface module.

contains a GET/LESS THAN OR EQUAL TO logic instruction and rung 3 contains the equivalent of a GET/GREATER THAN OR EQUAL TO logic instruction. Rung 4 contains the logic for switching the heaters on and off according to the high and low set-points. Rung 1 contains the logic that allows the thermocouple temperature to be monitored by the LED display board. Figure 10-51 shows the same operation programmed using Allen-Bradley PLC-5 or SLC-500 protocol.

Several common set-point control schemes can be performed by different PLC models. These include on/off control, proportional (P) control, proportional-integral (PI) control, and proportional-integral-derivative

(PID) control. Each involves the use of some form of *closed-loop control* to maintain a process characteristic such as a temperature, pressure, flow, or level at a desired value.

A typical block diagram of a closed-loop control system is shown in Fig. 10-52 on p. 290. A measurement is made of the variable to be controlled. This measurement is then compared to a reference point, or set-point. If a difference (error) exists between the actual and desired levels, the PLC control program will take the necessary corrective action. Adjustments are made continuously by the PLC until the difference between the desired and actual output is as small as is practical.

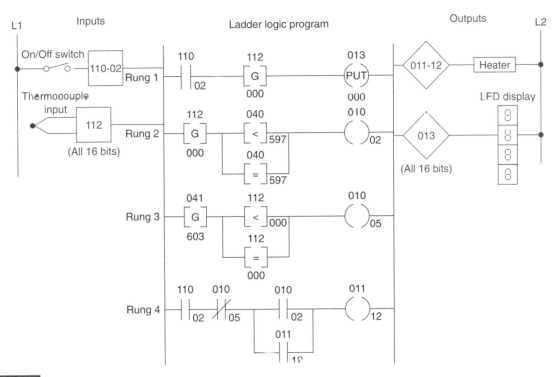

Fig. 10-50

Set-point temperature control program using Allen-Bradley PLC-2 protocol.

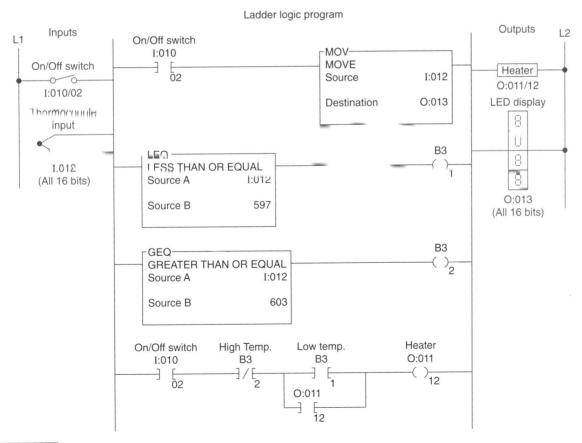

Fig. 10-51

Set-point temperature control program using Allen-Bradley PLC-5 or SLC-500 protocol.

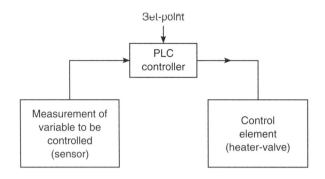

Set-point

PLC
controller

Measurement of
variable to be
controlled
(sensor)

Control
element
(heater-valve)

Fig. 10-52

Block diagram of closed-loop control system.

With *on/off* control (also known as *two-position* and *bang-bang control*), the output or final control element is either on or off—one for the occasion when the value of the measured variable is above the set-point, and the other for the occasion when the value is below the set-point. The controller will never keep the final control element in an intermediate position.

On/off control is inexpensive but not accurate enough for many process and machine control applications. On/off control almost always means overshoot and resultant system cycling. A *deadband* is usually required around the set-point to prevent relay chatter at set-point. On/off control does not adjust to the time constants of a particular system.

Proportional controls are designed to eliminate the hunting or cycling associated with on/off control. They allow the final control element to take intermediate positions between on and off. This permits *analog control* of the final control element to vary the amount of energy to the process, depending on how much the value of the measured variable has shifted from the desired value.

The process illustrated in Fig. 10-53 is an example of a proportional control process. The PLC analog output module controls the amount of fluid placed in the holding tank by adjusting the percentage of valve opening. The valve is initially open 100%. As the fluid level in the tank approaches the preset point, the processor modifies the output to degrade

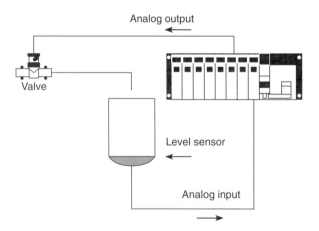

Analog output

Valve

Level sensor

Analog input

Fig. 10-53

Proportional control process.

closing the valve by different percentages, adjusting the valve to maintain a set-point.

Proportional-integral derivative (PID) control is the most sophisticated and widely used type of process control. PID operations are more complex and mathematically based (Fig. 10-54). PID controllers produce outputs that depend on the *magnitude, duration,* and *rate of change* of the system error signal. Sudden system disturbances are met with an aggressive attempt to correct the condition. A PID controller can reduce the system error to 0 faster than any other controller.

Either programmable controllers are equipped with PID I/O modules that produce PID control or they have sufficient mathematical functions of their own to allow PID control to be carried out. In a PID module, the proportional mode produces an output signal proportional to the difference between the measurement being taken and the set-point entered in the PLC. The integral control function produces an output proportional to the amount and length of time that the error signal is present. The *derivative* section produces an output signal proportional to the rate of change of the error signal.

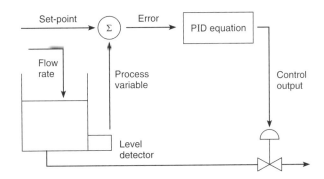

Fig. 10-54

Typical PID control loop.

Questions

1. Explain the difference between a register or word and a table or file.

2. What do data manipulation instructions allow the PLC to do?

3. Into what two broad categories can data manipulation instructions be placed?

4. What is involved in a data transfer instruction?

5. Explain what the logic rung in Fig. 10-55 is telling the processor to do.

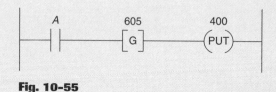

Fig. 10-55

6. The MOVE instruction will be used to copy the information stored in word N7:20 to N7:35. What address should be entered into the source and destination?

7. Explain the purpose of the MOVE WITH MASK instruction.

8. Explain the purpose of the BIT DISTRIBUTE instruction.

9. List three types of data manipulation used with file instructions.

10. List the six parameters and addresses that must be entered into the *file arithmetic and logic* (FAL) instruction.

11. Assume the *all mode* has been entered as part of a FAL instruction. How will this affect the transfer of data?

12. What is the advantage of using the *file copy* (COP) or *fill file* (FLL) instruction rather than the FAL instruction for the transfer of data?

13. What is involved in a DATA COMPARE instruction?

14. Name and draw the symbol for five different types of DATA COMPARE instructions.

15. Explain how the *limit test* (LIM) instruction compares data.

16. How are multibit I/O interfaces different from the discrete type?

17. Assume a thumbwheel switch is set for the decimal number 3286.

    a. What is the equivalent BCD value for this setting?
    b. What is the equivalent binary value for this setting?

18. Assume that a thermocouple is connected to an analog input module. Explain how the temperature of the thermocouple is communicated to the processor.

19. Name two typical analog output field devices.

20. With the aid of a block diagram, explain the basic operation of a closed-loop control system.

21. Explain what each of the logic rungs in Fig. 10-56 is telling the processor to do.

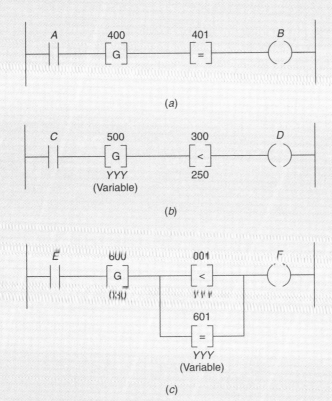

(a)

(b)

(c)

Fig. 10-56

22. Compare the operation of the final control element in on/off and proportional control systems.

23. Explain the type of output produced by each of the following sections of a PID module:

    a. Proportional control
    b. Integral control
    c. Derivative control

Problems

1. Study the data transfer program in Fig. 10-57 and answer the following questions:

    a. When S1 is off, what number will appear in the XXX position below the PUT instruction?

    b. When S1 is on, what number will appear in the XXX position below the PUT instruction?

    c. When S1 is on, what number will appear in the LED display?

    d. What is required for the number 216 to appear in the LED display?

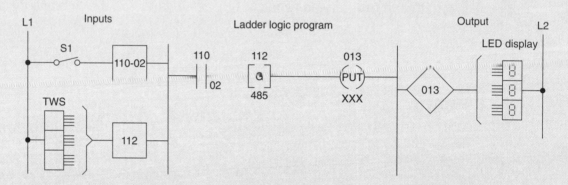

Fig. 10-57

2. Study the data transfer counter program in Fig. 10-58 and answer the following questions:

    a. Where does the three-digit number XXX below the GET instruction in rung 2 come from?

    b. Outline the steps to follow to operate the program so that output 010/01 is energized after 25 off-to-on transitions of input 110/17.

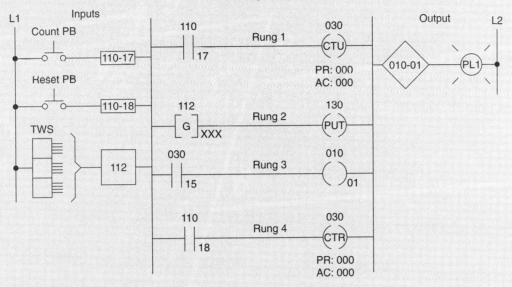

Ladder logic program

Fig. 10-58

3. Construct a nonretentive timer program that will turn on a pilot light after a time-delay period. Use a thumbwheel switch to vary the preset time-delay value of the timer.

4. Study the data compare program in Fig. 10-59 and answer the following questions:

   a. Will the pilot light come on whenever switch S1 is closed?

   b. Must switch S1 be closed to change the number XXX below the GET instruction?

   c. What number (XXX) must be below the GET instruction to have the pilot light turn on?

   d. What is word address 055?

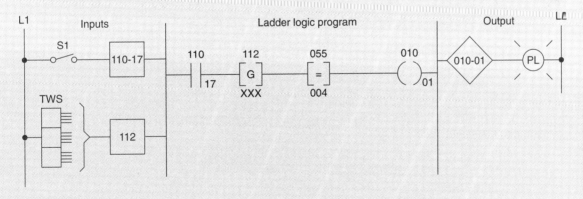

Fig. 10-59

5. Study the data compare program in Fig. 10-60 and answer the following questions:

   a. List the values that could be found below the GET instruction that would allow the pilot light to turn on.

   b. If the value in the word 112 is 003 and switch S1 is open, will the pilot light turn on?

   c. Since word 061 is a storage word, could you enter a number other than 012 into the three-digit position below the [<] instruction?

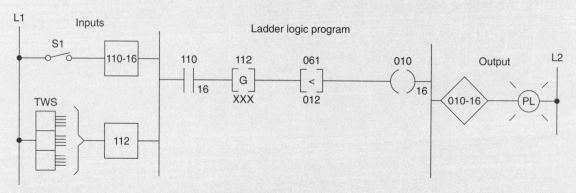

Fig. 10–60

6. Write a program to perform the following:

   a. Turn on pilot light 1 (PL1) if the thumbwheel switch value is less than 4.
   b. Turn on pilot light 2 (PL2) if the thumbwheel switch value is equal to 4.
   c. Turn on pilot light 3 (PL3) if the thumbwheel switch value is greater than 4.
   d. Turn on pilot light 4 (PL4) if the thumbwheel switch value is less than or equal to 4.
   e. Turn on pilot light 5 (PL5) if the thumbwheel switch value is greater than or equal to 4.

7. Write a program that will copy the value stored at address N7:56 into address N7:60.

8. Write a program that uses the mask move instruction to move only the upper 8 bits of the value stored at address I:2.0 to address O:2.1 and ignore the lower 8 bits.

9. Write a program that uses the FAL instruction to copy 20 words of data from the integer data file, starting with N7:40, into the integer data file, starting with N7:80.

10. Write a program that uses the COP instruction to copy 128 bits of data from the memory area, starting at B3:0, to the memory area, starting at B3:8.

11. Write a program that will cause a light to come on only if a PLC counter has a value of 6 or 10.

12. Write a program that will cause a light to come on if a PLC counter value is less than 10 or more than 30.

13. The temperature reading from a thermocouple is to bo read and stored in a memory location every 5 min for 4 h. The temperature reading is brought in continuously and stored in address N7:150. File #7:200 is to contain the data from the last full 4-hour period.

Math Instructions

After completing this chapter, you will be able to:

◆ Analyze and interpret math instructions as they apply to a PLC program

◆ Create PLC programs involving math instructions

◆ Apply combinations of PLC arithmetic functions to processes

Most PLCs have arithmetic function capabilities. The PLC can do many arithmetic operations per second for fast updating when needed. The usual interval between PLC arithmetic function updates is 1 or 2 scan times. This chapter covers the basic mathematical functions performed by PLCs and their applications.

11·1 MATH INSTRUCTIONS

Math instructions, like data manipulation instructions, enable the programmable controller to take on some of the qualities of a computer system. The PLC's math functions capability is not intended to allow it to replace a calculator, but rather to allow it to perform arithmetic functions on values stored in memory words.

Depending on what type of processor is used, various math instructions can be programmed. The basic mathematical functions performed by PLCs are: addition (+), subtraction (−), multiplication (×), and division (÷). These instructions use the contents of two words or registers and perform the

desired function. The PLC instructions for data manipulation (data transfer and data compare) are used with the math symbols to perform math functions. Math instructions are all output instructions.

11·2 ADDITION INSTRUCTION

The ADD instruction (+) performs the addition of two values stored in the referenced memory locations. How these values are accessed depends on the controller. Figure 11-1 shows the format used by the Allen-Bradley PLC-2 family of PLCs. Again, even though the format and instructions vary with each manufacturer, the concepts remain the same. In this controller, addition is accomplished by reporting the values stored in two GET instructions, immediately followed by the ADDITION instruction. When input device A is true, the value of word 030 (105) is added to the value of word 031 (080), and the sum (185) is stored in word 032.

Allen-Bradley PLC-5 and SLC-500 controllers use a block-formatted ADD instruction for addition. Figure 11-2 shows an example of the ADD instruction. In this example, when the rung is true, the value stored at the source A address, N7:0 (25), is added to the value stored at the source B address, N7:1 (50), and

```
Input A    030      031           032
 ┤ ├      ─┤G├─    ─┤G├─          ─(+)─
           105      080            185

                          (105 + 080 = 185)
```

Fig. 11-1

Allen-Bradley PLC-2 ADD instruction.

Ladder logic program

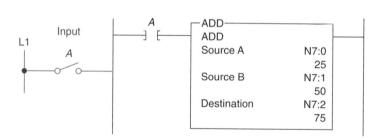

Fig. 11-2

Allen-Bradley PLC-5 and SLC-500 ADD instruction.

the answer (75) is stored at the destination address, N7:2. A constant may be entered either in source A or in source B.

The program of Fig. 11-3 shows how the ADD instruction can be used to add the accumulated counts of two up-counters. This application requires a light to come on when the sum of the counts from the two counters is equal to or greater than 350. GET instruction 050 is addressed to get the accumulated value from counter 050. GET instruction 051 is addressed to get the accumulated value from counter 051. These two values are added together by the addition address 020 of rung 3. Rung 4 is a GET/GREATER THAN OR EQUAL TO logic rung. The output in rung 4 will be logic true whenever the accumulated values in the two counters are equal to or greater than the referenced value, 350. A reset button is provided to reset the accumulated count of both counters to zero.

Figure 11-4 on p. 302 shows the same operation programmed using block-formatted PLC-5 or SLC-500 instructions.

When performing math functions, care must be taken to ensure that values remain in the range that the data-table can store or the overflow bit will be set. For example, with the Allen-Bradley PLC-5 controller, you cannot store a value larger than 32,767 at an integer address. For an Allen-Bradley PLC-2, the maximum value is 999.

11·3 SUBTRACTION INSTRUCTION

The SUBTRACT (−) instruction is similar to the ADD instruction. However, it uses a store word or register to reflect the difference between, rather than the sum of, two GET

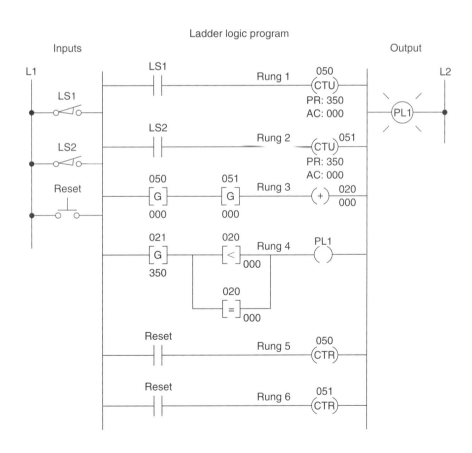

Fig. 11-3

PLC-2 counter program that uses the ADD instruction.

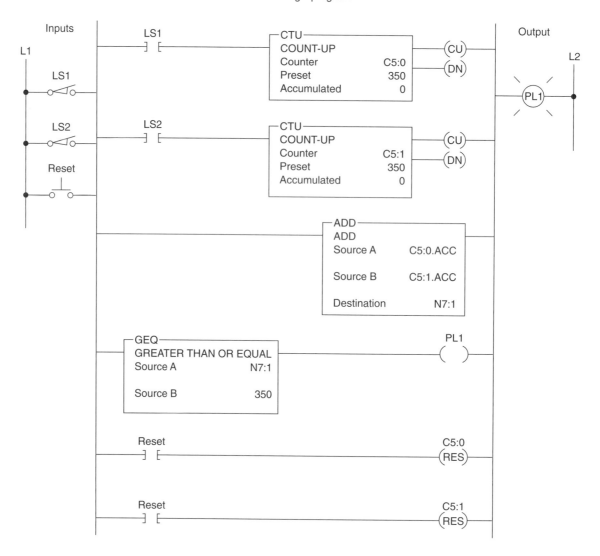

Fig. 11-4

PLC-5 and SLC-500 counter program that uses the ADD instruction.

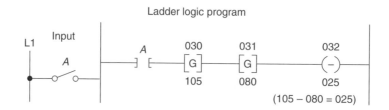

Fig. 11-5

Allen-Bradley PLC-2 SUBTRACT instruction.

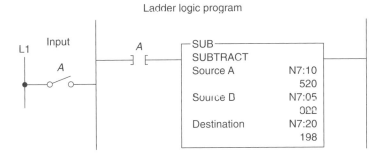

Ladder logic program

Fig. 11-6

Allen-Bradley block-formatted SUBTRACT instruction.

operation, inaccurate results may occur. Some PLCs flag an error if the number is negative.

Figure 11-6 shows an example of the block-formatted SUBTRACT instruction used as part of the Allen-Bradley PLC-5 and SLC-500 controllers instruction set. In this example, when the rung is true, the value stored at the source B address, N7:05 (322), is subtracted from the value stored at the source A address, N7:10 (520), and the answer (198) is stored at the destination address, N7:20.

values. Figure 11-5 shows an Allen-Bradley PLC-2 SUBTRACT instruction rung. When input device *A* is true, the value of word 031 (080) is subtracted from the value of word 030 (105), and the difference (025) is stored in word 032. Only positive values can be used with some PLCs. When the difference is a negative number and it is used for a subsequent

The program of Fig. 11-7 shows how the SUBTRACT function can be used to indicate a vessel overfill condition. This application requires an alarm to sound when a supply system leaks 5 lb or more of raw material into

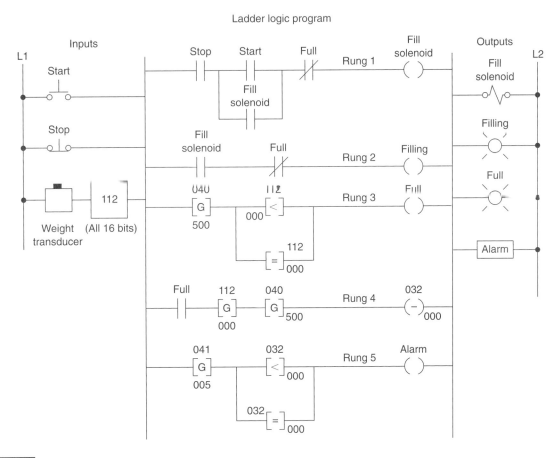

Fig. 11-7

Allen-Bradley PLC-2 overfill alarm program.

the vessel after a preset weight of 500 lb has been reached. When the start button is pressed, the fill solenoid (rung 1) and filling indicating light (rung 2) are turned on and raw material is allowed to flow into the vessel. The vessel has its weight monitored continuously by the PLC program (rung 3) as it fills. When the weight reaches 500 lb, the fill solenoid is de-energized and the flow is cut off. At the same time, the filling pilot light indicator is turned off and the full pilot light indicator (rung 3) is turned on. Should the fill solenoid leak 5 lb or more of raw material into the vessel, the alarm (rung 5) will

energize and stay energized until the over-flow level is reduced below the 5-lb overflow limit. Figure 11-8 shows the same operation programmed using Allen-Bradley PLC-5 or SLC-500 block-formatted instructions.

11·4 MULTIPLICATION INSTRUCTION

The Allen-Bradley PLC-2 MULTIPLY (×) (×) instruction operates on *two* stored memory words instead of one. Two memory words

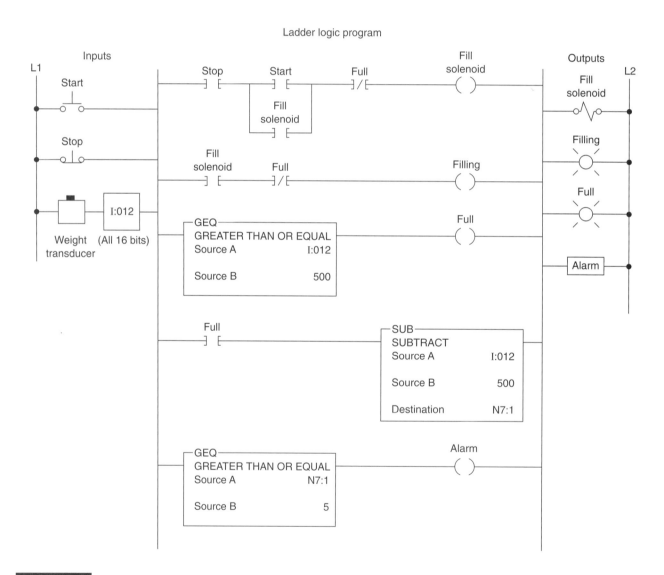

Ladder logic program

Fig. 11-8

Allen-Bradley PLC-5 and SLC-500 overfill alarm program.

are used so that a number larger than 999 can be displayed and stored. The first word contains the *most significant digit* (MSD) and the second word contains the *least significant digit* (LSD). If the product is less than six digits, leading zeros appear in the product. For good documentation habits, the manufacturer recommends using *consecutive* word addresses for the two addresses of the MULTIPLY instruction.

Figure 11-9 shows a simple MULTIPLY program. When input device *A* is true, the value of word 030 (123) is multiplied by the value of word 031 (061), and the product (7 503) is stored in word 032 (007) and word 033 (503). As a result, rung 2 will become

true, turning output PL1 on. Figure 11-10 shows the same operation programmed using Allen-Bradley PLC-5 or SLC-500 block-formatted instructions.

Figure 11-11 on p. 306 shows how the Allen-Bradley PLC-2 MULTIPLY function is used as part of an oven temperature control program. In this program, the PLC calculates the upper and lower *deadband* or off/on limits about the set point. The upper and lower limits are set automatically at ±1% regardless of the set-point value. The set-point temperature is adjustable by means of thumbwheel switches, and an analog thermocouple interface module is used to monitor the current temperature of the oven. In this example, the

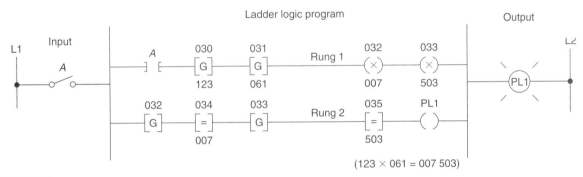

Fig. 11-9

Allen-Bradley PLC-2 MULTIPLY instruction.

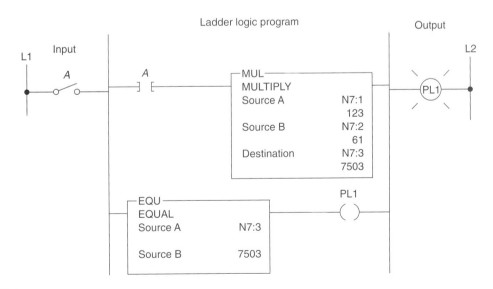

Fig. 11-10

Allen-Bradley PLC-5 and SLC-500 MULTIPLY instruction.

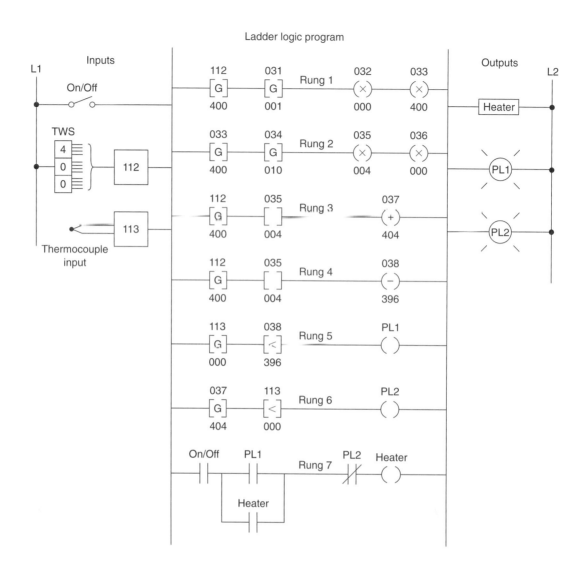

Ladder logic program

Inputs

L1

On/Off

TWS

4
0
0

112

113

Thermocouple input

Rung 1

112 [G] 400 — 031 [G] 001 — 032 (×) 000 — 033 (×) 400

Rung 2

033 [G] 400 — 034 [G] 010 — 035 (×) 004 — 036 (×) 000

Rung 3

112 [G] 400 — 035 [ ] 004 — 037 (+) 404

Rung 4

112 [G] 400 — 035 [ ] 004 — 038 (−) 396

Rung 5

113 [G] 000 — 038 [<] 396 — PL1 ( )

Rung 6

037 [G] 404 — 113 [<] 000 — PL2 ( )

Rung 7

On/Off ⊢⊢ — PL1 ⊢⊢ — PL2 ⫫ — Heater ( )

Heater ⊢⊢

Outputs

L2

Heater

PL1

PL2

Fig. 11-11

Automatic control of upper and lower set-point limits using Allen-Bradley PLC-2 instruction set.

set-point temperature is 400°F: therefore, the electric heaters will be turned on when the temperature of the oven drops to less than 396°F and stay on until the temperature rises above 404°F. If the set point is changed to 100°F, the deadband remains at ±1%, with the lower limit being 99°F and the upper limit being 101°F. The number stored in word 037 represents the upper temperature limit, while the number stored in word 038 represents the lower temperature limit. Figure 11-12 shows the same operation programmed using Allen-Bradley PLC-5 or SLC-500 block-formatted instructions.

11•5 DIVISION INSTRUCTION

Operation of the DIVIDE (÷) (÷) instruction is very similar to the operation of the MULTIPLY instruction. The Allen-Bradley PLC-2 DIVIDE takes the number stored in the first GET instruction of the division rung and divides it by the number stored in the second GET instruction. The result of the division is held in two words or registers as referenced by the output coils. The first output holds the *integer,* while the second output holds the *decimal fraction.*

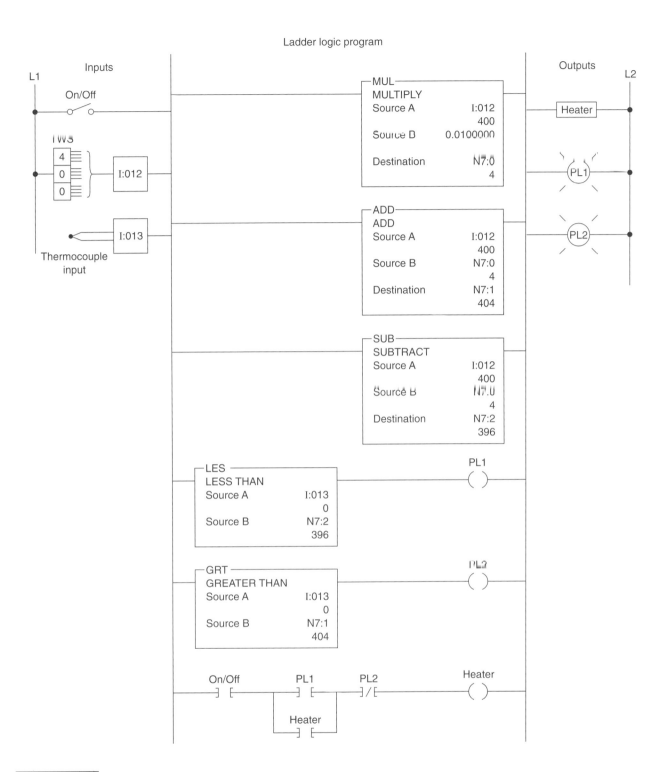

Ladder logic program

Fig. 11-12

Automatic control of upper and lower set-point limits using
Allen-Bradley PLC-5 and SLC-500 instruction set.

Figure 11-13 shows a simple Allen-Bradley PLC-2 DIVIDE instruction program. When input device *A* is true, the value of word 030 (150) is divided by the value of word 031 (040), and the quotient is stored in words 032 and 033 (003.750, or 3.75). As a result, rung 2 will become true, turning output PL1 on. Figure 11-14 shows the same operation programmed using Allen-Bradley PLC-5 or SLC-500 block-formatted instructions. Note that the destination address is the floating-point data file F0.0. If this destination were changed to an integer file, the result would be rounded off. If the remainder is 0.5 or above, the result is rounded up; if the remainder is less than 0.5, the answer is rounded down.

Figure 11-15 shows how the DIVIDE function is used as part of a program to convert Celsius temperature to Fahrenheit. In this application, the thumbwheel switch connected to the input module indicates Celsius temperature. The program is designed to convert the recorded Celsius temperature in the data table to Fahrenheit values for display. The formula

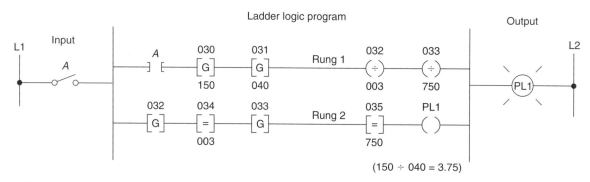

$$°F = \left(\frac{9}{5} \cdot °C \right) + 32$$

forms the basis for the program. In this example, a current temperature reading of

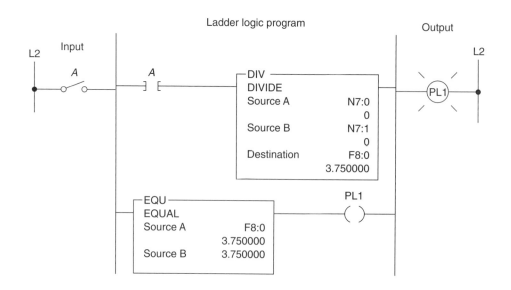

Fig. 11-13

Allen-Bradley PLC-2 DIVIDE instruction.

Fig. 11-14

Allen-Bradley PLC-5 and SLC-500 DIVIDE instruction.

Ladder logic program

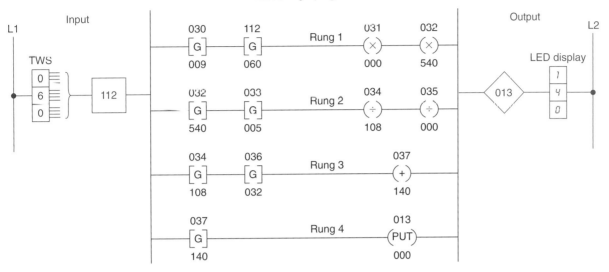

Fig. 11-15

Converting Celsius temperature to Fahrenheit using
Allen-Bradley PLC-2 instruction set.

Ladder logic program

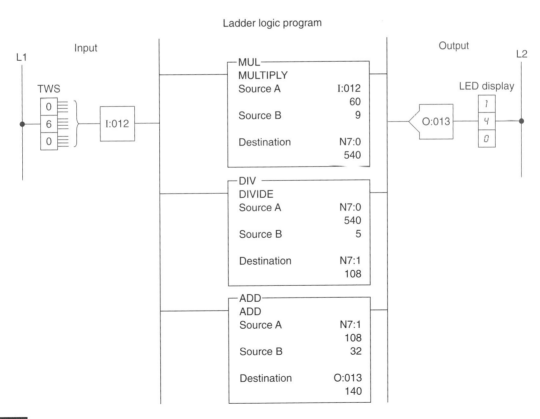

Fig. 11-16

Converting Celsius temperature to Fahrenheit using
Allen-Bradley PLC-5 or SLC-500 instruction set.

60°C is assumed. In rung 1, the GET instruction at address 030 multiplies the temperature (60°C) by 9 and stores the product (540) in address 032. In rung 2, the GET instruction at address 033 divides 5 into 540 and stores the quotient 108 in address 034. In rung 3, the GET instruction at address 036 adds 32 to the value of 108 and stores the sum 140 in address 037. Thus, 60°C = 140°F. In rung 4, the GET/PUT instruction pair transfers the converted temperature reading, 140°F, to the LED display. Figure 11-16 on p. 309, shows the same operation programmed using Allen-Bradley PLC-5 or SLC-500 block-formatted instructions.

•6 OTHER WORD LEVEL MATH INSTRUCTIONS

Figure 11-17 shows an example of the *square root* (SQR) instruction. The number whose square root we want to determine is placed in the source. When the instruction is true, the function calculates the square root and places it in the destination. If the value of the source is negative, the instruction will store the square root of the absolute value of the source at the destination.

Figure 11-18 shows an example of the *negate* (NEG) instruction. This instruction negates (changes the sign of) the value stored at the source address, N7:52, and stores the result at the destination address, N7:53, when the instruction is true. Positive numbers will be stored in straight binary format, and negative numbers will be stored in two's complement.

The *clear* (CLR) instruction is used to set the destination value of a word to 0. The example shown in Fig. 11-19 changes the value stored in the destination address, N7:22, to 0 when the instruction is true.

The *convert to BCD* (TOD) instruction is used to convert 16-bit integers into *binary coded decimal* (BCD) values. This instruction could be used when transferring data from the processor (which stores data in binary format) to an external device, such as an LED display, that functions in BCD format. The example shown in Fig. 11-20 will convert the binary bit pattern at the source address, N7:23, into a BCD bit pattern of the same decimal value at the destination address, O:20. The source displays the value 10, which is the correct decimal value; however,

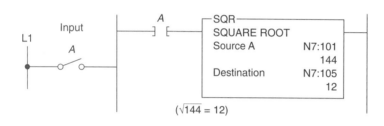

Fig. 11-17

Square root instruction (SQR).

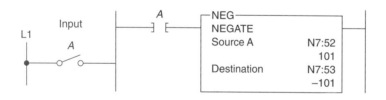

Fig. 11-18

Negate instruction (NEG).

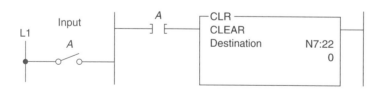

Fig. 11-19

Clear instruction (CLR).

Ladder logic program

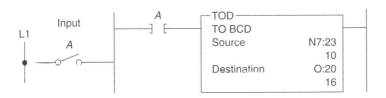

Fig. 11-20

Convert to BCD instruction.

Ladder logic program

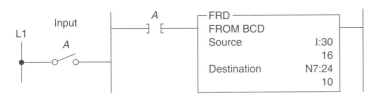

Fig. 11-21

Convert from BCD instruction.

the destination displays the value 16. Since the processor interprets all bit patterns as binary, the value 16 is the binary interpretation of the BCD bit pattern. The bit pattern for 10 BCD is the same as the bit pattern for 16 binary.

The *convert from BCD* (FRD) instruction is used to convert *binary coded decimal* (BCD) values to integer values. This instruction could be used to convert data from a BCD

Ladder logic program

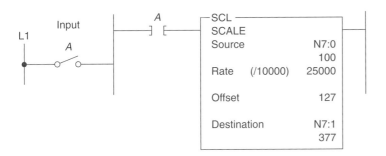

Fig. 11-22

Scale data instruction.

external source, such as a BCD thumbwheel switch, to the binary format in which the processor operates. The example shown in Fig. 11-21 will convert the BCD bit pattern stored at the source address, I:30, into a binary bit pattern of the same decimal value at the destination address, N7:24.

The *scale data* (SCL) instruction is used to allow very large or very small numbers to be enlarged or reduced by the rate value. In the example shown in Fig. 11-22, the number 100 stored at the source address, N7:0, is multiplied by 25,000, divided by 10,000, and added to 127. The result, 377, is placed in the destination address, N7:1.

11·7 FILE ARITHMETIC OPERATIONS

File arithmetic functions include file add, file subtract, file multiply, file divide, file square root, file convert from BCD, and file convert to BCD. The *file arithmetic and logic* (FAL) instruction can combine an arithmetic operation with file transfer. The arithmetic operations that can be implemented with the FAL are ADD, SUB, MULT, DIV, and SQR.

The *file add* function of the FAL instruction can be used to perform addition operations on multiple words. In the example shown in Fig. 11-23 on p. 312, when the rung goes true, the expression tells the processor to add the data in file address N7:25 to the data stored in file address N7:50 and stores the result in file address N7:100. The rate per scan is set at all, so the instruction goes to completion in one scan.

Figure 11-24 shows an example of the *file sub-tract* function of the FAL instruction. In this example, when the rung goes true, the processor subtracts a program constant (255) from each word of file address N10:0 and stores the result at the destination file address, N7:255. The rate per scan is set at 2, so it will take 2 scans from the moment the instruction goes true to complete its operation.

Figure 11-25 shows an example of the *file multiply* function of the FAL instruction. In this example, when the rung goes true, the data in file address N7:330 is multiplied by

Ladder logic program

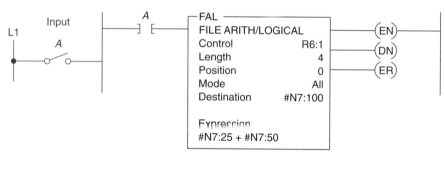

Fig. 11-23

File add function of the FAL instruction.

Ladder logic program

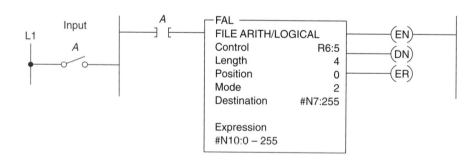

Fig. 11-24

File subtract function of the FAL instruction.

the data in element address N7:23, with the result stored in file address N7:500.

Figure 11-26 shows an example of the *file divide* function of the FAL instruction. In this example, the data in file address F8:20 is divided by the data in file address F8:100, with the result stored in element address F8:200. The mode is incremental, so the instruction operates on one set of elements for each false-to-true transition of the instruction.

Ladder logic program

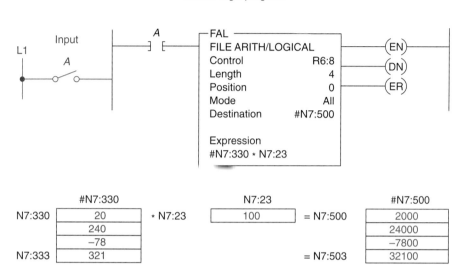

Fig. 11-25

File multiply function of the FAL instruction.

Ladder logic program

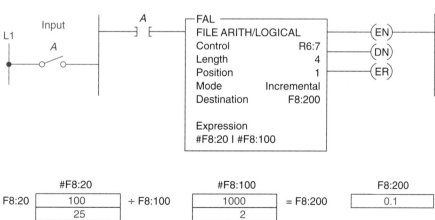

Fig. 11-26

File divide function of the FAL instruction.

Questions

1. Explain the purpose of math instructions as applied to the PLC.

2. What are the four basic math functions that can be performed on some PLCs?

3. What standard format is used for PLC math instructions?

4. Explain what each of the logic rungs in Fig. 11-27 is telling the processor to do.

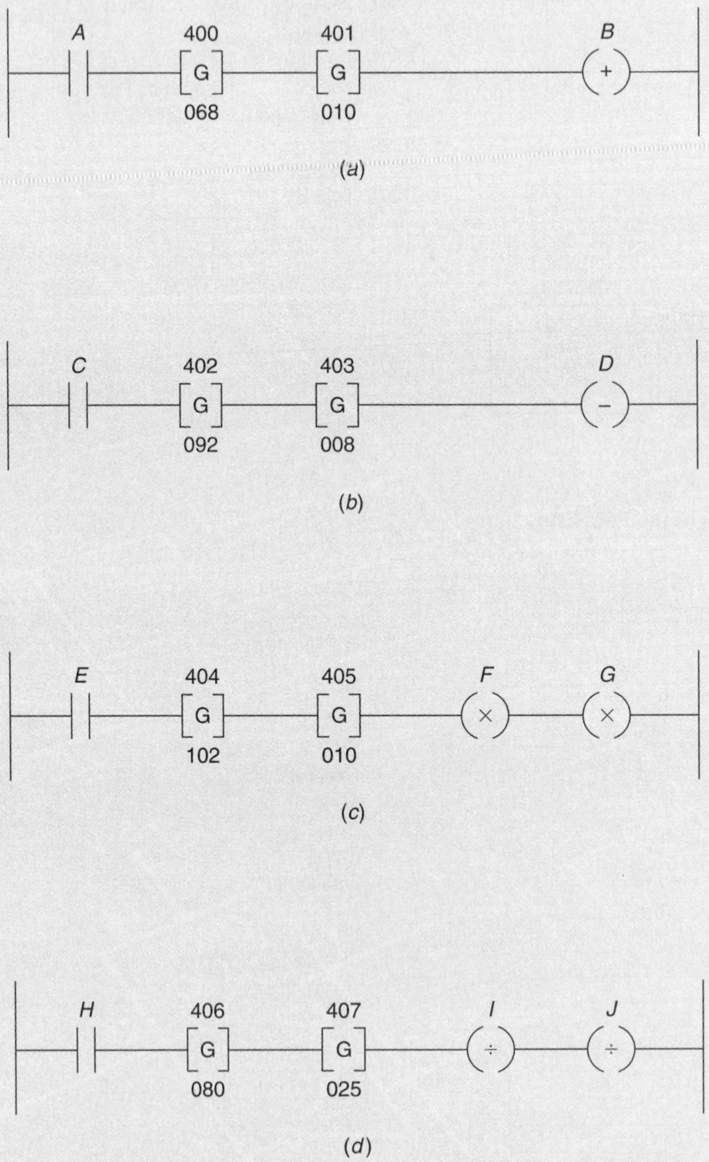

Fig. 11-27

5. With reference to Fig. 11-28:

    a. What is the value of the number stored in word 402?
    b. When will output *C* be energized?

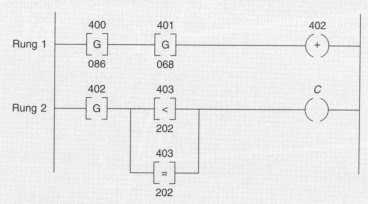

Fig. 11-28

6. With reference to Fig. 11-29:

    a. What is the value of the number stored in word 032?
    b. When will output *D* be energized?

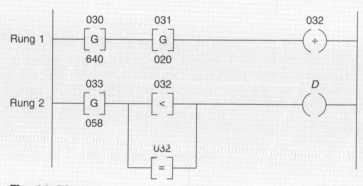

Fig. 11-29

7. With reference to Fig. 11-30:

    a. What is the value of the number stored in word 402?

    b. What is the value of the number stored in word 403?

    c. For rung 2 to have logic continuity, what numbers must be stored in addresses 404 and 405?

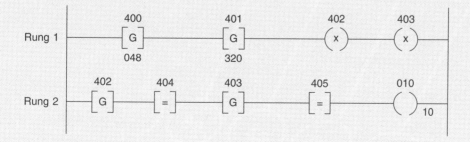

Fig. 11-30

Math Instructions

8. With reference to Fig. 11-31:

    a. What is the value of the number stored in word 402?

    b. What is the value of the number stored in word 403?

    c. For rung 2 to have logic continuity, what numbers must be stored in words 404 and 405?

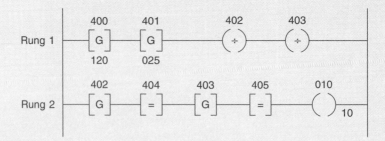

Fig. 11-31

9. Would math instructions be classified as input or output instructions?

10. With reference to Fig. 11-32, what is the value of the number stored at source B if N7:3 contains a value of 60 and N7:20 contains a value of 80?

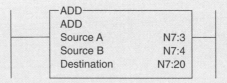

Fig. 11-32

11. With reference to Fig. 11-33, what is the value of the number stored at destination if N7:3 contains a value of 500?

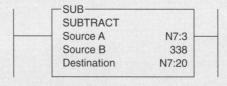

Fig. 11-33

12. With reference to Fig. 11-34, what is the value of the number stored at destination if N7:3 contains a value of 40 and N7:4 contains a value of 3?

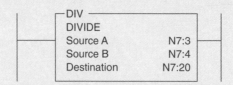

Fig. 11-34

Chapter 11

13. With reference to Fig. 11-35, what is the value of the number stored at the destination if N7:3 contains a value of 15 and N7:4 contains a value of 4?

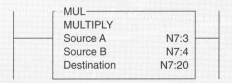

```
┌─ MUL ──────────────────┐
│ MULTIPLY               │
│ Source A        N7:3   │
│ Source B        N7:4   │
│ Destination     N7:20  │
└────────────────────────┘
```

Fig. 11-35

14. With reference to Fig. 11-36, what is the value of the number stored at N7:20 if N7:3 contains a value of −345?

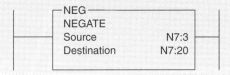

```
┌─ NEG ──────────────────┐
│ NEGATE                 │
│ Source          N7:3   │
│ Destination     N7:20  │
└────────────────────────┘
```

Fig. 11-36

15. With reference to Fig. 11-37, what will be the value of each of the bits in word B3:3 when the rung goes true?

```
          ┌─ CLR ──────────────────────┐
──┤ ├─────│ CLEAR                       │
          │ Destination         B3:3    │
          │         0000111100001111    │
          └─────────────────────────────┘
```

Fig. 11-37

16. With reference to Fig. 11-38, what is the value stored at N7:101 when the rung goes true?

```
          ┌─ SQR ──────────────────────┐
──┤/├─────│ SQUARE ROOT                 │
          │ Source A           N7:101   │
          │ Destination        N7:105   │
          │                         4   │
          └─────────────────────────────┘
```

Fig. 11-38

17. With reference to Fig. 11-39, list the values that will be stored in file #N7:10 when the rung goes true.

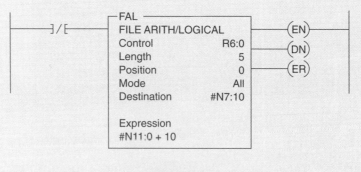

File #N11:0

328
150
10
32
0

Fig. 11-39

Problems

1. Answer each of the following with reference to the *counter program* shown in Fig. 11-3:

   a. Assume the accumulated count of counters 050 and 051 to be 148 and 036, respectively. State the value of the number stored in each of the following words at this point: (i) 050, (ii) 051, (iii) 020, (iv) 021.

   b. Will output PL1 be energized at this point? Why?

   c. Assume the accumulated count of counters 050 and 051 to be 250 and 175, respectively. State the value of the number stored in each of the following words at this point: (i) 050, (ii) 051, (iii) 020, (iv) 021.

   d. Will output PL1 be energized at this point? Why?

2. Answer each of the following with reference to the *overfill alarm program* shown in Fig. 11-7:

   a. Assume that the vessel is filling and has reached the 300-lb point. State the status of each of the logic rungs (true or false) at this point.

   b. Assume that the vessel is filling and has reached the 480-lb point. State the value of the number stored in each of the following words at this point: (i) 040, (ii) 112, (iii) 032, (iv) 041.

   c. Assume that the vessel is filled to a weight of 502 lb. State the status of each of the logic rungs (true or false) for this condition.

   d. Assume that the vessel is filled to a weight of 510 lb. State the value of the number stored in each of the following words for this condition: (i) 040, (ii) 112, (iii) 032, (iv) 041.

   e. With the vessel filled to a weight of 510 lb, state the status of each of the logic rungs (true or false).

3. Answer the following with reference to the *upper and lower temperature control program* shown in Fig. 11-11:

    a. Assume that the set-point temperature is 600°F. At what temperature will the electric heaters be turned on and off?

    b. Assume that the set-point temperature is 600°F and the thermocouple input module indicates a temperature of 590°F. What is the value of the number stored in each of the following words at this point: (i) 112, (ii) 031, (iii) 032, (iv) 033, (v) 034, (vi) 035, (vii) 036, (viii) 037, (ix) 038, (x) 113?

    c. Assume that the set-point temperature is 600°F and the thermocouple input module indicates a temperature of 608°F. What is the status (energized or not energized) of each of the following outputs: (i) PL1, (ii) PL2, (iii) heater?

4. With reference to the *converting Celsius temperature to Fahrenheit program* shown in Fig. 11-15, state the value of the number stored in each of the following words for a thumbwheel setting of 035: (i) 030, (ii) 112, (iii) 031, (iv) 032, (v) 033, (vi) 034, (vii) 035, (viii) 036, (ix) 037, (x) 013.

5. Design a program that will add the values stored at N7:23 and N7:24 and store the result in N7:30 whenever input *A* is true, and then, when input *B* is true, will copy the data from N7:30 to N7:31.

6. Design a program that will take the accumulated value from TON timer T4:1 and display it on a 4-digit, BCD format set of LEDs. Use address O:023 for the LEDs. Include the provision to change the preset value of the timer from a set of 4-digit BCD thumbwheels when input *A* is true. Use address I:012 for the thumbwheels.

7. Design a program that will implement the following arithmetic operation:

    ⊙ Use a MOVE instruction and place the value 45 in N7:0 and 286 in N7:1.

    ⊙ Add the values together and store the result in N7:2.

    ⊙ Subtract the value in N7:2 from 785 and store the result in N7:3.

    ⊙ Multiply the value in N7:3 by 25 and store the result in N7:4.

    ⊙ Divide the value in N7:3 by 35 and store the result in F8:0.

8.   **A.** There are three part conveyor lines (1-2-3) feeding a main conveyor. Each of the three conveyor lines has its own counter. Construct a PLC program to obtain the total count of parts on the main conveyor.

    B. Add a timer to the program that will update the total count every 30 s.

Math Instructions

9. With reference to Fig. 11-40, when the input goes true, what value will be stored at each of the following:

    a. N7:3
    b. N7:5
    c. F8:1

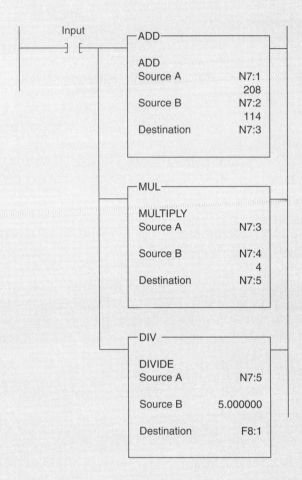

Fig. 11-40

10. With reference to Fig. 11-41, when the input goes true, what value will be stored at each of the following?

    a. N7:3
    b. N7:4
    c. N7:5
    d. N7:6

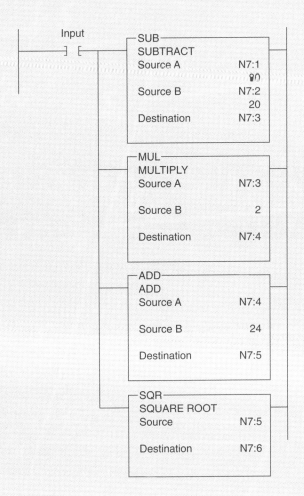

Input

SUB
SUBTRACT
Source A   N7:1
                     90
Source B   N7:2
                     20
Destination   N7:3

MUL
MULTIPLY
Source A   N7:3

Source B   2

Destination   N7:4

ADD
ADD
Source A   N7:4

Source B   24

Destination   N7:5

SQR
SQUARE ROOT
Source   N7:5

Destination   N7:6

Fig. 11-41

11. Two part conveyor lines, A and B, feed a main conveyor line M. A third conveyor line, R, removes rejected parts a short distance away from the main conveyor. Conveyors A, B, and R have parts counters connected to them. Construct a PLC program to obtain the total parts output of main conveyor M.

12. A main conveyor has two conveyors, A and B, feeding it. Feeder conveyor A puts six-packs of canned soda on the main conveyor. Feeder conveyor B puts eight-packs of canned soda on the main conveyor. Both feeder conveyors have counters that count the number of *packs* leaving them. Construct a PLC program to give a *total can* count on the main conveyor.

Math Instructions

12

Sequencer and Shift Register Instructions

After completing this chapter, you will be able to:

◆ Identify and describe the various forms of mechanical sequencers and explain the basic operation of each

◆ Interpret and explain information associated with PLC sequence input, output, and load instructions

◆ Compare the operation of an event-driven and a time-driven sequencer

◆ Describe the operation of bit and word shift registers

◆ Interpret and develop programs that use shift registers

This chapter explains (1) how the PLC sequencer and shift register functions operate and (2) how they can be applied to control problems. The sequencer instruction evolved from the mechanical drum switch, and it can handle complex sequencing control problems more easily than does the drum switch. Shift registers are often used to track parts on automated manufacturing lines by shifting either status or values through data files.

Sequencer instructions are named after the mechanical sequencer switches they replace. These switches are often referred to as *drum switches, rotary switches, stepper switches,* or *cam switches* (Fig. 12-1), in addition to the *sequencer switch* identification. Figure 12-2 illustrates the operation of a cam-operated sequencer switch. An electric motor is used to drive the cams. A series of leaf-spring–mounted contacts interacts with the cam so that, in different degrees of rotation of the cam, various contacts are closed and opened to energize and de-energize various electrical devices.

Figure 12-3 illustrates a typical mechanical drum-operated sequencer switch. The switch consists of series of contacts that are operated by pegs located on a motor-driven drum. The pegs can be placed at random locations around the circumference of the drum to operate contacts. When the drum is rotated, contacts that align with the pegs will close, while the contacts where there are no pegs will remain open. In this example, the presence of a peg can be thought of as logic 1, or on, while the absence of a peg can be logic 0, or off.

Figure 12-3 also shows an equivalent sequencer data table for the drum cylinder. If the first five steps on the drum cylinder were removed and flattened out, they would appear as illustrated in the table. Each horizontal location where there was a peg is now represented by a 1 (on), and the positions where there were no pegs are each represented by a 0 (off).

Sequencer switches are used whenever a repeatable operating pattern is required. An excellent example is the sequencer switch used in dishwashers to pilot the machinery through a

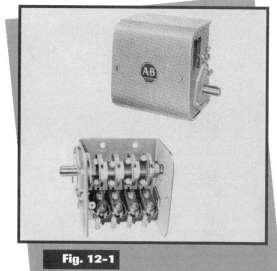

wash cycle (Fig. 12-4). The cycle is always the same, and each step occurs for a specific time. The domestic washing machine is another example of the use of a sequencer, as are dryers and similar time-clock–controlled devices.

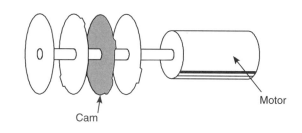

Motor

Cam

(a)

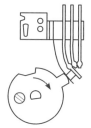 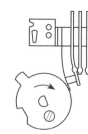 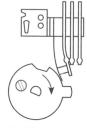

Position (1)
Switch 1 open
Switch 2 open

Position (2)
Switch 1 closed
Switch 2 open

Position (3)
Switch 1 closed
Switch 2 closed

Position (4)
Switch 1 closed
Switch 2 open

(b)

Fig. 12-2

Mechanical cam-operated sequencer. *(a)* Cam-driving mechanism. *(b)* Cam and contact operation.

Peg locations in cylinder

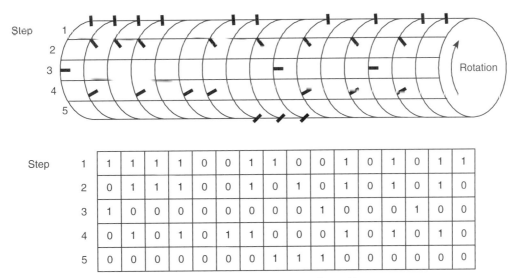

Step																
1	1	1	1	1	0	0	1	1	0	0	1	0	1	0	1	1
2	0	1	1	1	0	0	1	0	1	0	1	0	1	0	1	0
3	1	0	0	0	0	0	0	0	0	1	0	0	0	1	0	0
4	0	1	0	1	0	1	1	0	0	0	1	0	1	0	1	0
5	0	0	0	0	0	0	0	1	1	1	0	0	0	0	0	0

Equivalent sequencer data table

Fig. 12-3

Mechanical drum-operated sequencer.

Figure 12-5 on pp. 326-327 shows the wiring diagram and data table for a dishwasher that uses a cam-operated sequencer commonly known as the *timer*. In this unit, a synchronous motor drives a mechanical train that, in turn, drives a series of cam wheels. The cam advances in increments of about 45 s in duration. Normally, the timer motor operates continuously throughout the cycle of operation. The data table in Fig. 12-5

outlines the sequence of operation of the timer. Each time increment is 45 s. A total of sixty 45 s steps are used to complete the 45-min operating cycle. The numbers in the "Active Circuits" column refer to the circled numbers found on the schematic diagram.

•2 SEQUENCER INSTRUCTIONS

Sequencer instructions are used typically to control automatic assembly machines that have a consistent and repeatable operation. Sequencer instructions can make programming many applications a much easier task. For example, the on/off operation of 16 discrete outputs can be controlled, using a sequencer instruction, with only one ladder rung. By contrast, the equivalent contact-coil ladder control arrangement would need 16 rungs in the program.

The sequencer instruction is a powerful instruction found on most PLCs today. A programmed sequencer can replace a

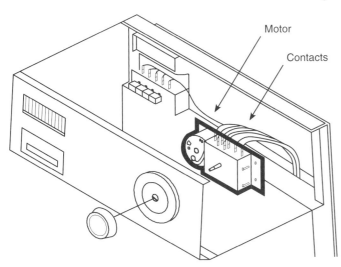

Fig. 12-4

Dishwasher sequencer switch.

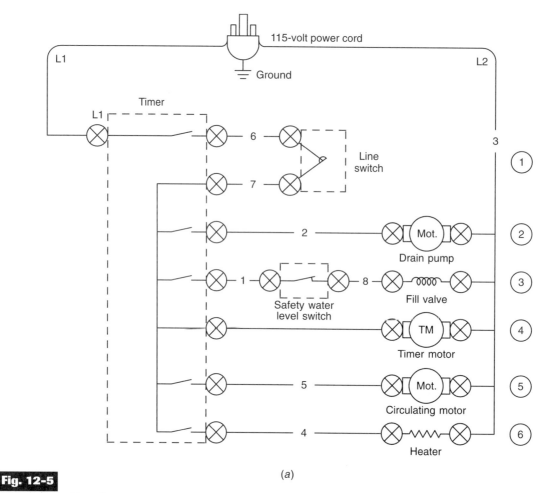

Fig. 12-5

Dishwasher wiring diagram and data table.

mechanical drum switch. To program a sequencer, binary information is entered into a series of consecutive memory words. These consecutive memory words are referred to as a *word file.* Data are first entered into a word file for each sequencer step. As the sequencer advances through the steps, binary information is transferred from the word file to the output word(s).

The *sequencer output* (SQO) instruction can be used to control output devices sequentially. The desired sequence of operation is stored in a data file, and this information is then transferred sequentially to the outputs. Figure 12-6 on p. 328 illustrates how the sequencer output instruction works. In this example, 16 lights are used for outputs. Each light represents one bit address (1 through 16) of output word

050. The lights are programmed in a four-step sequence to simulate the operation of two-way traffic lights.

Data are entered into a word file for each sequencer step, as illustrated in Fig. 12-7 on p. 328. In this example, words 60, 61, 62, and 63 are used for the four-word file. Using the programmer, binary information (1's and 0's) that reflects the desired light sequence is entered into each word of the file. For ease of programming, some PLCs allow the word file data to be entered using the octal, hexadecimal, BCD, or similar number system. When this is the case, the required binary information for each sequencer step must first be converted to whatever number system is employed by the PLC. This information is then entered with the programmer into the word file.

Machine function		Timer increment	Active circuits					
Off		0–1						
First prerinse	Drain	2	1	2		4		
	Fill	3	1		3	4	5	
	Rinse	4–5	1			4	5	6
	Drain	6	1	2		4	5	
Prewash	Fill	7	1		3	4	5	
	Wash	8–10	1			4	5	6
	Drain	11	1	2		4	5	
Second prerinse	Fill	12	1		3	4	5	
	Rinse	13–15	1			4	5	6
	Drain	16	1	2		4		
Wash	Fill	17	1		3	4		
	Wash	18–30	1			4	5	6
	Drain	31	1	2		4	5	
First rinse	Fill	32	1		3	4	5	
	Rinse	33–34	1			4	5	6
	Drain	35	1	2		4	5	
Second rinse	Fill	36	1		3	4	5	
	Rinse	37–41	1			4	5	6
	Drain	42	1	2		4	5	
Dry	Dry	43–58	1			4		6
	Drain	59	1	2		4		6
	Dry	60	1			4		6

(b)

Fig. 12-5 (continued)

Dishwasher wiring diagram and data table.

Once the data have been entered into the word file of the sequencer, the PLC is ready to control the lights. When the sequencer is activated and advanced to *step 1,* the binary information in word 060 of the file is transferred into word 050 of the output. As a result, lights 1 and 12 will be switched on and all the rest will remain off. Advancing the sequencer to *step 2* will transfer the data from word 061 into word 050. As a result, lights 1, 8, and 12 will be on and all the rest will be off. Advancing the sequencer to *step 3* will transfer the data from word 062 into word 050. As a result, lights 4 and 9 will be on and all the rest will be off. Advancing the sequencer to *step 4* will transfer the data from word 063 into word 050. As a result,

lights 4, 5, and 9 will be on and all the rest will be off. When the last step is reached, the sequencer is either automatically or manually reset to step 1.

When a sequencer operates on an entire output word, there may be outputs associated with the word that do *not* need to be controlled by the sequencer. In our example, bits 2, 3, 5, 7, 10, 11, 13, 14, 15, and 16 of output word 050 are not used by the sequencer but could be used elsewhere in the program.

To prevent the sequencer from controlling these bits of the output word, a *mask* word (040) is used. The use of a mask word is

Step 1: Bits 1 and 12 are on.

Step 2: Bits 1, 8, and 12 are on.

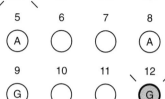
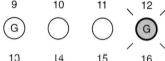
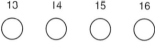

Step 3: Bits 4 and 9 are on.

Step 4: Bits 4, 5, and 9 are on.

Fig. 12-6

Sequencer steps (R-Red light, G-Green light, A-Amber light).

	16	15	14	13	12	11	10	9	8	7	6	5	4	3	2	1	
Word 050	0	0	0	0	0	0	0	0	0	0	0	0	0	0	0	0	Output
Word 060	0	0	0	0	1	0	0	0	0	0	0	0	0	0	0	1	Step 1
Word 061	0	0	0	0	1	0	0	0	1	0	0	0	0	0	0	1	Step 2
Word 062	0	0	0	0	0	0	0	1	0	0	0	0	1	0	0	0	Step 3
Word 063	0	0	0	0	0	0	0	1	0	0	0	1	1	0	0	0	Step 4

Four word file

Fig. 12-7

Binary information for each sequencer step.

	16	15	14	13	12	11	10	9	8	7	6	5	4	3	2	1	
Word 050	0	0	0	0	0	0	0	0	0	0	0	0	0	0	0	0	Output
Word 040	0	0	0	0	1	0	0	1	1	0	0	1	1	0	0	1	Mask
Word 060	0	0	0	0	1	0	0	0	0	0	0	0	0	0	0	1	Step 1
Word 061	0	0	0	0	1	0	0	0	1	0	0	0	0	0	0	1	Step 2
Word 062	0	0	0	0	0	0	0	1	0	0	0	0	1	0	0	0	Step 3
Word 063	0	0	0	0	0	0	0	1	0	0	0	1	1	0	0	0	Step 4

Fig. 12-8

Using a mask word.

illustrated in Fig. 12-8. The mask word selectively screens out data from the sequencer word file to the output word. For each bit of output word 050 that the sequencer is to control, the corresponding bit of mask word 040 must be set to 1. All other bits of output word 050 are set to 0 and thus can be used independently of the sequencer.

Sequencers, like other PLC instructions, are programmed differently with each PLC, but again the concepts are the same. The advantage of sequencer programming over the conventional program is the large savings of memory words. Typically, the sequencer program can do in 20 words or less what a standard program can do in 100 words.

Sequencer output instructions can be either block- or coil-formatted and contain a counter or timer and a file. The instructions require the entry of more than one address. The PLC block-formatted sequencer instruction used as part of the Allen-Bradley PLC-2 controller instruction set is illustrated in Fig. 12-9. Sequencer instructions are usually retentive and there can be an upper limit to the number of external outputs and steps

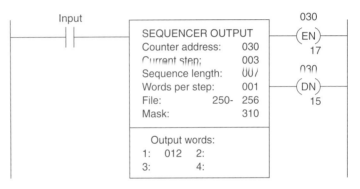

Counter address: The address of the counter controlling the sequencer operation.

Current step: The step of the sequencer being transferred to the output (accumulated value of counter).

Sequence length: The total number of steps required in the sequence operation (preset value of the counter).

Words per step: The width of the sequencer table in terms of the output words.

File: The starting and finishing data table word addresses of the file that will contain the data for the operation of the sequencer instruction.

Mask: The starting and finishing data table word addresses of the mask file.

Output words: Words that will be controlled by the sequencer instruction.

Fig. 12-9

Allen-Bradley PLC-2 controller sequencer output function block.

that can be operated on by a single instruction. Many sequencer instructions reset the sequencer automatically to step 1 on completion of the last sequence step. Other instructions provide an individual reset control line or a combination of both.

12•3 SEQUENCER PROGRAMS

A sequencer program can be *event-driven* or *time-driven.* An event-driven sequencer operates similarly to a mechanical stepper switch that increments by one step for each pulse applied to it. A time-driven sequencer operates similarly to a mechanical drum switch that increments automatically after a preset time period.

A sequencer chart, similar to that shown in Fig. 12-10, is a table that lists the sequence of operation of the outputs controlled by the *sequencer* instruction. These tables use a *matrix-style* chart format. A matrix is a two-dimensional, rectangular array of quantities. A time-driven sequencer chart usually indicates outputs on its horizontal axis and the time duration on its vertical axis. An event-driven sequencer indicates outputs on its horizontal axis and the input, or event, on its vertical axis.

A sequencer data worksheet is also used to document the different sequencer steps before they are programmed into the PLC. This form allows you to document sequencer data in an orderly, systematic way, reducing the chances for programming errors.

Figure 12-11 shows the program for a coil-formatted, time-driven sequencer used with Allen-Bradley SLC-150 controllers. This four-step, time-driven sequencer controls six external outputs. The status of the outputs, for each step, is as recorded in the data worksheet form. Outputs 015 and 016 are masked so they can be used elsewhere in the program. The program entry code data are simply the mask and output status data converted from binary to hexadecimal code.

With time-driven sequencers, each step has a function similar to a timer instruction because it involves an accumulated time value and a programmed, preset time value. The preset time values for each step are as recorded in the data worksheet form.

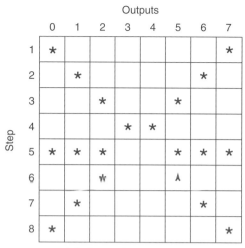

★ Indicates that output is energized

(a) Matrix-style chart

0000000010000001
0000000001000010
0000000000100100
0000000000011000
0000000011100111
0000000000100100
0000000001000010
0000000010000001

(b) Sequencer output file words

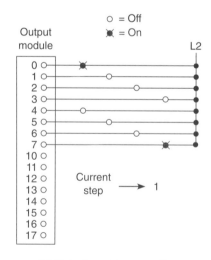

(c) Output module connection

Fig. 12-10
Sequencer chart.

This particular program (Fig. 12-11) simulates the operation of two-way traffic lights. Beginning with the sequencer reset, when the SQO instruction goes true, step 1 is initiated. As a result, outputs 009 and 014 are on and all the rest are off. After a preset time of 30 s (assuming SQO remains true), step 2 is initiated. As a result, outputs 009, 012, and 014 are on and all the rest are off. After 5 s, step 3 begins and outputs 010 and 013 are on and all the rest are off. After 30 s, step 4 begins and outputs 010, 011, and 013 are on and all the rest are off. After 5 s, the cycle repeats automatically with step 1. The sequence can be stopped at any step by opening the on/off switch S1. The sequence can be reset to step 1 at any time by pressing reset button PB1.

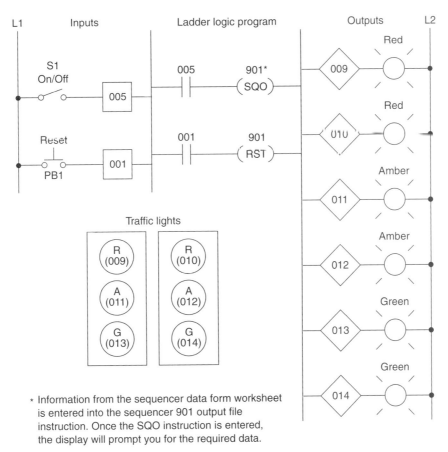

Address _SQO 901_   Time driven [X]   Event driven [ ]

Sequencer data										Program entry code		Preset (Time)
	Sequencer data group →	Second				First				2nd Grp	1st Grp	
	Module group number →	4				3						
	I/O terminal address →	016	015	014	013	012	011	010	009			
Event description	Mask data →	0	0	1	1	1	1	1	1	3	F	
Red – Green	Step no. 1	0	0	1	0	0	0	0	1	2	1	30 s
Red – Green and Amber	2	0	0	1	0	1	0	0	1	2	9	5 s
Green – Red	3	0	0	0	1	0	0	1	0	1	2	30 s
Green and Amber – Red	4	0	0	0	1	0	1	1	0	1	6	5 s

(a) Sequencer data worksheet used to document the different sequencer steps

* Information from the sequencer data form worksheet is entered into the sequencer 901 output file instruction. Once the SQO instruction is entered, the display will prompt you for the required data.

(b) PLC program

Fig. 12-11

Time-driven sequencer.

In Fig. 12-12, you can see the traffic light circuit of Fig. 12-11 programmed as an event-driven sequencer. The event-driven sequencer functions similarly to the counter instruction because it involves an accumulated counter value and a programmed, preset counter value. If the preset counter value is set to 1, closing and opening step pushbutton PB1 will step you manually through the different sequencer steps. Beginning with the sequencer reset, step 1 is initiated and remains so until step pushbutton PB1 is actuated. Momentarily closing PB1 produces a false-to-true transition that increments the sequencer to step 2. Thus, the event-driven sequencer counts false-to-true transitions of the sequencer rung. When the accumulated count value reaches the preset count value, the sequencer advances to the next step and the accumulated value increments from zero again. If the preset count value were set for 2, it would take two false-to-true transitions of the sequencer rung to advance the sequence one step.

Address _____SQO 901_____   Time driven ☐   Event driven ☒

Sequencer data								Program entry code		Preset (Count)
Sequencer data group →	Second				First					
Module group number →	4				3					
I/O terminal address →	016	015	014	013	012	011	010	009	2nd Grp	1st Grp	
Event description — Mask data →	0	0	1	1	1	1	1	1	3	F	
Red – Green — Step no. 1	0	0	1	0	0	0	0	1	2	1	1
Red – Green and Amber — 2	0	0	1	0	1	0	0	1	2	9	1
Green – Red — 3	0	0	0	1	0	0	1	0	1	2	1
Green and Amber – Red — 4	0	0	0	1	0	1	1	0	1	6	1

(a) Sequencer data worksheet used to document the different sequencer steps

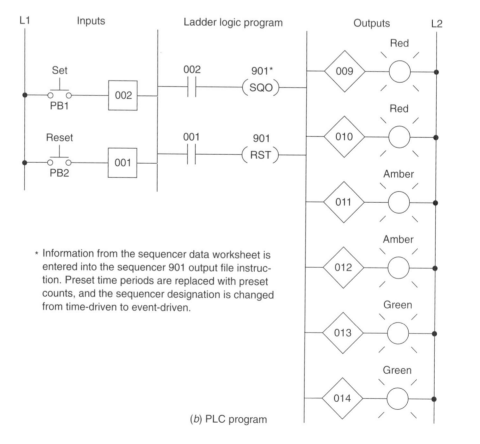

* Information from the sequencer data worksheet is entered into the sequencer 901 output file instruction. Preset time periods are replaced with preset counts, and the sequencer designation is changed from time-driven to event-driven.

(b) PLC program

Fig. 12-12

Event-driven sequencer.

The programming of sequencers will vary between PLC models and manufacturers, but the operational concepts are the same. The sequence of events controlled by the sequencer is determined by the bit pattern of each consecutive word and the number of words in the sequence. An example of the *sequencer output* (SQO) instruction available as part of the Allen-Bradley PLC-5 and SLC-500 instruction sets is shown in Fig. 12-13. The following parameters and addresses must be entered into the instruction block:

◇ **File** is the address of the sequencer file. You must use the file indicator (#) for this address. The file contains the data that will be transferred to the destination address when the instruction undergoes a false-to-true transition. Each word in the file represents a position, starting with position 0 and continuing to the file length. The actual file length will be one word longer than that indicated by the length in the instruction.

◇ **Mask** is a hexadecimal code or the address of the mask word or file through which the instruction moves data. Set mask bits to 1 to pass data, and reset mask bits to 0 to mask data. Use a mask word or file if you want to change the mask according to application requirements. If the mask is a file, its length will be equal to the length of the sequencer file. The two files track automatically.

◇ **Destination** is the address of the output location where the data is written to when it is copied from the file. The destination may be a word address or a file address. If it is a file address, the position of the file and the position of the destination will automatically be the same. In most applications, the destination will be a word address.

◇ **Control,** which is the instruction's address and control element, will be from an R data file. The control stores the status byte of the instruction, the length of the sequencer file, and the instantaneous position in the file as follows:

	15	13	11	08	00
Word 0	EN	DN	ER	FD	
Word 1	Length of sequencer file				
Word 2	Position				

◇ The **enable bit (EN**; bit 15) is set by a false-to-true rung transition and indicates the instruction is enabled. It follows the rung condition.

◇ The **done bit (DN**; bit 13) is set by the instruction after it has operated on the last word in the sequencer file. The done bit will reset on the true-to-false transition once the instruction has operated on the last position.

◇ The **error bit (ER**; bit 11) is set when the processor detects a negative position value, or a negative or zero length value.

◇ **Length** is the number of steps of the sequencer file, starting at position 1. Position 0 is the start-up position. The instruction resets (wraps) to position 1 at each cycle completion. The actual file length will be 1 plus the file length entered in the instruction.

◇ **Position** is the word location or step in the sequencer file from which the instruction moves data. Any value up to the file length may be entered, but the instruction will always reset to 1 on the true-to-false transition after the instruction has operated on the last position.

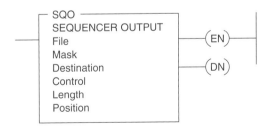

Fig. 12-13

Allen-Bradley PLC-5 and SLC-500 *sequencer output* (SQO) instruction.

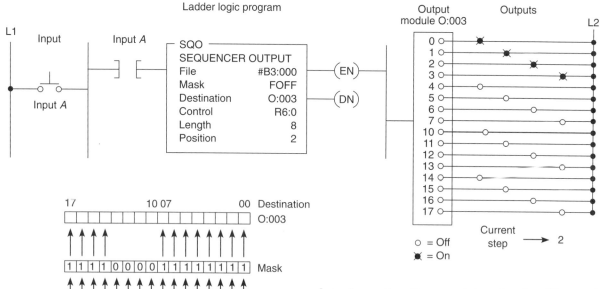

Ladder logic program

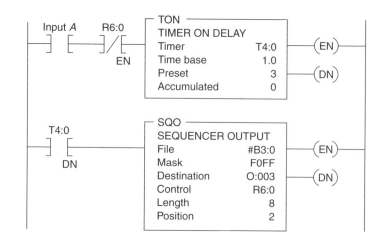

Fig. 12-14

Event-driven *sequencer output* instruction program.

The event-driven sequencer output program of Fig. 12-14 can be used to explain the operation of the sequencer output instruction. Data are copied from file #B3:000 at the bit locations where there is a 1 in the mask, to the destination O:003 on a false-to-true transition of input *A*. The position indexes one position, and the data are then copied. Once the position reaches the last position, then on the true-to-false transition of the instruction, the position will reset to 1. Position 0 is executed under the following conditions: the position is at 0, the instruction is true, and the processor goes from the program to the run mode. Position 0 is often used as a home or starting position, with a 0 loaded into this position through the program. When the instruction sees a false-to-true transition, it indexes to position 1 and copies the

data from the first position in the file to the destination. On subsequent sequences, it will reset to position 1. Note that the data in O:003 matches the data in position 2 in the file, except for the data in bits 10 through 13. These bits may be controlled from elsewhere in the program since they are not affected by the sequencer instruction because of the 0 in these bit positions in the mask. For the sequencer to be incremented automatically through each step, it must have a timer incorporated into its ladder logic program. Figure 12-15 shows a timer with a preset of 3 s, which is used to pulse the input for the sequencer. The enable bit of the sequencer is used to reset the timer after each increment

Fig. 12-15

Time-driven *sequencer output* instruction program.

occurs. This circuit increments automatically through the eight steps of the sequencer at 3-s intervals when input *A* is closed.

The *sequencer input* (SQI) instruction is also common on PLCs. This sequencer allows input data to be compared for equality against data stored in the sequencer file. For example, it can make comparisons between the states of input devices and their desired states: if the conditions match, the instruction is true.

The sequencer input program of Fig. 12-16 can be used to explain the operation of the *sequencer input* instruction. The entries in the instruction are similar to those in the *sequencer output* instruction, except the destination is replaced by the source. The SQI instruction compares a file of input image data (I:031), through a mask (FFF0), to a file of reference data (#N7:11) for equality. When the status of all nonmasked bits of the word at that particular step match those of the corresponding reference word, the instruction goes true; otherwise, the instruction is false. The input data can indicate the state of an input device, such as the combination of input switches shown in this example program. When the combination of opened and closed switches are equal to the combination of 1's and 0's on a step in the sequencer reference file, the PL1 output of the sequencer becomes energized.

The program of Fig. 12-17 on p. 336 illustrates the use of the *sequencer input* (SQI) and *sequencer output* (SQO) instructions in pairs to monitor and control, respectively, a sequential operation. When programming sequencer input and output instructions in pairs, use the same control address, length value, and position value in each instruction. The sequencer input instruction is therefore indexed by the sequencer output instruction because both control elements have the same address, R6:5. This type of programming technique allows input and output sequences to function in unison, causing a specific output sequence to occur when a specific input sequence takes place.

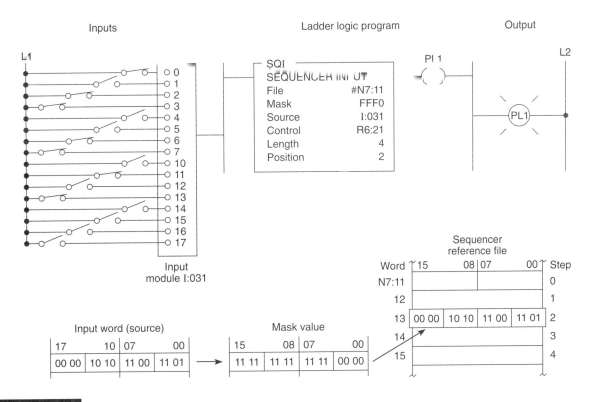

Fig. 12-16

Sequencer input (SQI) instruction program.

Ladder logic program

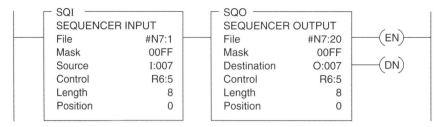

Fig. 12-17

Sequencer input instruction programmed in the same rung as a *sequencer output* instruction.

Allen-Bradley PLCs also have a *sequencer load* (SQL) instruction that functions like a word-to-file move. It can be used as a teaching tool to load data into a sequencer file, one step at a time. For example, a machine may be jogged manually through its sequence of operation, with its input devices read at each step. At each step, the status of the input devices is written to the data file in the *sequencer input* instruction. As a result, the file is loaded with the desired input status at each step, and these data are then used for comparison with the input devices when the machine is run in automatic mode.

The sequencer load program of Fig. 12-18 can be used to explain the operation of the sequencer load instruction. The sequencer load instruction is used to load the file and does *not* function during the machine's normal operation. It replaces the manual loading of data into the file with the programming

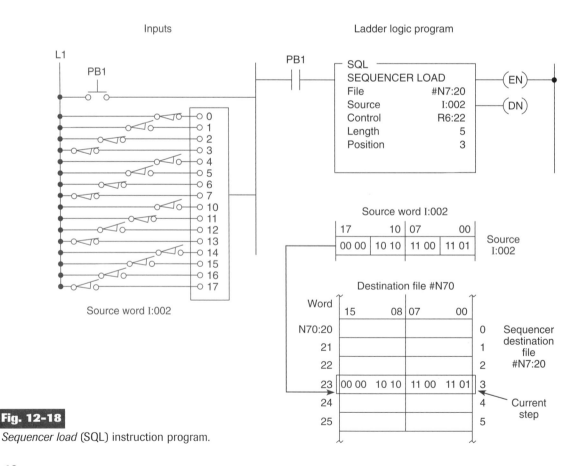

Fig. 12-18

Sequencer load (SQL) instruction program.

Chapter 12

terminal. The parameters entered in the instruction are similar to those entered in the SQI and SQO instructions. The *sequencer load* instruction does not use a mask. It copies data from the source address to the file. When the instruction goes from false to true, the instruction indexes to the next position and copies the data. When the instruction has operated on the last position and has a true-to-false transition, it resets to position 1. It transfers data in position 0 only if it is at position 0 and the instruction is true, and the processor goes from the program to run mode. By manually jogging the machine through its cycle, the switches connected to input I:002 of the source can be read at each position and written into the file by momentarily pressing PB1.

12•4 SHIFT REGISTERS

The PLC not only uses a fixed pattern of register (word) bits, but it can easily manipulate and change individual bits. A bit *shift register* is a register that allows the shifting of bits through a single register, or group of registers. The bit shift register shifts bits serially (from bit to bit) through an array in an orderly fashion.

A shift register can be used to simulate the movement or *track* the flow of parts and information. This is accomplished by shifting either status or values through data files. Common applications for shift registers include the following:

◆ Tracking parts through an assembly line

◆ Controlling machine or process operations

◆ Inventory control

◆ System diagnostics

Figure 12-19 illustrates the basic concept of a shift register. A common shift pulse or clock causes each bit in the

shift register to move 1 position to the right. At some point, the number of data bits fed into the shift register will exceed the register's storage capacity. When this happens, the first data bits fed into the shift register by the shift pulse are lost at the end of the shift register. Typically, data in the shift register could represent the following:

◆ Part types, quality, and size

◆ The presence or absence of parts

◆ The order in which events occur

◆ Identification numbers or locations

◆ A fault condition that caused a shutdown

You can program a shift register to shift status data either right or left by shifting either status or values through data files. When tracking parts on a status basis, *bit* shift registers are used. Bit shift instructions will shift bit status from a source bit address, through a data file, and out to an unload bit, one bit at a time. There are two bit shift instructions: *bit shift left* (BSL), which shifts bit status from a lower address number to a higher address number through a data file; and *bit shift right* (BSR), which shifts data from a higher address number to a lower address number through a data file (Fig. 12-20 on p. 338). Some PLCs provide a *circulating shift register* function, which allows you to repeat a pattern again and again.

When working with a bit shift register, each bit is identified by its position in the register. Therefore, working with any bit in the register

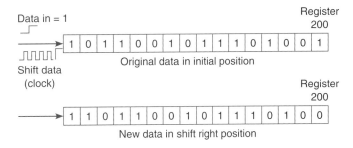

Fig. 12-19

Basic concept of a shift register.

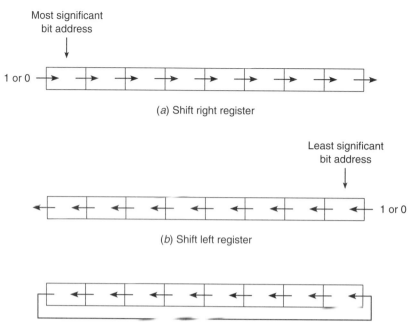

Most significant
bit address

(a) Shift right register

Least significant
bit address

1 or 0

(b) Shift left register

(c) Wraparound or circulating shift register

Fig. 12-20

Bit shift right and bit shift left registers.

becomes a matter of identifying the position it occupies rather than the conventional word number/bit number addressing scheme.

Figure 12-21 shows examples of the *bit shift left* (BSL) and *bit shift right* (BSR) instructions available as part of the instruction set for Allen-Bradley PLC-5 and SLC-500 controllers.

```
┌ BSL ──────────────┐
│ BIT SHIFT LEFT     │
│ File               │
│ Control            │
│ Bit address        │
│ Length             │
└────────────────────┘
```

(a) Bit shift left instruction

```
┌ BSR ──────────────┐
│ BIT SHIFT RIGHT    │
│ File               │
│ Control            │
│ Bit address        │
│ Length             │
└────────────────────┘
```

(b) Bit shift right instruction

Fig. 12-21

Allen-Bradley PLC-5 and SLC-500 *bit shift left* and *bit shift right* instructions.

The data file used for a shift register usually is the bit file because its data are displayed in binary format, making it easier to read. BSL and BSR are output instructions that load data into a bit array one bit at a time. The data are shifted through the array, then unloaded one bit at a time.

To program a *bit shift* instruction, you need to provide the processor with the following information:

◆ **File** is the address of the bit array you want to manipulate. You must start the array at a 16-bit word boundary. For example, use bit 0 of word 1, 2, 3, etc. You can end the array at any bit number up to the file maximum. The instruction invalidates all bits beyond the last bit in the array (as defined by the length), up to the next word boundary.

◆ **Control** is the address of the control structure, which controls the operation of the instruction. It is reserved for the instruction and cannot be used to control any other instruction. The three words that make up the control element are listed in the following table:

Bit:	15	13	11	10
Control Word	EN	DN	ER	UL
Length Word	Stores the length of the file, in bits			
Position Word	Points to the current bit and toggles between 0 and a value equal to the length of the file			

The enable bit, bit 15, is set when the instruction is true. The done bit, bit 13, is set when the instruction has shifted all the bits in the file 1 position. It resets when the instruction goes false. The error bit, bit 11, is set when the instruction has detected an error, which can happen when a negative number is entered in the length. The unload bit, bit 10, is the bit location into which the status from the last bit in the file shifts when the instruction goes from false to true. When the next shift occurs, these data are lost, unless additional programming is done to retain the data.

◇ **Bit address** is the address of the source bit. The instruction inserts the status of this bit in either the first (lowest) bit position (for the BSL instruction) or the last (highest) bit position (for the BSR instruction) in the array.

◇ **Length** is the number of bits in the bit array, or the file length, in bits.

The *bit shift left* program of Fig. 12-22 can be used to describe the operation of the BSL instruction. A shift pulse is generated by a false-to-true transition of limit switch LS1. When the rung goes from false to true, the enable bit is set and the data block is shifted to the left (to a higher bit number) one bit position. The specified bit, at sensor bit address I:002/00, is shifted into the first bit position, B3:010/00. The last bit is shifted out of the array and stored in the unload bit, R6:0/UL. The status that was previously in the unload bit is lost. Note that all the bits in the unused portion of the last word of the file are invalid and should not be used elsewhere in the program. For wraparound operation, set the position of the bit address to the last bit of the array or to the UL bit, whichever applies.

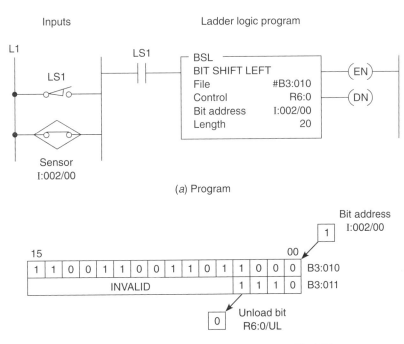

(a) Program

(b) Data block array before shift pulse generated by LS1

Fig. 12-22

Bit shift left program.

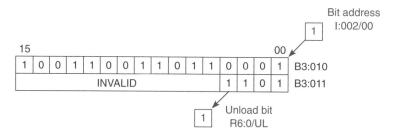

Bit address I:002/00

```
15                                    00
1 0 0 1 1 0 0 1 1 0 1 1 0 0 0 1   B3:010
        INVALID           1 1 0 1   B3:011
```

Unload bit R6:0/UL

(c) Data block array after shift pulse generated by LS1

Fig. 12-22 (continued)

Bit shift left program.

The *bit shift right* program of Fig. 12-23 can be used to describe the operation of the BSR instruction. Before the rung goes from false to true, the status of bits in words B3:100 and B3:101 are as shown. The status of the bit address, I:002/05, is a 0, and the status of the unload bit, R6:1/UL, is a 1. When limited switch LS1 closes, the status of the bit

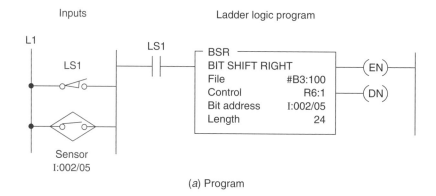

(a) Program

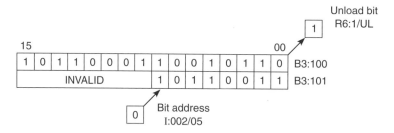

Unload bit R6:1/UL

```
15                                    00
1 0 1 1 0 0 0 1 1 0 0 1 0 1 1 0   B3:100
        INVALID           1 0 1 1 0 0 1 1   B3:101
```

Bit address I:002/05

(b) Data block array before shift pulse generated by LS1

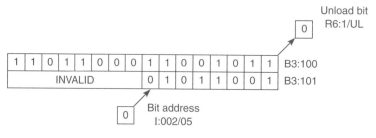

Unload bit R6:1/UL

```
1 1 0 1 1 0 0 0 1 1 0 0 1 0 1 1   B3:100
        INVALID           0 1 0 1 1 0 0 1   B3:101
```

Bit address I:002/05

(c) Data block array after shift pulse generated by LS1

Fig. 12-23

Bit shift right program.

address, I:002/05, is shifted into B3:101/07, which is the twenty-fourth bit in the file. The status of all the bits in the file are shifted one position to the right, through the length of 24 bits. The status of B3:100/00 is shifted to the unload bit, R6:1/UL. The status that was previously in the unload bit is lost.

The program of Fig. 12-24 illustrates a spray-painting operation controlled by a shift register. Each file bit location represents a station on the line, and the status of the bit indicates whether or not a part is present at that station. The bit address, I:001/02, detects whether or not a part has come on the line. The shift register's function is used to keep track of the items to be sprayed. A bit shift left instruction is used to indicate a forward motion of the line. As the parts pass along the production line, the shift register bit patterns represent the items on the conveyor hangers to be painted. LS1 is used to detect the hanger and LS2 detects the part. The logic of this operation is such that when a part to be painted and a part hanger occur together (indicated by the simultaneous operation of LS1 and LS2), a logic 1 is input into the shift register. The logic 1 will cause the undercoat spray gun to operate and, five steps later, when a 1 occurs in the shift register, the topcoat spray gun is operated. Limit switch 3 counts the parts as they exit the oven. The count obtained by limit switch 2 and limit switch 3 should be equal at the end of the spray-painting run (PL1 is energized) and is an indication that the parts commencing the spray-painting run equal the parts that have completed it. A logic 0 in the shift register indicates that the conveyor has no parts on it to be sprayed, and it therefore inhibits the operation of the spray guns.

The program of Fig. 12-25 on p. 343 illustrates a bit shift operation used to keep track of carriers flowing through a 16-station machine. Proximity switch #1 senses a carrier, while proximity switch #2 senses a part on the carrier. Pilot lights connected to output module O:004 turn on as carriers with parts move through the machine. They turn off as empty carriers move through. Station #4 is an inspection station. If the part fails, the inspectors push PB1 as they remove the part from the system, which turns output O:004/04 off. Rework is added back into the system at station #6. When the operator puts a part on an empty carrier, he or she pushes PB2, turning output O:004/06 on.

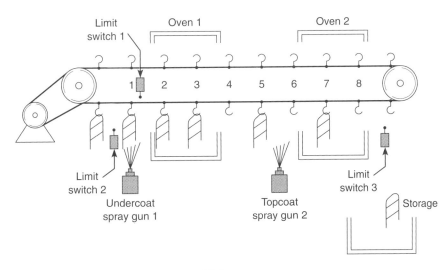

(a) Process

Fig. 12-24

Shift register spray-painting application.

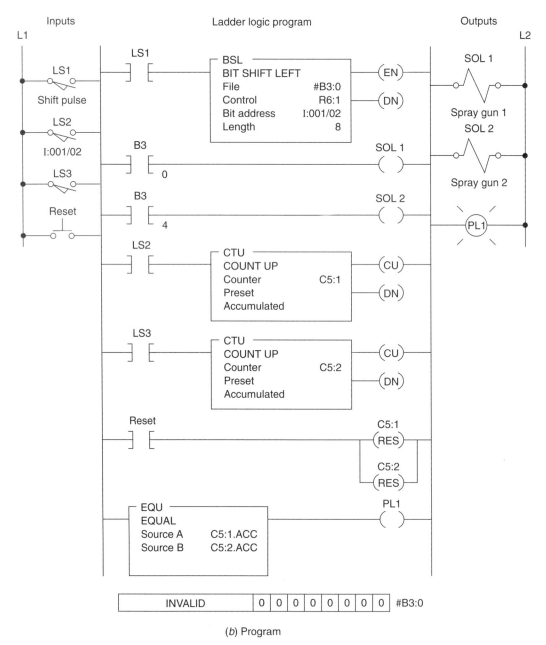

Inputs   Ladder logic program   Outputs

(b) Program

Fig. 12-24 (continued)

Shift register spray-painting application.

12.5 WORD SHIFT REGISTERS

Bit shift registers are classified as *synchronous* registers because information is shifted one bit at a time within a word, or from one word to another. The synchronous shift register may also be referred to as a serial shift register.

Asynchronous word shift registers permit stacking of data in a file. Two separate shift pulses are required: one to shift data into the file (LOAD), and one to shift data out of the

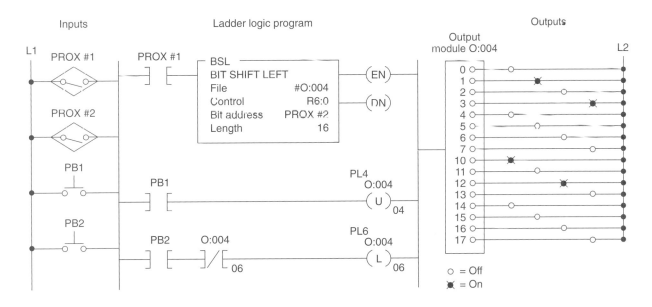

Fig. 12-25

Bit shift operation used to keep track of carriers flowing through a 16-station machine.

file (UNLOAD). These two shift pulses operate independently (asynchronously) of each other. There are two basic types of PLC word shift registers:

◆ FIFO (first in, first out)
◆ LIFO (last in, first out)

An example of the FIFO instruction pair is shown in Fig. 12-26. Both of the FIFO instructions are output instructions, and they are used as a pair. The *FIFO load* (FFL) instruction loads data into a file from a source element; the *FIFO unload* (FFU) instruction unloads data from a file to a destination word. When used in pairs, these instructions establish an asynchronous shift register (stack) that stores and retrieves data in a prescribed order.

When you program a FIFO stack, use the *same* file and control addresses, length, and position values for *both* instructions in the pair. You need to provide the processor with the following information:

◆ **Source** is the word address location from which data that are entered into the FIFO file comes. The load instruction retrieves the value from this address and loads it into the next word in the stack.

◆ **Destination** is the address that stores the value that exits from the stack. This is where the data go as they are indexed from the FIFO file on a false-to-true transition of the FFU instruction. Any data currently in the destination are written over by the new data when the FFU is indexed.

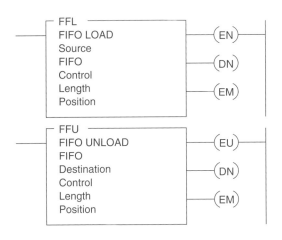

Fig. 12-26

FIFO load (FFL) and *FIFO unload* (FFU) instruction pair.

Sequencer and Shift Register Instructions

343

- **FIFO** is the address of the stack. It must be an indexed word address in the input, output, status, bit, or integer file. The same address is programmed for the FFL and FFU instructions.

- **Control** is the file address of the control structure. The status bits, stack length, and position value are stored in this element. The FIFO load and FIFO unload instructions share the same control element, which may not be used to control any other instructions. The control bits in the FIFO control element are:

Bit:	15	14	13	12
	EN	EU	DN	EM

The EN bit is the *FIFO load* enable bit and follows the status of the FFL instruction. The EU bit is the *FIFO unload* enable bit and follows the status of the FFU instruction. The DN bit (the done bit) indicates that the position has reached the FIFO length, that is, that the FIFO file is full. When the DN bit is set, it inhibits the transfer of any additional data from the source to the FIFO file. The EM bit is set when the last piece of data entered from the source has been transferred to the destination

and the position is 0. If the FFU has a false-to-true transition after the EM bit is set, 0's will be loaded into the destination.

- **Length** lets you specify the maximum number of words in the stack.

- **Position** is the pointer in the FIFO file. It indicates where the next piece of data from the source will be entered and also how many pieces of data are currently entered in the FIFO. Enter a position value only if you want the instruction to start at an offset at power-up; otherwise, enter 0. Your ladder program can change the position if necessary.

The program of Fig. 12-27 can be used to describe how data are indexed in and out of a FIFO file using the FFL and FFU instruction pair. Data enter the FIFO file from the source address, N7:10, on a false-to-true transition of input *A*. Data are placed at the position indicated in the instruction on a false-to-true transition of the FFL instruction, after which the position indicates the current number of data entries in the FIFO file. The FIFO file fills from the beginning address of the FIFO file and indexes to one higher address for each false-to-true transition of input *A*. A false-to-true

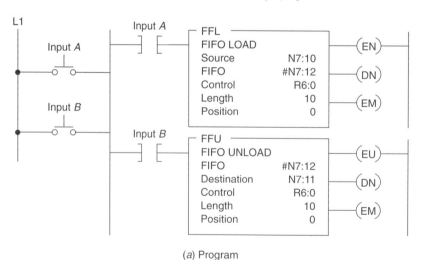

(a) Program

Fig. 12-27

How data are indexed in and out of a FIFO file.

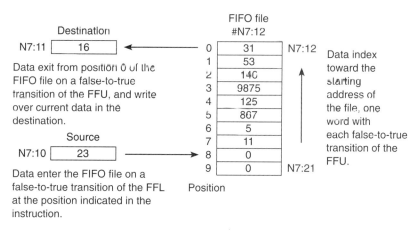

(b) Transfer of data

Fig. 12-27 (continued)

How data are indexed in and out of a FIFO file.

transition of input *B* causes all data in the FIFO file to shift one position toward the starting address of the file, with the data from the starting address of the file shifting to the destination address, N7:11.

The FIFO instruction is often used for inventory control. For example, if a number of parts are to be manufactured, each part would be assigned a different code. Once the PLC begins the manufacturing process, the different coded parts can be pulled out automatically in the order prescribed by the FIFO instruction.

The LIFO (last in, first out) instruction inverts the order of the data it receives by outputting the last data received first and the first data received last. Essentially, a LIFO is a stack that allows data to be added without disturbing the data already contained in the stack. An example of the LIFO instruction pair is shown in Fig. 12-28. The difference between FIFO and LIFO stack operation is that the LIFO instruction removes data in the *reverse* of the order they are loaded (last in, first out). Otherwise, LIFO instructions operate the same as FIFO instructions.

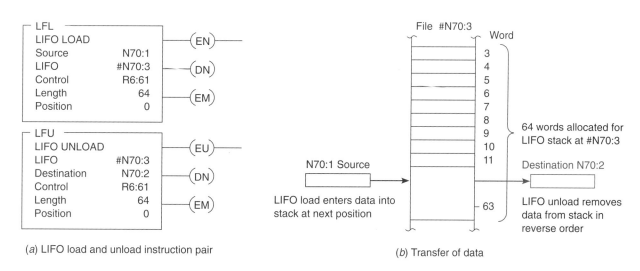

(a) LIFO load and unload instruction pair

(b) Transfer of data

Fig. 12-28

LIFO instruction pair.

Sequencer and Shift Register Instructions

Questions

1. Explain the basic operation of a cam-operated sequencer switch.

2. What type of operations are sequencer switches most suitable for?

3. What is the advantage of sequencer programming over conventional programming methods?

4. With reference to a PLC sequencer output instruction:

   a. Where is the information for each sequencer step entered?

   b. What is the function of the output word?

   c. Explain the transfer of data that occurs as the sequencer is advanced through its various steps.

5. Explain the purpose of a mask word when used in conjunction with the sequencer instruction.

6. What are the two limits placed on some sequencer output instructions?

7. Sequencer instructions are usually retentive. Explain what this means.

8. Explain the difference between an event-driven and a time-driven sequencer.

9. Explain the function of a sequencer input instruction.

10. Explain the function of a sequencer load instruction.

11. How does a bit shift register manipulate individual bits?

12. List four common applications for shift registers.

13. Name two types of bit shift instructions.

14. Name the two types of shift pulses used with asynchronous word shift registers and state the function of each.

15. Explain the difference between a FIFO register and a LIFO register.

Problems

1. Answer each of the following with reference to the *dishwasher circuit* shown in Fig. 12-5 on pp. 326-327 (the time per step is 45 s):

    a. How many cam switches can be found in the timer?

    b. How many timed steps are there for one complete operating cycle?

    c. What is the value of the time interval for each step?

    d. State the five output devices operated by the timer.

    e. What is the total length of time that the heater is on for one complete cycle?

    f. What output devices would normally be on when the timer is at the 20-min point in the cycle?

    g. What is the greatest length of time that the fill valve stays energized?

    h. Explain the function of the safety water level switch.

    i. Outline the sequence in which the outputs are energized for the first rinse portion of the cycle.

    j. Why is the timer motor off for only one step in the entire operating cycle?

2. Construct an equivalent sequencer data table for the six steps of the drum-operated sequencer drawn in Fig. 12-29.

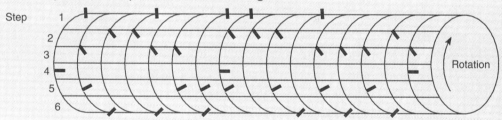

Fig. 12-29

3. Answer the following with reference to the sequencer word file of Fig. 12-30:

    a. Assume that output bit addresses 1 through 16 are controlling lights 1 through 16. State the status of each light for each of the steps.

    b. What output bit addresses could be masked?

    c. State the status of each bit of output word 25 for step 3 of the sequencer cycle.

    d. If word 31 is to be entered into the PLC using the hexadecimal code, how would it be written?

	16	15	14	13	12	11	10	9	8	7	6	5	4	3	2	1	
Word 25	0	0	0	0	0	0	0	0	0	0	0	0	0	0	0	0	Output

	16	15	14	13	12	11	10	9	8	7	6	5	4	3	2	1	
Word 30	1	1	1	1	1	1	1	1	1	1	1	1	1	1	1	1	Step 1
Word 31	1	0	1	0	1	0	1	0	1	0	1	0	1	0	1	0	Step 2
Word 32	0	1	0	1	0	1	0	1	0	1	0	1	0	1	0	1	Step 3
Word 33	0	0	0	0	0	0	0	0	0	0	0	0	0	0	0	0	Step 4

Fig. 12-30

Sequencer and Shift Register Instructions

4. Answer each of the following with reference to the time-driven sequencer traffic light program shown in Fig. 12-11 on p. 331:

    a. How many bit addresses are controlled by this sequencer?

    b. What is the mask data program code for this sequencer?

    c. Assume normal operation. When input 005 is turned on, the sequencer will perform the steps as they have been programmed. If input 005 is turned off during normal operation, what happens to the sequencer at that point and when input 005 is turned back on again?

    d. Suppose input 005 were turned off at step 3 in the run mode. What outputs would be energized?

    e. If you wanted to control outputs 010, 011, 014, 015, and 016, what would the program entry code for your mask data be?

    f. What is the total time required for one complete cycle of the sequencer?

    g. State for what step(s) of the sequencer each of the following outputs would be on:

        1. 009   3. 011   5. 013   7. 015

        2. 010   4. 012   6. 014   8. 016

    h. Pressing the reset button resets the sequencer to what step?

    i. What outputs not used by the sequencer can be used elsewhere in the program?

5. Answer each of the following with reference to the event-driven sequencer traffic light program shown in Fig. 12-12 on p. 332:

    a. Which input condition is made true to reset the sequencer to step 1?

    b. When does the sequencer advance to the next step?

    c. Assume power to the sequencer is lost and then returned. In what way will the operation of the sequencer be affected?

    d. If the preset value of step 3 were changed from 1 to 2, how would this affect the operation of the sequencer?

6. Using whatever PLC sequencer output instruction you are most familiar with, develop a program that will operate the cylinders in the desired sequence. The time between each step is to be 3 s. The desired sequence of operation will be as follows:

- All cylinders to retract.
- Cylinder 1 advance.
- Cylinder 1 retract and cylinder 3 advance.
- Cylinder 2 advance and cylinder 5 advance.
- Cylinder 4 advance and cylinder 2 retract.
- Cylinder 3 retract and cylinder 5 retract.
- Cylinder 6 advance and cylinder 4 retract.
- Cylinder 6 retract.
- Sequence to repeat.

7. Using whatever PLC sequencer output instruction you are most familiar with, develop a program to implement an automatic car-wash process. The process is to be event-driven by the vehicle, which activates various limit switches (LS1 through LS6) as it is pulled by a conveyor chain through the car-wash bay. Design the program to operate the car wash in the following manner:

- The vehicle is connected to the conveyor chain and pulled inside the car-wash bay.
- LS1 turns the water input valve on.
- LS2 turns on the soap release valve, which mixes with the water input valve to provide a wash spray.
- LS3 shuts off the soap valve, and the water input valve remains on to rinse the vehicle.
- LS4 shuts off the water input valve and activates the hot wax valve, if selected.
- LS5 shuts off the hot wax valve and starts the air-blower motor.
- LS6 shuts off the air blower. The vehicle exits the car wash.

8. Using whatever PLC sequencer input and output instructions you are most familiar with, develop a program that contains sequencer input and output instructions on the same rung and that meets the following criteria:

Inputs True to Cause Outputs to Index at Indicated Output Step	Output Step	Outputs True at Indicated Output Step
I:002/00, I:002/10	1	O:015/15, O:015/17
I:002/11, I:002/15	2	O:015/04
I:002/11	3	O:015/03, O:015/13
I:002/05, I:002/07	4	O:015/10
I:002/04	5	O:015/11, O:015/16

Mask out all unused input and output bits. Construct a chart to show the data that must be entered in the *sequencer input* and *sequencer output* files.

9. A product moves continuously down an assembly line that has 4 stations, as shown in Fig. 12-31. The product enters the inspection zone, where its presence is sensed by the proximity switch. The inspector examines it and activates a reject button if the product fails inspection. If the product is defective, reject status lights come on at stations 1, 2, and 3 to tell the assembler to ignore the part. When a defective part reaches station 4, a diverter gate is activated to direct that part to a reject bin. Using whatever PLC bit shift register you are most familiar with, develop a program to implement this process.

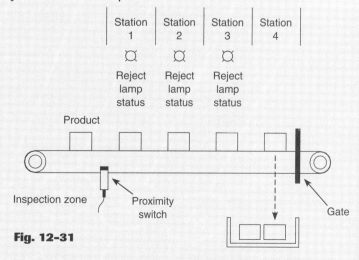

Fig. 12-31

PLC Installation Practices, Editing, and Troubleshooting

After completing this chapter, you will be able to:

- ◆ Outline and describe requirements for a PLC enclosure

- ◆ Identify and describe the functions of bleeder resistors in PLCs

- ◆ Differentiate between off-line and on-line programming

- ◆ Describe proper grounding practices and preventive maintenance tasks associated with PLC systems

- ◆ List and describe specific PLC troubleshooting procedures

This chapter discusses guidelines for the installation, maintenance, and troubleshooting of a PLC-controlled system. Information on proper grounding that ensures personal safety as well as correct operation of equipment is included. Unique troubleshooting procedures that apply specifically to PLCs are listed and explained.

13.1 PLC ENCLOSURES

A PLC system, if installed properly, should give years of trouble-free service. The design nature of PLCs includes a number of rugged design features that allow them to be installed in almost any industrial environment. However, problems can occur if the system is not installed properly.

Programmable logic controllers are generally placed within an *enclosure* (Fig. 13-1). An enclosure is the chief protection from atmospheric conditions. The National Electrical Manufacturers Association (NEMA) has defined enclosure types, based on the degree of protection an enclosure will provide. For most solid-state control devices, a NEMA 12 enclosure is recommended. This type of enclosure is for general-purpose areas and is designed to be dust-tight. In addition, metal enclosures also help to minimize the effects of electromagnetic radiation that may be generated by surrounding equipment.

Every PLC installation will dissipate heat from its power supplies, local I/O racks, and processor. This heat accumulates in the enclosure and must be dissipated from it into the surrounding air. For many applications, normal convection cooling will keep the controller components within the specified temperature operating range. Proper spacing of components within the enclosure is usually sufficient for heat dissipation. The temperature inside the enclosure must not exceed the maximum operating temperature of the controller (typically 60°C maximum). Additional cooling provisions, such as a fan

Fig. 13-1

PLC system mounted within an enclosure.
(Courtesy of Klockner-Moeller Ltd.)

or blower, may be required where high ambient temperatures are encountered. Figure 13-2 shows the typical layout of components for a PLC installation.

The enclosure should have a power disconnect so that, when required, the PLC can be worked on with the power off. Also, a viewing window is desirable so that the indicators and the PLC and modules can be viewed without having to open the enclosure during normal operation. Viewing windows installed in the doors of control panels help facilitate maintenance and troubleshooting chores for factory personnel.

A hardwired *master control relay* (MCR) is normally included as part of the enclosure layout. The hardwired MCR is used to interrupt power to the I/O rack in the event of a system failure, but it will still allow power to be maintained at the CPU. In addition, an isolation transformer provides the following:

◆ Physical isolation from the main power distribution

◆ Voltage transformation, if required, to provide 110 or 240 V to the ac distribution system.

When the PLC is operated in a noise-polluted industrial environment, special consideration should be given to possible electrical interference. Malfunctions resulting from noise are temporary occurrences of operating errors that can result in hazardous machine operation in certain applications. Noise usually enters through input, output, and power supply lines. Noise may be coupled into these lines by an electrostatic field or through electromagnetic induction. The following reduces the effect of electrical interference:

◆ PLC design features

◆ Proper mounting of the controller within an enclosure

◆ Proper equipment grounding

◆ Proper routing of wiring

◆ Proper suppression added to noise-generating devices

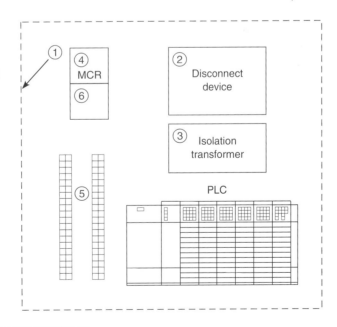

(1) NEMA rated enclosure suitable for your application and environment that shields your controller from electrical noise and airborne contaminants

(2) Disconnect, to remove power from the system

(3) Fused isolation transformer or a constant voltage transformer, as your application requires

(4) Master control relay/emergency-stop circuit

(5) Terminal blocks or wiring ducts

(6) Suppression devices for limiting electromagnetic interference (EMI) generation

(a) Typical component layout

Fig. 13-2

Typical PC installation.

PLC Installation Practices, Editing, and Troubleshooting

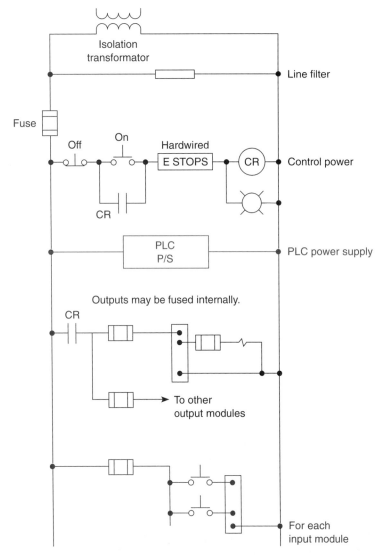

Isolation transformer

Line filter

Fuse

Off On Hardwired
E STOPS CR Control power

CR

PLC
P/S PLC power supply

Outputs may be fused internally.

CR

To other
output modules

For each
input module

(b) Typical wiring layout. In the configuration shown, the emergency stop (ESTOP) relay contacts are in series with the output module power, but not with the PLC power supply and the input modules. For an emergency stop with the input values still there and the PLC still running, valuable troubleshooting data are made available.

Fig. 13-2 (continued)
Typical PC installation.

To increase the operating noise margin, the controller should be located away from noise-generating devices such as large ac motors and high-frequency welders. Potential noise generators include relays, solenoids, motors, and motor starters, especially when operated by hard contacts such as pushbuttons or selector switches. Suppression for noise generation may be necessary when these types of loads are connected as output devices, or when they are connected to the same supply line that powers the PLC. Figure 13-3 shows typical noise-suppression methods.

Lack of surge suppression on inductive loads may contribute to processor faults and sporadic operation. RAM can be corrupted (lost), and I/O modules can appear faulty or can reset themselves. When inductive devices are energized or de-energized, they can cause an electrical pulse to be backfed into the PLC system. The backfed pulse, when entering the PLC system, can be mistaken by the PLC for a computer pulse. It takes only one false pulse to create a malfunction of the orderly flow of PLC operational sequences. For extremely noisy environments, use a memory module and program it for automatic loading on processor fault or power cycle for quick recovery.

Careful wire routing also helps to cut down on electrical noise. Within the PLC enclosure, input power to the processor module should be routed separately from the wiring to I/O modules. *Never* run signal wiring and power wiring in the same conduit. Segregate I/O wiring by signal type, and bundle wiring with similar electrical characteristics together. Wiring with different signal characteristics should be routed into the enclosure by separate paths whenever possible. A fiberoptic system, which is totally immune to all kinds of electrical interference, can also be used for signal wiring.

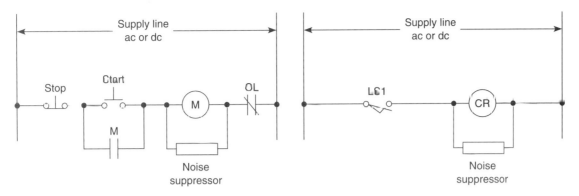

(a) Circuit connection

(b) Typical devices. *(Courtesy of Allen-Bradley Company, Inc.)*

Fig. 13-3

Typical noise-suppression methods.

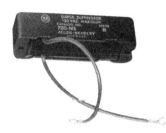

·3 LEAKY INPUTS AND OUTPUTS

Many field input devices, such as proximity switches, used with PLC-based systems are of a solid-state design. Any electronically based input sensor that uses a solid-switch silicon controlled rectifier (SCR), triac, or transistor will have a small leakage current even when in the off state. Often, the leaky input will only cause the module's input indicator to flicker. The leakage may, however, result in a falsely activated PLC input. To correct the problem, a bleeder resistor is connected across or in parallel with the input, as shown in Fig. 13-4.

This leakage may also occur with the solid-state switch used in many output modules. A similar problem can be created when a high-impedance output load device is used with these modules. Figure 13-5 on p. 356 shows how a bleeder resistor is connected to bleed off this unwanted leakage current.

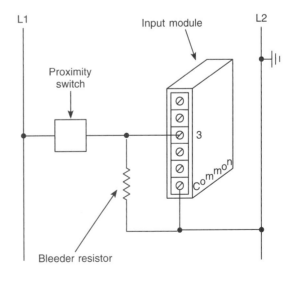

Fig. 13-4

Connection for leaky input devices.

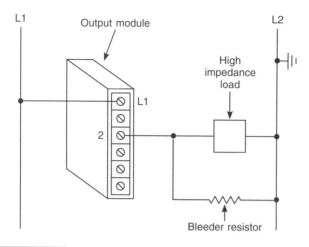

Fig. 13-5

Connection for leaky output devices.

13•4 GROUNDING

Proper grounding is an important safety measure in all electrical installations. The authoritative source on grounding requirements for a PLC installation is the National Electrical Code. The code specifies the type of conductors, color codes, and connections necessary for safe grounding of electrical components. According to the code, the grounding path must be permanent (no solder), continuous, and able to conduct safely the ground-fault current in the system with minimal impedance. In the event of a high value of ground current, the temperature of the conductor could cause the solder to melt, resulting in interruption of the ground connection. In addition to the grounding required for the controller and its enclosure, you must also provide proper grounding for all controlled devices in your application. Most manufacturers provide detailed information on the proper grounding methods to use in an enclosure. Figure 13-6 shows typical grounding connections for an enclosure.

With solid-state control systems, grounding helps to limit the effects of noise due to electromagnetic induction. The following grounding practices will help reduce electrical noise interference:

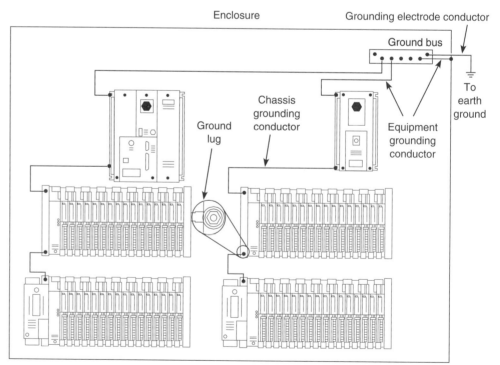

Fig. 13-6

PLC ground connections. (*Courtesy of Allen-Bradley Company, Inc.*)

Note: When using this grounding configuration, make no connections to EQUIP GND on the power supply terminal strips. This can cause ground loops.

◆ All PLC equipment and enclosure backplates should be grounded individually to a central point on the enclosure frame.

◆ Ground wires should be separated from power wiring at the point of entry to the enclosure.

◆ All ground connections should be made with star washers between the grounding wire and lug and metal enclosure surface.

◆ Paint or other nonconductive material should be scraped away from the area where a chassis makes contact with the enclosure.

◆ The minimum ground wire size should be No. 12 AWG stranded copper for PLC equipment grounds and No. 8 AWG stranded copper for enclosure backplate grounds.

◆ The enclosure should be grounded properly to the ground bus.

◆ The machine ground should be connected to the enclosure and to earth ground.

◆ The ground connection should have a very low resistance. A rule of thumb would be less than 0.1 Ω dc resistance between the device and ground.

Ground loops can also cause noise problems and are often difficult to find. They generally occur when *multiple* grounds exist (Fig. 13-7). The farther the grounds are apart, the more likely is the noise problem. A potential can exist between the power supply earth and the remote earth. Ground-loop interference results from multiple grounds that form loops ideal for picking up interference. If a varying magnetic field passes through one of these loops, a voltage is produced and current flows in the loop.

•5 VOLTAGE VARIATIONS AND SURGES

The power supply section of the PLC system is built to sustain line fluctuations and still allow the system to function within its operating range. Where line voltage variation is excessive, a constant voltage transformer can be used to solve the problem. The constant voltage transformer stabilizes the input voltage by compensating for voltage changes at the primary to maintain a steady voltage at the secondary.

When current in an inductive load is interrupted or turned off, a very high voltage spike is generated. If not suppressed, these voltage spikes can reach several thousand volts and produce surges of damaging high currents. To avoid this situation, a suppression network

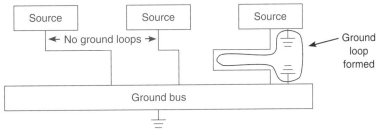

Certain connections require shielded cables to help reduce the effects of electrical noise coupling. Ground each shield at one end only. A shield grounded at both ends forms a ground loop, which can cause a processor to fault.

Fig. 13-7

Formation of ground loops.

should be installed to limit the voltage spike as well as the rate of change of current through the inductor. Generally, output modules designed to drive inductive loads include suppression networks built in as part of the module circuit.

An additional external suppression device is recommended if an output module is used to control devices such as relays, solenoids, motor starters, or motors. The suppression device is wired in parallel (directly across) and as close as possible to the load device. Figure 13-8 illustrates different methods of suppressing dc and ac inductive loads. The suppression components used must be rated appropriately to suppress the switching transient characteristic of the particular inductive device. Check the installation manual for the PLC you are using for the proper type and rating of the suppression device.

The *metal oxide varistor* (MOV) (Fig. 13-9) surge suppressor functions in the same manner as back-to-back zener diodes. Each zener diode acts as an open circuit until the reverse voltage across it exceeds its rated value. Any greater voltage peak instantly makes the diode act like a short circuit that bypasses this voltage away from the rest of the circuit. Additional suppression is especially important if your inductive device is in series or parallel to a hard contact, such as a pushbutton or selector switch. Switching inductive loads without surge suppression can significantly reduce the lifetime of contacts. It is recommended that you locate the suppression device as close as possible to the load device.

13•6 PROGRAM EDITING

After you have entered the rungs for your program, you may need to modify them. *Editing* is simply the ability to make changes to an existing program through a variety of editing functions. Using the editing function, instructions and rungs can be added or

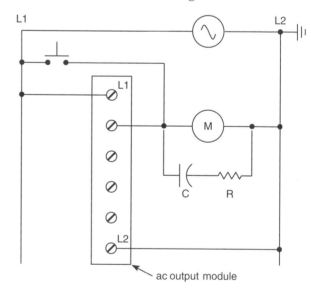

(a) For inductive dc load devices, a clamping diode connected in reverse-bias is suitable. A diode conducts only when current to the solenoid is switched off. When the inductive load switches off, it tries to maintain the same current and the voltage across the inductive load reverses. The diode provides a path for this current so that the field in the inductor collapses.

(b) Surge suppression is also known as *snubbing*. An RC circuit can be used for suppression of ac load devices. The resistor and capacitor connected in series slows the rate of rise of the transient voltage.

Fig. 13-8

Suppressing dc and ac inductive loads.

deleted: addresses, data, and bits can be changed. Again, the editing format varies with different manufacturers and PLC models.

When editing a processor's logic, the use of the *search* function can be extremely helpful. This function is used to search the program for specific addressed instructions. Activation of the search function locates the specified addressed instruction in the processor memory. The circuit containing the searched instruction is then displayed automatically on the screen for user inspection. If desired, the user can then modify the instruction itself, or the circuit containing the instruction. Depending on the PLC, the search function can be used to search for:

◆ An instruction with an address
◆ An instruction type
◆ An address
◆ A rung of logic

Another important editing technique is the use of the *cursor control* keys. There are usually four keys that are used for cursor control. These keys are often engraved with arrows pointing up, down, right, and left. A blinking instruction in a ladder rung of the operator terminal display indicates the cursor position. The cursor controls allow you to move through your program from instruction to instruction or from rung to rung.

13•7 PROGRAMMING AND MONITORING

When programming a PLC, several instruction entry modes are available, depending on the manufacturer and the model of the unit. A personal computer, with appropriate software, is often used to program and monitor the program in the PLC. Additionally, it makes possible *off-line programming*, which involves writing and storing the program in the personal computer without its

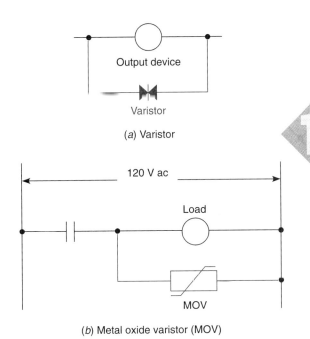

(a) Varistor

(b) Metal oxide varistor (MOV)

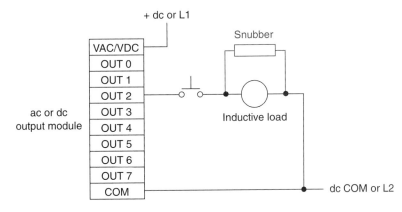

(c) Adding a snubber circuit across the inductive load can significantly increase the lifetime of the pushbutton contacts.

Fig. 13-9

Snubbing circuits.

Fig. 13-10

Off-line programming. This powerful feature allows you to develop and edit processor control programs off-line (without the need of a programmable controller) by utilizing the memory of the personal computer. Programs can be developed in a comfortable (nonfactory) environment, documented, stored on disk, and downloaded into the processor memory. *(Courtesy of Square D Company.)*

An *on-line programming mode* permits the user to change the program during machine operation. As the PLC controls its equipment or process, the user can add, change, or delete control instructions and data values as desired. Any modification made is executed immediately on entry of the instruction. Therefore, the user should assess all possible sequences of machine operation resulting from the change in advance. On-line programming should be done only by experienced personnel who understand fully the operation of the PLC they are dealing with and the machinery being controlled. If at all possible, changes should be made off-line to provide a safe transition from existing programming to new programming.

Data monitor is a feature that allows you to display data from any place in the data table. Depending on the PLC, the data monitor function can be used to do the following:

◇ View data within an instruction

◇ Store data or values for an instruction prior to use

◇ Set or reset values and/or bits during a debug operation for control purposes

◇ Change the radix or data format

Always be careful when manipulating data using the data monitor function. Changing data could affect the program and turn output devices on or off.

The *contact histogram* function allows you to view the transition history (the on and off states) of a data table value. The status of the bit(s) (on or off) and the length of time the bit(s) remained on or off (in hours, minutes, seconds, hundredths of a second) is displayed. When viewing a contact histogram file, the accumulated time indicates the total time that the histogram function was running. The delta time of the contact histogram indicates the elapsed time between the changes in states.

being connected to the PLC (Fig. 13-10), and later downloading it to the PLC. In contrast, *on-line programming* involves programming or entering ladder logic directly into the PLC. *Off-line programming* is the safest manner in which to edit a program because additions, changes, and deletions do not affect the operation of the system until downloaded to the PLC.

Often a *continuous test mode* is provided that causes the processor to operate from the user program without energizing any outputs. This allows the control program to be executed and debugged while the outputs are disabled. A check of each rung can be done by monitoring the corresponding output rung on the programming device. A *single-scan* test mode may also be available for debugging the control logic. This mode causes the processor to complete a single scan of the user program each time the single scan key is pressed with no outputs being energized.

13·8 PREVENTIVE MAINTENANCE

The biggest deterrent to PLC system faults is a proper preventive maintenance program. Although PLCs have been designed to minimize maintenance and provide trouble-free operation, there are several preventive measures that should be looked at regularly.

Many control systems operate processes that must be shut down for short periods for product changes. The following preventive maintenance tasks should be carried out during these short shutdown periods:

◆ Any filters that have been installed in enclosures should be cleaned or replaced to ensure that clear air circulation is present inside the enclosure.

◆ Dust or dirt accumulated on PLC circuit boards should be cleaned. If dust is allowed to build up on heat sinks and electronic circuitry, an obstruction of heat dissipation could occur and cause circuit malfunction. Furthermore, if conductive dust reaches the electronic boards, a short circuit could result and cause permanent damage to the circuit board. Ensuring that the enclosure door is kept closed will prevent the rapid buildup of these contaminants.

◆ Connections to the I/O modules should be checked for tightness to ensure that all plugs, sockets, terminal strips, and module connections are making connections, and that the module is installed securely. Loose connections may result not only in improper function of the controller, but also in damage to the components of the system.

◆ All field I/O devices should be inspected to ensure that they are adjusted properly. Circuit boards dealing with process control analogs should be calibrated every 6 months. Other devices, such as sensors, should be done on a monthly basis. End devices

in the environment, which have to translate mechanical signals into electrical, may gum up, get dirty, crack, or break—and then they will no longer trip at the correct setting.

◆ Care should be taken to ensure that heavy noise- or heat-generating equipment is not moved too close to the PLC.

◆ Check the condition of the battery that backs up the RAM memory in the CPU. Most CPUs have a status indicator that shows whether the battery's voltage is sufficient to back up the memory stored in the PLC. If a module is to be replaced, it must be replaced with exactly the same type of module.

◆ Stock commonly needed spare parts. Input and output modules are the PLC components that fail most often.

◆ Keep a master copy of operating programs used.

To avoid injury to personnel and to prevent equipment damage, all connections should be checked with power removed from the system. In addition to disconnected electrical power, all other sources of power (pneumatic and hydraulic) should be de-energized before working on a machine or process controlled by a PLC. Most companies use lockout and tag procedures to make sure that equipment does not operate while maintenance and repairs are conducted. A personnel protection tag is placed on the power source for the equipment and the PLC, and can be removed only by the person who originally placed the tag. In addition to the tag, a lock is also attached so that equipment cannot be energized.

13·9 TROUBLESHOOTING

In the event of a PLC fault, a careful, systematic approach should be used when troubleshooting the system to resolve the problem. The ability to monitor the activity of equipment control logic on a CRT screen is a big advantage in many troubleshooting situations (Fig. 13-11).

When a problem does occur, the first step in the troubleshooting procedure is to identify the problem and its source. The source of a problem can generally be narrowed down to the processor module, I/O hardware, wiring, machine inputs or outputs, or the ladder logic program. Once a problem is recognized, it is usually quite simple to deal with. The following sections will deal with troubleshooting these potential problem areas.

Processor Module

The processor is responsible for the *self-detection* of potential problems. It performs error checks during its operation and sends status information to indicators that are normally located on the front of the processor module. Typical diagnostics include memory OK, processor OK, battery OK, and power supply OK.

The processor then monitors itself continually for any problems that might cause the controller to execute the user program improperly. Depending on the controller, a

Fig. 13-11
Technician monitoring a PLC system on a CRT screen.
(Courtesy of PSI Repair Services, Inc.)

set of fault relay contacts may be available. The fault relay is controlled by the processor and is activated when one or more specific fault conditions occur. The fault relay contacts are used to disable the outputs and signal a failure.

Most PLCs incorporate a *watchdog timer* to monitor the scan process of the system. The watchdog timer is usually a separate timing circuit that must be set and reset by the processor within a predetermined period. The watchdog timer circuit will *time-out* if a processor hardwire malfunction occurs, and will immediately halt the operation of the PLC. For example, if the program scan value equals the watchdog value, a watchdog major error will be declared. Operation manuals show how to apply this function. Errors in memory data are also detected through various built-in diagnostic routines.

The PLC processor hardware is not likely to fail because today's microprocessors and microcomputer hardware are very reliable when operated within the stated limits of temperature, moisture, etc. The PLC processor chassis is typically designed to withstand harsh environments.

Input Malfunctions

If the controller is operating in the run mode but output devices do not operate as programmed, the most likely problem source is one of the following:

◆ I/O devices

◆ Wiring between I/O modules, I/O devices, and user power

◆ User power

◆ I/O modules

Narrowing down to one of the above as the problem source can usually be accomplished by comparing the actual status of the suspect I/O with controller status indicators. Usually each input or output device has at least two status indicators. One of these indicators is on the I/O module; the other indicator is provided by the programming device monitor.

If input hardware is suspected to be the source of a problem, the first check is to see if the status indicator on the input module illuminates when it is receiving power from its corresponding input device (e.g., pushbutton, limit switch). If the status indicator on the input module does *not* illuminate when the input device is on, take a voltage measurement across the input terminal to check for the proper voltage level (Fig. 13-12). If the voltage level is correct,

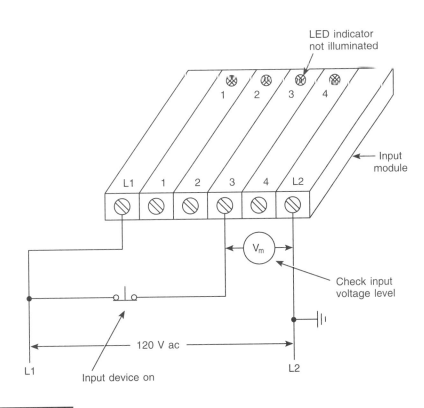

Fig. 13-12

Checking for input malfunctions.

then the input module should be replaced. If the voltage level is not correct, then check for faults with power to the input device; wiring among input device, input module, and user power; and the input device itself.

If the programming device monitor does not show the correct status indication for a condition instruction, the input module may not be converting the input signal properly to the logic level voltage required by the processor module. In this case, the input module should be replaced. If a replacement module does *not* eliminate the problem and wiring is assumed to be correct, then the I/O rack, communication cable, or processor should be suspected. Figure 13-13 shows a typical input device troubleshooting guide. This guide reviews condition instructions and how their true/false status relates to external input devices. Status indication is provided by displaying all logically true instructions in a rung in reverse video (darkened).

Output Malfunctions

When an output does not energize as expected, first check the output module blown fuse indicator. Usually this indicator will illuminate only when the output circuit corresponding to the blown fuse is energized. If this indicator is illuminated, correct the cause of the malfunction and replace the blown fuse in the module.

If the blown fuse indicator is not illuminated (fuse OK), then check to see if the output device is responding to the LED status indicator. If an output rung is energized, the module status indicator is on, and the output device is not responding, then the wiring to the output device or the output device itself should be suspected. If, according to the programming device monitor, an output device is commanded to turn on but the status indicator is off, then the module should be replaced. Figure 13-14 shows a typical output device troubleshooting guide.

Ladder Logic Program

The ladder logic program is also not likely to fail, assuming that the program was at one time working correctly. A hardware fault in the memory IC that holds the ladder logic program could alter the program, but this is a PLC hardware failure. If all other possible sources of trouble have been eliminated, the ladder logic program should be reloaded into the PLC from the master copy of the program.

While the ladder logic program is not likely to fail, the process may be in a state that was unaccounted for in the original program and thus is not controlled properly. In this case, the program needs to be modified to include this new state. A careful examination of the description of the control system and the ladder logic program can help identify this type of fault.

The *force on* and *force off* instructions allow you to turn specific bits on or off for testing purposes. Forcing lets you simulate operation or control an output device. For example, forcing a solenoid valve on will tell you immediately whether the solenoid is functional when the program is bypassed. If it is, the problem must be related to the software and not the hardware. If the output fails to respond when forced, it is either the actual output module causing the problem, or the solenoid itself is malfunctioning. *Take all necessary precautions to protect personnel and equipment during forcing.*

Certain diagnostic instructions may be included as part of a PLC's instruction set for troubleshooting purposes. The *temporary end* (TND) instruction (Fig. 13-15*a* on p. 366) is used when you want to change the amount of logic scanned to progressively debug your program. It operates only when its rung conditions are true and stops the processor from scanning any logic beyond the TND instruction. When the processor encounters the TND instruction, it resets the watchdog timer (to 0), performs an I/O update, and begins running the ladder program at the first instruction in the main program.

Input device troubleshooting guide				
Input device condition	Input module status indicator	Operator terminal status indicator		Possible problem source(s)
		⊣⊢	⊣/⊢	
—o⌿o— Closed — on	On	True Dark ⊣■⊢	False Normal ⊣/⊢	None, correct status indication.
—o⟍o— Open — off	Off	False Normal ⊣⊢	True Dark ■/⊢	None, correct status indication.
—o⌿o— Closed — on	On	False Normal ⊣⊢	True Dark ■/⊢	1. I/O module. 2. Processor/operator terminal communication.
—o⌿o— Closed — on	Off	False Normal ⊣⊢	True Dark ■/⊢	1. Wiring/power to I/O module. 2. I/O module.
—o⟍o— Open — off	Off	True Dark ■	False Normal /⊢	1. Programming error. 2. Processor/operator terminal communication.
—o⟍o— Open — off	On	True Dark ⊣■⊢	False Normal ⊣/⊢	1. Short circuit in input device or wiring. 2. Input module.

Fig. 13-13

Typical input device troubleshooting guide. *(Courtesy of Allen-Bradley Company, Inc.)*

Output device troubleshooting guide			
Output device condition	Output module status indicator	Operator terminal status indicator	Possible problem source
—⋀— Energized — on	On	True Dark ■	None, correct status indication.
—⋀— De-energized — off	Off	False Normal ◯	None, correct status indication.
—⋀— De-energized — off	On	True Dark ■	1. Wiring to output device. 2. Output device.
—⋀— De-energized — off	Off	True Dark ■	1. Blown fuse — output module. 2. Output module malfunction.

Fig. 13-14

Typical output device troubleshooting guide. *(Courtesy of Allen-Bradley Company, Inc.)*

The *suspend* (SUS) instruction (Fig. 13-15b) is used to trap and identify specific conditions for program debugging and system troubleshooting. When the rung is true, this instruction places the controller in the *suspend idle* mode. Operation is suspended and the suspend ID number is placed in word 7 (S:7) of the status file so that you can track down the error in the logic. The program number containing the executed SUS instruction is placed in word 8 (S:7) of the status file. All outputs are de-energized.

The wiring between the field devices and the terminals of the I/O modules is a likely place for problems to occur. Faulty wiring and mechanical connection problems can interrupt or short the signals sent to and from the I/O modules.

The sensors and actuators connected to the I/O of the process can also fail. Mechanical switches can wear out or be damaged during normal operation. Motors, heaters, lights,

Input

—] [——————(TND)—

(a) Temporary end (TND) instruction

Input   Input   Input
A   B   C

Fig. 13-15

Diagnostic instructions.

(b) Suspend (SUS) instruction

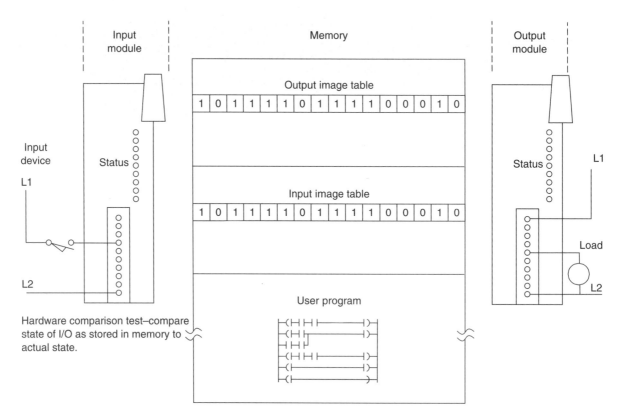

Logic observation–determine validity of decisions made by processor.

Fig. 13-16

General methods of troubleshooting.

and sensors can also fail. Input and output field devices must be compatible with the I/O module to ensure proper operation.

Some PLC manufacturers allow the same output coil to be used more than once in the ladder program. As a result, multiple rung conditions can control the same output coil, making troubleshooting more difficult. In the case of duplicate outputs, the monitored rung may be true; but if a rung farther down in the ladder diagram is false, the PLC will keep the output off. Some software allows for checking multiple coil use.

When a problem occurs, the best way to proceed is to try to logically identify the devices or connections that could be causing the problem rather than arbitrarily checking every connection, switch, motor, sensor, I/O module, etc. First, observe the system in operation and try to describe the problem. Using these observations and the description of the control system, you should identify the possible sources of trouble. Compare the logic status of the hardwired inputs and outputs to their actual state (Fig. 13-16). Any disagreements indicate malfunctions as well as their approximate location.

Most of your troubleshooting can be accomplished by interpreting the status indicators on the I/O modules. The key is to know if the status indicators are telling you there is a fault or the system is normal. Quite often you will be given a troubleshooting guide, map, or tree that presents a list of observed problems and their possible sources. The troubleshooting tree (Fig. 13-17) and input/output troubleshooting guides (Tables 13-1 and 13-2 on pp. 368-369) are typical of the types used.

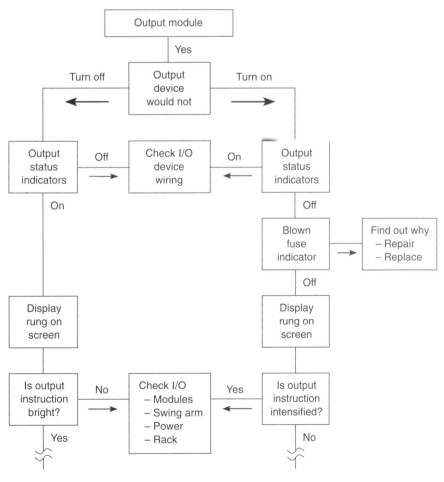

Fig. 13-17

Typical troubleshooting tree.

Table 13-1

INPUT TROUBLESHOOTING GUIDE

If Your Input Circuit LED Is...	And Your Input Device Is...	And	Probable Cause
On	On/Closed/Activated	Your input device will not turn off.	Device is shorted or damaged.
		Your program operates as though it is off.	Input circuit wiring or module.
			Input is forced off in program.
	Off/Open/Deactivated	Your program operates as though it is on and/or the input circuit will not turn off.	Input device off-state leakage current exceeds input circuit specification.
			Input device is shorted or damaged.
			Input circuit wiring or module.
Off	On/Closed/Activated	Your program operates as though it is off and/or the input circuit will not turn on.	Input circuit is incompatible.
			Low voltage across the input.
			Input circuit wiring or module.
			Input signal turn-on time too fast for input circuit.
	Off/Open/Deactivated	Your input device will not turn on.	Input device is shorted or damaged.
		Your program operates as though it is on.	Input is forced on in program.
			Input circuit wiring or module.

Table 13-2

OUTPUT TROUBLESHOOTING GUIDE

If Your Output Circuit LED Is...	And Your Output Device Is...	And	Probable Cause
On	On/Energized	Your program indicates that the output circuit is off or the output circuit will not turn off.	Programming problem: - Check for duplicate outputs and addresses. - If using subroutines, outputs are left in their last state when not executing subroutines. - Use the force function to force output off. If this does not force the output off, output circuit is damaged. If the output does force off, then check again for logic/programming problem.
			Output is forced on in program.
			Output circuit wiring or module.
	Off/De-energized	Your output device will not turn on and the program indicates that it is on.	Low or no voltage across the load.
			Output device is incompatible: - Check specifications and sink/source compatibility (if dc output).
			Output circuit wiring or module.
Off	On/Energized	Your output device will not turn off and the program indicates that it is off.	Output device is incompatible.
			Output circuit off state leakage current may exceed output device specification.
			Output circuit wiring or module.
			Output device is shorted or damaged.
	Off/De-energized	Your program indicates that the output circuit is on or the output circuit will not turn on.	Programming problem: - Check for duplicate outputs and addresses. - If using subroutines, outputs are left in their last state when not executing subroutines. - Use the force function to force output on. If this does not force the output on, output circuit is damaged. If the output does force on, then check again for logic/programming problem.
			Output is forced off in program.
			Output circuit wiring or module.

Questions

1. Why are PLCs generally placed within an enclosure?

2. What methods are used to keep enclosure temperatures within allowable limits?

3. State two ways in which electrical noise may be coupled into a PLC control system.

4. List four potential noise-generating devices.

5. Describe two ways in which careful wire routing can help cut down on electrical noise.

6. A. What type of input field devices and output modules are most likely to have a small leakage current flow when they are in the off state? Why?

   B. How can leakage currents be reduced?

7. When line voltage variations to the PLC power supply are excessive, what can be done to solve the problem?

8. A. Under what condition will an inductive load generate a very high voltage spike?

   B. What can be used to suppress a dc load?

   C. What can be used to suppress an ac load?

9. A. What is the purpose of PLC editing functions?

   B. What is the purpose of the search function as part of the editing process?

   C. What is the purpose of the cursor control keys as part of the editing process?

10. A. Explain the difference between off-line and on-line programming.

    B. Which method is safer? Why?

11. List four preventive maintenance tasks that should be carried out on the PLC installation regularly.

12. A. State two important reasons for grounding a PLC installation properly.

    B. List four important grounding practices to follow when installing a PLC system.

13. **A.** List four types of diagnostic fault indicators that often operate from the processor's self-detection circuits.

   B. When a processor comes equipped with a fault relay, how does this circuit usually operate?

   C. What is the prime function of a watchdog timer circuit?

14. What causes ground-loop interference?

15. Explain the operation of a MOV snubber.

16. List four common uses for the data monitor function.

17. What information is provided by the contact histogram function?

18. How do most companies ensure that equipment does not operate while maintenance and repairs are conducted?

19. While the ladder logic program is not likely to fail, what type of ladder program fault is possible?

20. Explain how forcing is used as part of the troubleshooting process.

21. What happens when the processor encounters a *temporary end* instruction?

22. Explain the function of the *suspend* instruction.

23. In what negative ways can faulty wiring and connections affect signals sent to and from the I/O modules?

24. How can you determine the compatibility of field devices and I/O modules?

Problems

1. The enclosure door of a PLC installation is not kept closed. What potential problem could this create?

2. A fuse is blown in an output module. Suggest two possible reasons why the fuse blew.

3. Whenever a crane located over a PLC installation is started from a standstill, temporary malfunction of the PLC system occurs. What is one likely cause of the problem?

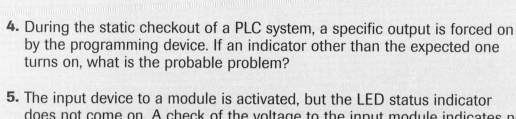

4. During the static checkout of a PLC system, a specific output is forced on by the programming device. If an indicator other than the expected one turns on, what is the probable problem?

5. The input device to a module is activated, but the LED status indicator does not come on. A check of the voltage to the input module indicates no voltage is present. Suggest two possible causes of the problem.

6. An output is forced on. The module logic light comes on, but the field device does not work. A check of the voltage on the output module indicates the proper voltage level. Suggest two possible causes of the problem.

7. A specific output is forced on, but the LED module indicator does not come on. A check of the voltage at the output module indicates a voltage far below the normal on level. What is the first thing to check?

8. An electronic-based input sensor is wired to a high-impedance PLC input and is falsely activating the input. How can this problem be corrected?

9. An LED logic indicator is illuminated, and according to the programming device monitor, the processor is not recognizing the input. If a replacement module does not eliminate the problem, what two other items should be suspected?

10. **A.** An NO field limit switch examined for an on state normally cycles from on to off five times during one machine cycle. How could you tell by observing the LED status light that the limit switch is functioning properly?

   B. How could you tell by observing the programming device monitor that the limit switch is functioning properly?

   C. How could you tell by observing the LED status light whether the limit switch was stuck open?

   D. How could you tell by observing the programming device monitor whether the limit switch was stuck open?

   E. How could you tell by observing the LED status light if the limit switch was stuck closed?

   F. How could you tell by observing the programming device monitor if the limit switch was stuck closed?

11. Assume that, prior to putting a PLC system into operation, you want to verify that each *input device* is connected to the correct input terminal and that the input module or point is functioning properly. Outline the safest method of carrying out this test.

12. Assume that, prior to putting a PLC system into operation, you want to verify that each *output device* is connected to the correct output terminal and that the output module or point is functioning properly. Outline the safest method of carrying out this test.

13. With reference to the ladder logic program of Fig. 13-18, add instructions to modify the program to ensure that a second pump (pump 2) does not run while pump 1 is running. If this condition occurs, the program should suspend operation and enter code identification number 6549 into S2:7.

Ladder logic program

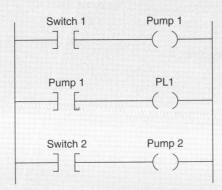

Fig. 13-18

14

Process Control and Data Acquisition Systems

This chapter introduces the kinds of industrial processes that can be PLC controlled. The acquisition of data is included as a type of process. Open-loop and closed-loop control systems are defined, and the fundamental characteristics of each are discussed.

After completing this chapter, you will be able to:

- ◆ Discuss the operation of continuous process, batch production, and individual products production

- ◆ Compare individual, centralized, and distributive control systems

- ◆ Explain the function of the major components of a process control system

- ◆ Describe the difference among on/off, proportional, derivative, and integral types of control

- ◆ Outline the function of the different parts of a data acquisition system

- ◆ Discuss common terms associated with the selection, operation, and connection of a data acquisition system

14•1 TYPES OF PROCESSES

Process control involves the automatic regulation of a control system. The ability of a PLC to perform math functions and utilize analog signals makes it ideally suited for this type of operation. Typical applications of process control systems include automobile assembly, petrochemical production, oil refining, power generation, and food processing. Any operation that requires the manipulation and control of one or more variables is a type of process control system. Commonly controlled variables in a process include temperature, speed, position, flow rate, pressure, and level. The types of processes carried out in modern manufacturing industries can be grouped into three general areas (in terms of the kind of operation that takes place):

◇ Continuous process

◇ Batch production

◇ Individual, or discrete, products production

A *continuous process* is one in which raw materials enter one end of the system and the finished product comes out the other end of the system; the process itself runs continuously. Once the process commences, it is continuous for a relatively long time. The period may be measured in minutes, days, or even months, depending on the process.

Figure 14-1 shows a continuous process engine assembly line. Engine blocks are fed into one end of the system and completed engines exit at the other end. In the continuous-type process, the product material is subjected to different treatments as it flows through the process (in this case, assembly, adjustment, and inspection). Auto assembly involves the use of automated machines or robots. Parts are supplied as needed at each station.

In *batch processing* there is no flow of product material from one section of the process to another. Instead, a set amount of each of

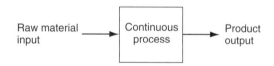

(a) Block diagram: the process runs for a relatively long time

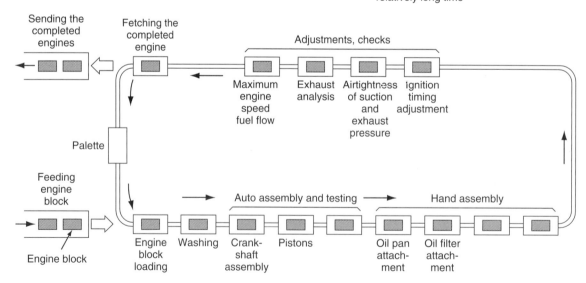

(b) Engine assembly line: involves assembly, adjustment, and inspection

Fig. 14-1

Continuous process.

the inputs to the process is received in a batch, and then some operation is performed on the batch to produce a finished product or an intermediate product that needs additional processing. The process is carried out, the finished product is stored, and another batch of product is produced. Each batch of product may be different. Many chemically based products are manufactured by using batch processes. Figure 14-2 shows a batch process system. Two ingredients are added together, mixed, and heated; a third ingredient is added; and all three are processed and then stored. Each batch may have different characteristics by design.

The *individual, or discrete, product production process* is the most common of all processing systems. With this manufacturing process, a series of operations produces a useful output product. The item produced may be bent, drilled, welded, and so on, at different steps in the process. The workpiece is normally a discrete part that must be handled on an individual basis. Figure 14-3 on p. 378 shows an individual product production process.

In the modern automated industrial plant, the operator merely sets up the operation and initiates a start, and the operations of the machine are accomplished automatically. These automatic machines and processes were developed to mass-produce products, control very complex operations, or operate machines accurately for long periods. They replaced much human decision, intervention, and observation.

Machines were originally mechanically controlled, then they were electromechanically controlled, and today they are often controlled by purely electrical or electronic means through programmable logic controllers (PLCs) or computers. The control of machines or processes can be divided into the following categories:

◆ Electromechanical
◆ Hardwired electronic
◆ Programmable hardwired electronic
◆ Programmable logic (PLC)
◆ Computer

Possible control configurations include individual, centralized, and distributed. *Individual control* is used to control a single machine. This type of control does not normally require communication with other controllers. Figure 14-4 on p. 378 shows an

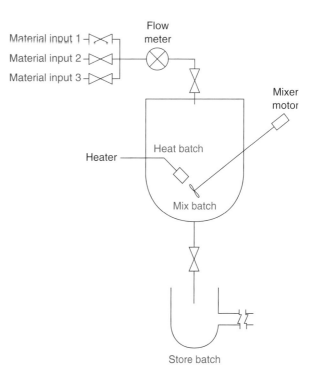

(a) Block diagram: a given quantity of material is processed through its manufacturing steps as a unit, each step being completed before the unit passes on to the next step.

Fig. 14-2

Batch process.

(b) Multicomponent/multiformula batching system

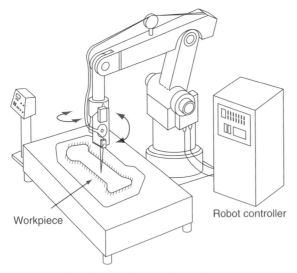

Workpiece

Robot controller

Industrial robot used in an individual item process

Fig. 14-3

Individual product production.

individual control application for the manufacture of aluminum handrails for indoor and outdoor applications. The operator enters the feed length and batch count via the interface control panel and then presses the start button to initiate the process. Rail lengths vary widely. The operator needs to select the rail length and the number of rails to cut.

Centralized control is used when several machines or processes are controlled by one central controller (Fig. 14-5). The control layout uses a *single,* large control system to control many diverse manufacturing processes and operations. Each individual step

in the manufacturing process is handled by a central control system controller. No exchange of controller status or data is sent to other controllers. Some processes require central control because of the complexity of decentralizing the control tasks into smaller ones. One disadvantage of centralized control is that, if the main controller fails, the whole process stops. A central control system is especially useful in a large, interdependent process plant where many different processes must be controlled for efficient use of the facilities and raw materials. It also provides a central point where alarms can be monitored and processes can be changed, thus eliminating the need to travel constantly throughout the plant.

The *distributive control system (DCS)* differs from the centralized system because each machine is handled by a *dedicated control system.* Each dedicated control is totally independent and could be removed from the overall control scheme if it were not for the manufacturing functions it performs. Distributive control involves two or more computers communicating with each other to accomplish the complete control task. This type of control typically employs *local area networks* (LANs), in which several computers control different stages or processes locally and are constantly exchanging information and reporting the status on the process. Communication among the computers is done through single coaxial cables or fiberoptics at very high speeds.

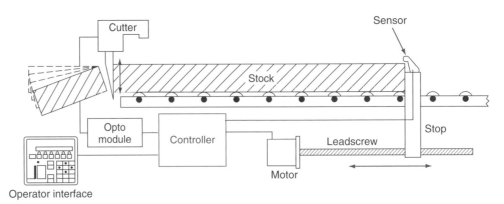

Fig. 14-4

Individual control.

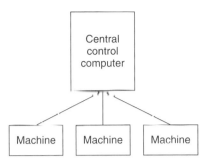

Fig. 14-5

Centralized control can be interpreted as a large individual control applied to a large process or to several machines.

Because of their flexibility, distributive control systems have emerged as the system of choice for numerous batch and continuous process automation requirements. Distributive control permits the distribution of the processing tasks among several control elements. Instead of just one computer located at a central control point doing all the processing, each local loop controller, placed very close to the point being controlled, has processing capability.

Figure 14-6 shows a DCS supervised by a host computer. The host computer could be a personal computer and could handle tasks such as program up- and download, alarm reporting, data storage, and operator interaction

facilities. Remote computer operators overseeing the system can be located some distance from the industrial environment in dust-free and temperature-controlled conditions. Each PLC controls its associated machine or process. Depending on the process, one PLC failure would not stop the complete system most of the time.

•2 STRUCTURE OF CONTROL SYSTEMS

A *process control system* can be defined as the functions and operations necessary to change a material either physically or chemically. Process control normally refers to the manufacturing or processing of products in industry. In the case of a programmable controller, the process or machine is operated and supervised under the control of the user program. Figure 14-7 on p. 380 shows the major components of a process control system, which includes the following:

Sensors

◆ Provide inputs from the process and from the external environment

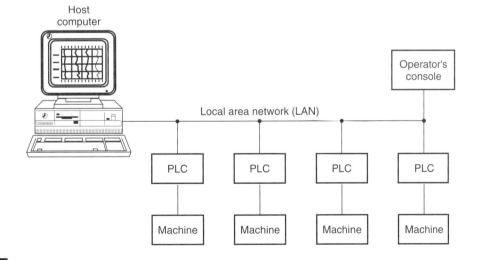

Fig. 14-6

Distributive control system (DCS). This system is a design approach in which factory or machine control is divided into several subsystems, each managed by a unique programmable logic controller (PLC), yet all are interconnected to form a single entity. Various types of communication busses are used to connect the controllers.

- Convert physical information such as pressure, temperature, flow rate, and position into electrical signals

- Are related to a physical variable so that their electrical signal can be used to monitor and control a process

Operator-Machine Interface

- Allows inputs from a human to set up the starting conditions or alter the control of a process

- Allows human inputs through various types of switches, controls, and keypads

- Operates using supplied input information that may include emergency shutdown or changing the speed, the type of process to be run, the number of pieces to be made, or the recipe for a batch mixer

Signal Conditioning

- Involves converting input and output signals to a usable form

- Includes signal-conditioning techniques such as amplification, attenuation, filtering, scaling, A/D and D/A converters

Actuators

- Convert system output electrical signals into physical action

- Have process actuators that include flow control valves, pumps, positioning drives, variable speed drives, clutches, brakes, solenoids, stepping motors, and power relays

- Indicate the state of the process variables (machine-operator interface) through external actuators such as meters, cathode-ray tube (CRT) monitors, printer, alarms, and pilot lights

- Can send outputs directly from the controller to a computer for storage of data and analysis of results (machine–machine interface)

Controller

- Makes the system's decisions based on the input signals

- Generates output signals that operate actuators to carry out the decisions

Control systems can be classified as open-loop or closed-loop. Figure 14-8 depicts a typical *open-loop control system.* The process is controlled by inputting to the controller the desired *set point* (also known as *command,* or *reference*) necessary to achieve the ideal operating point for the process and accepting whatever output results. Since the

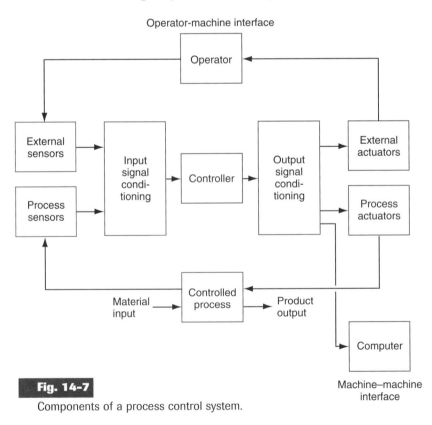

Fig. 14-7

Components of a process control system.

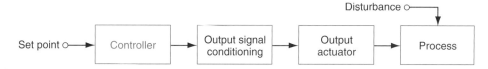

Fig. 14-8

Open-loop control system.

only input to the controller is the set point, it is apparent that an open-loop system controls the process blindly; that is, the controller receives no information concerning the present status of the process or the need for any corrective action. Open-loop control reduces system complexity and costs less when compared to closed-loop control. However, open-loop control has an element of uncertainty and is not considered to be as accurate as closed-loop control.

Open-loop control is still found in many industrial applications. Figure 14-9 shows an example of an open-loop *stepper motor* control system operated by a PLC. Stepper motors are often used to control position in low-power, low-speed applications. A stepper motor is basically a permanent magnet motor with several sets of coils, termed *phases* (*A* and *B*), located around the rotor. The phases are wired to the PLC output assembly and are energized, in turn, under the control of the user program. The rotor turn is determined by which phases are turned on. The programmable controller does not receive feedback from the motor to indicate that rotation has occurred, but it is assumed that the motor has responded correctly because the motor phases are driven in a sequence by step pulses. Open-loop, or nonfeedback, control is only as stable as the load and the individual components of the system.

A *closed-loop control system* is one in which the output of a process affects the input control signal. The system measures the actual output of the process and compares it to the desired output. Adjustments are made continuously by the control system until the difference between the desired and actual output is as small as is practical. Closed-

loop systems are more difficult to adjust or calibrate than open-loop systems, but they have a higher certainty that the process output is occurring and a more accurate indication of the output accuracy. Figure 14-10 p. 382 depicts a typical closed-loop control system. The actual output is sensed and fed back (hence, the name *feedback control*) to be subtracted from the set-point input that indicates what output is desired. If a difference occurs, a signal to the controller causes it to take action to change the actual output until the difference is 0. The operation of the component parts are as follows:

Set Point

◆ The input that determines the desired operating point for the process.

◆ Normally provided by a human operator, although it may also be supplied by another electronic circuit.

Process Variable

◆ The signal that contains information about the current process status.

◆ Refers to the feedback signal.

◆ Ideally, matches the set point (indicating that the process is operating exactly as desired).

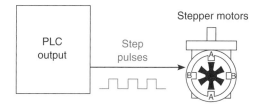

Fig. 14-9

Open-loop stepper motor control.

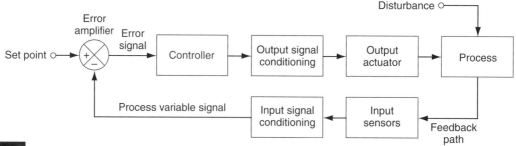

Fig. 14-10

Closed-loop control system.

Error Amplifier

◆ Determines whether the process operation matches the set point.

◆ Quite often a differential amplifier circuit provides output, referred to as the *error signal* or the *system deviation signal.*

◆ The magnitude and polarity of the error signal will determine how the process will be brought back under control.

Controller

◆ Produces the appropriate corrective output signal based on the error signal input.

Output Actuator

◆ The component that directly affects a process change.

◆ Has motors, heaters, fans, and solenoids that are all examples of output actuators.

The container-filling process illustrated in Fig. 14-11 is an example of a continuous control process. An empty box is moved into position and filling begins. The weight of the box and contents is monitored. When the actual weight equals the desired weight, filling is halted. In this operation:

◆ A sensor attached to the scale weighing the container generates the voltage signal or digital code that represents the weight of the container and contents.

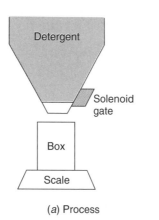

(a) Process

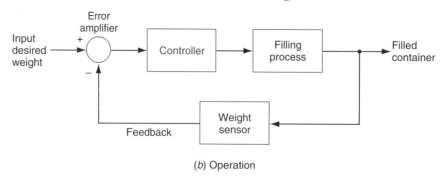

(b) Operation

Fig. 14-11

Container-filling closed-loop process.

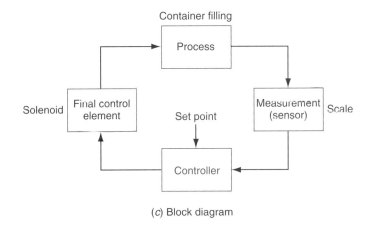

(c) Block diagram

Fig. 14-11 (continued)

Container-filling closed-loop process.

- ◆ The sensor signal is subtracted from the voltage signal or digital code that has been input to represent the desired weight.

- ◆ As long as the difference between the input signal and feedback signal is greater than 0, the controller keeps the solenoid gate open.

- ◆ When the difference becomes 0, the controller outputs a signal that closes the gate.

•3 CONTROLLERS

Controllers may be classified in several different ways. For example, they may be classified according to the type of power they use. Two common types in this category are electric and pneumatic controllers. *Pneumatic* controllers are decision-making devices that operate on air pressure. *Electric* (or *electronic*) controllers operate on electric signals.

Controllers are also classified according to the type of control they provide. In this section we will discuss four types of control in this category:

- ◆ On/off
- ◆ Proportional (P)
- ◆ Integral (I)
- ◆ Derivative (D)

With *on/off control* (also known as *two-position* and *bang-bang control*), the final control element is either on or off—one for the occasion when the value of the measured variable is above the set point, and the other for the occasion when the value is below the set point. The controller will never keep the final control element in an intermediate position. Controlling activity is achieved by the period of on-off cycling action.

Figure 14-12 shows a system using on/off control, in which a liquid is heated by steam. If the liquid temperature goes below the set point, the steam valve opens and the steam is turned on. When the liquid temperature goes above the set point, the steam valve closes and the steam is shut off. The on/off cycle will continue as long as the system is operating.

Figure 14-13 on p. 384 illustrates the control response for an on/off temperature controller. The output turns on when the temperature falls below the set point and turns off when the temperature reaches the set

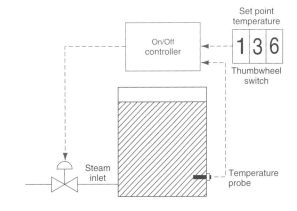

Fig. 14-12

On/off liquid heating system.

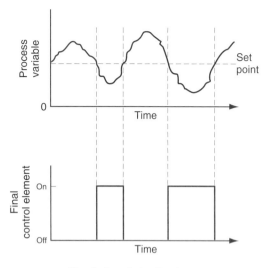

Simplest control action is on/off.

Fig. 14-13

On/off control.

point. Control is simple, but overshoot and cycling about the set point can be disadvantageous in some processes. The measured variable will oscillate around the set point at an amplitude and frequency that depend on the capacity and time response of the process. This oscillation is typical of the on/off controller. Oscillations may be reduced in amplitude by increasing the sensitivity of the controller. This increase will cause the controller to turn on and off more often, a possibly undesirable result.

A *deadband* is usually established around the set point. The deadband of the controller is usually a selectable value that determines the error range above and below the set point that will not produce an output as long as the process variable is within the set limits. The inclusion of deadband eliminates any hunting by the control device around the set point. Hunting occurs when minor adjustments of the controlled position are continually made due to minor fluctuations. On/off temperature control is usually used:

◆ When a precise control is unnecessary

◆ In systems that cannot handle the energy being turned on and off frequently

◆ When the mass of the system is so great that temperatures change extremely slowly

◆ For alarm systems

Proportional controls are designed to eliminate the hunting or cycling associated with on/off control. They allow the final control element to take intermediate positions between on and off. This permits *analog control* of the final control element to vary the amount of energy to the process, depending on how much the value of the measured variable has shifted from the desired value.

A proportional controller allows tighter control of the process variable because its output can take on any value between fully on and fully off, depending on the magnitude of the error signal. Figure 14-14 shows an example of a motor-driven analog proportional control valve used as a final control element. Typically, it receives an input current between 4 mA and 20 mA from the controller; in response, it provides linear movement of the valve. A 4–20 mA loop is a common method of transmitting analog data over long distances in industrial process control environments. A value of 4 mA corresponds to a minimum value (often 0) and 20 mA corresponds to a maximum value (full scale). The 4mA lower limit allows the system to detect opens. If the circuit is open, 0 mA would result, and the system can issue an alarm. Because the signal is a current, it is unaffected by reasonable variations in connecting wire resistance and is less susceptible to noise pickup from other signals than is a voltage signal. Twisted pairs of wires are normally used to minimize the effects of magnetic fields from motors, transformers, etc.

Proportioning action can also be accomplished by turning the final control element on and off for short intervals. This *time proportioning* (also known as *proportional pulsewidth modulation*) varies the ratio of on time to off. Figure 14-15 shows an example of time proportioning used to produce varying wattage from a 200 W heater element.

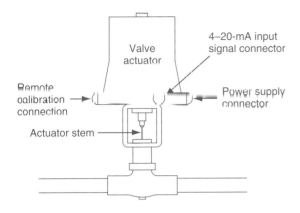

(a) Proportional control valve

Actuator current (mA)	Valve response (% open)
4	0
6	12.5
8	25
10	37.5
12	50
14	62.5
16	75
18	87.5
20	100

(b) Changes in valve response in reaction to increases in actuator current

Fig. 14-14

Motor-driven analog proportional control valve.

(a) To produce 100 W from a 200 W heater, the heater must be on half the time.

(b) To produce 50 W, the heater must be on one-quarter, or 25%, of the time.

(c) To produce 25 W, the heater must be on one-eighth, or 12.5%, of the time.

Fig. 14-15

Time proportioning of a 200 W heater element.

The proportioning action occurs within a "proportional band" around the set point, as illustrated in Fig. 14-16 on p. 386. Outside this band, the controller functions as an on/off unit, with the output either fully on (below the band) or fully off (above the band). Within the band, however, the output is turned on and off in the ratio of the measurement difference from the set point. At the set point (the midpoint of the proportional band), the output on:off ratio is 1:1; that is, the on time and off time are equal. If the temperature is further from the set point, the on and off times vary in proportion to the temperature difference. If the temperature is below the set point, the output will be on longer; if the temperature is too high, the output will be off longer.

In theory, a proportional controller should be all that is needed for process control.

Any change in system output is corrected by an appropriate change in controller output. Unfortunately, the operation of a proportional controller leads to process deviation known as *offset* or *droop*. This steady-state error is the difference between the attained value of the controller and the required value, as shown in Fig. 14-17 on p. 386. It may require an operator to make a small adjustment (manual reset) to bring the controlled variable to set point on initial start-up, or whenever the process conditions change significantly.

Proportional control is often used in conjunction with derivative control, integral control, or a combination of both. Processes with long time lags and a large maximum rate of rise (for example, a heat exchanger) require wide proportional bands to eliminate oscillation. The wide band can result in large offsets with changes in the load. To eliminate these offsets, automatic reset (integral) can be used. Derivative (rate) action can be used on processes with long time delays to speed recovery after a process disturbance.

Process Control and Data Acquisition Systems

Time proportional				4–20 mA proportional	
Percent on	On time (seconds)	Off time (seconds)	Temp. (°F)	Output level	Percent output
0.0	0.0	20.0	over 540	4 mA	0.0
0.0	0.0	20.0	540.0	4 mA	0.0
12.5	2.5	17.5	530.0	6 mA	12.5
25.0	5.0	15.0	520.0	8 mA	25.0
37.5	7.5	12.5	510.0	10 mA	37.5
50.0	10.0	10.0	500.0	12 mA	50.0
62.5	12.5	7.5	490.0	14 mA	62.5
75.0	15.0	5.0	480.0	16 mA	75.0
87.5	17.5	2.5	470.0	18 mA	87.5
100.0	20.0	0.0	460.0	20 mA	100.0
100.0	20.0	0.0	under 460	20 mA	100.0

Example: heating
Set point: 500°F
Proportional band: 80°F (±40°F)

Fig. 14-16

Proportional band.

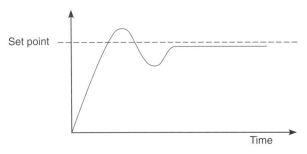

Set point

Time

(a) The net result of this action is an offset signal that is slightly lower than the set-point value. Depending on the PLC application, this offset may or may not be acceptable.

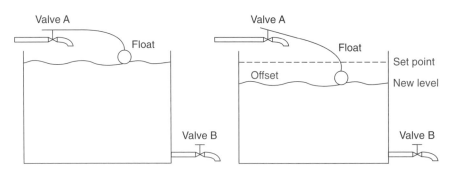

(b) In this example when valve B opens, liquid flows out and the level in the tank drops. This causes the float to lower opening valve A, allowing more liquid in. This process continues until the level drops to a point where the float is low enough to open valve A, thus allowing the same input as is flowing out. The level will stabilize at a new lower level, not at the desired set point.

Fig. 14-17

Proportional control steady-state error.

The *integral action,* sometimes termed *reset action,* responds to the size and time duration of the error signal; therefore, the output signal from an integral controller is the mathematical integral of the error. An error signal exists when there is a difference between the process variable and the set point, so the integral action will cause the output to change and continue to change until the error no longer exists. Integral action eliminates steady-state error. The amount of integral action is measured as minutes per repeat or repeats per minute, which is the relationship between changes and time.

The *derivative mode controller* responds to the speed at which the error signal is changing—that is, the greater the error change, the greater the correcting output. The derivative action is measured in terms of time.

Proportional plus integral (PI) *control* combines the characteristics of both types of control (Fig. 14-18). A step change in the measurement causes the controller to respond proportionally, followed by the integral response, which is added to the proportion response. Because the integral mode determines the output changes as a function of time, the more integral action found in the control, the faster the output changes. The PI control mode is used when changes in the process load do not happen very often, but when they do happen, changes are small.

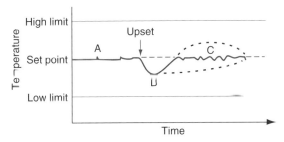

- To eliminate the offset error, the controller needs to change its output until the process variable error is zero.
- Reset (integral) control action changes the controller output by the amount needed to drive the process variable back to the set point value.
- The new equilibrium point after reset action is at point C.
- Since the proportional controller must always operate on its proportional band, the proportional band must be shifted to include the new point C. A controller with reset does this automatically.

Fig. 14-18

Proportional plus integral (PI) control action.

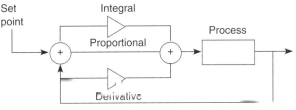

- During setup, the set point, proportional band, reset (integral), rate (derivative), and the output limits are specified.
- All these can be changed during operation to tune the process.
- The integral term improves accuracy, and the derivative reduces overshoot for transient upsets.
- The output can be used to control valve positions, temperature, flow metering equipment, and so on.

(*a*) PID control loop

Rate action acts on error just like reset does, but rate action is a function of the rate of change rather than the magnitude of error. Rate action is applied as a change in output for a selectable time interval, usually stated in minutes. Rate-induced change in controller output is calculated from the derivative of the change in input. Input change, rather than proportional control error change, is used to improve response. Rate action quickly positions the output where proportional action alone would eventually position the output. In effect, rate action puts the brakes on any offset or error by quickly shifting the proportional band.

Proportional plus derivative (PD) *control* is used in process-control systems with errors that change very rapidly. By adding derivative control to proportional control, we obtain a controller output that responds to the measurement's rate of change as well as to its size.

Proportional-integral-derivative (PID) *controllers* produce outputs that depend on the magnitude, duration, and rate of change of the system error signal (Fig. 14-19). Sudden system disturbances are met with an aggressive attempt to correct the condition. A PID controller can reduce the system error to zero faster than any other controller. Because it has an integrator and a differentiator, however, this controller must be *custom-tuned* to each process controlled. The PID controller is the most sophisticated and widely used type of process controller.

To reduce fluctuations quickly and accurately in a control system, it is often necessary to

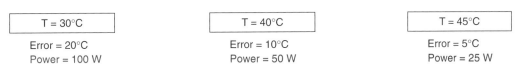

(*b*) PID allows the output "power" level to be varied. For example, assume that a furnace is set at 50°C. The heater power will increase as the temperature falls further below the 50°C set point. The lower the temperature, the higher the power. PID has the effect of gently turning the power down as the signal gets close to the set point.

Fig. 14-19

PID control.

solve for conditions by using mathematical equations. These equations are referred to as *algorithms*. An algorithm is any fixed set of instructions that will solve a particular type of problem in a finite number of steps. In a PLC-based system, the CPU implements these instructions in the process of calculating signals to be sent to process actuators.

The long-term operation of any system, large or small, requires a mass–energy balance between input and output. If a process were operated at equilibrium at all times, control would be simple. Since change does occur, the critical parameter in process control is time, that is, how long it takes for a change in any input to appear in the output. System time constants can vary from fractions of a second to many hours. The PID controller has the ability to tune its control action to specific process time constants and therefore to deal with process changes over time. PID control changes the amount of output signal in a mathematically specified way that accounts for the amount of error and the rate of signal change. A common PID equation for obtaining PID control is:

$$Co = \overbrace{K\,(E}^{\text{Proportional}} + \overbrace{1/Ti \int_0^t E\,dt}^{\text{Integral}}$$

$$+ \overbrace{KD\,[E - E\,(n-1)]/dt)}^{\text{Derivative}} + \text{bias}$$

where Co = control output
K = controller gain (no units)
$1/Ti$ = reset gain constant (repeats per min)
KD = rate gain constant (min)
dt = time between samples (min)
bias = output bias
E = error; equal to measured min set point or set point minus measured min
$E(n-1)$ = error from last sample

Either programmable controllers can be fitted with input/output assemblies that produce PID control, or they will already have sufficient mathematical functions to allow PID control to be carried out. PID is essentially an equation that the controller uses to evaluate the controlled variable. Figure 14-20 illustrates how a programmable logic controller can be used in the control of a PID loop. The controlled variable (pressure) is measured and feedback is generated. The PLC program compares the feedback to the set point and generates an error signal. The error is examined in three ways: with proportional, integral, and derivative methodology. The controller then issues a command (control output) to correct for any measured error by adjustment of the position of the outlet valve.

The *response* of a PID loop is the rate at which it compensates for error by adjusting the output. The PID loop is adjusted or *tuned* by changing the proportional gain, the integral gain, and/or the derivative gain. A PID loop is normally tested by making an abrupt change to the set point and observing the controller's response rate. Adjustments can then be made as follows:

◇ As the proportional gain is increased, the controller responds faster.

◇ If the proportional gain is too high, the controller may become unstable and oscillate.

◇ The integral gain acts as a stabilizer.

◇ Integral gain also provides power, even if the error is zero (e.g., even when an oven reaches its set point, it still needs power to stay hot).

◇ Without this base power, the controller will droop and hunt for the set point.

◇ The derivative gain acts as an anticipator.

◇ Derivative gain is used to slow the controller down when change is too fast.

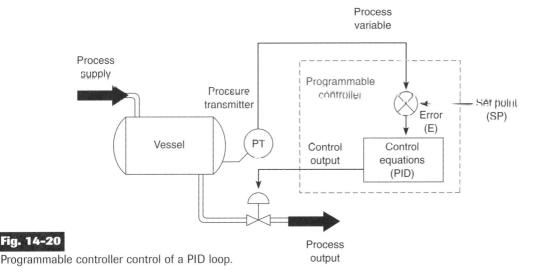

Fig. 14-20

Programmable controller control of a PID loop.

Basically, PID controller tuning consists of determining the appropriate values for the gain (proportional band), rate (derivative), and reset (integral) time tuning parameters (control constants) that will give the control required. Depending on the characteristics of the deviation of the process variable from the set point, the tuning parameters interact to alter the controller's output and produce changes in the value of the process variable. In general, three methods of controller tuning are used:

Manual

◆ The operator estimates the tuning parameters required to give the desired controller response.

◆ The proportional, integral, and derivative terms must be adjusted, or tuned, individually to a particular system using a trial and error method.

Semiautomatic or Autotune

◆ The controller takes care of calculating and setting PID parameters.
  - Measure sensor
  - Calculate error, sum of error, rate of change of error
  - Calculate desired power with PID equation
  - Update control output

Fully Automatic or Intelligent

◆ Also known in the industry as *fuzzy logic control.*

◆ Uses artificial intelligence to readjust PID tuning parameters continually as necessary.

◆ Rather than calculating an output with a formula, the fuzzy logic controller evaluates rules. The first step is to "fuzzify" the error and change-in-error from continuous variables into linguistic variables, like "negative large" or "positive small." Simple if-then rules are evaluated to develop an output. The resulting output must be de-fuzzified into a continuous variable such as valve position.

Temperature ramp and *soak programming* (Fig. 14-21 on p. 390) is set up by selecting a series of set points, with individual tuning of the control parameters. The set point changes linearly between the endpoints over a selected time period. You can set the number of repeat cycles for each ramp and soak sequence.

If the process does not require that the heat-up rate be controlled, a standard set-point controller may be used to control temperature. The oven load will reach the desired temperature as quickly as the product and

Process Control and Data Acquisition Systems

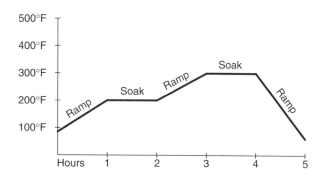

Fig. 14-21

Ramp and soak profile.

oven heating capacities will allow, but it will not necessarily be linear. If a controlled heat-up is required (e.g., heat-up at 1°F/min), a programmable, ramping controller is needed. Such a controller allows a specific, linear heat-up rate to be programmed.

Soak time refers to the point when the product has reached the desired process temperature for the desired length of time. A programmable controller can be programmed to remain at the desired temperature for a specified period then cool down to complete the cycle. For most applications, the soak cycle begins once the time specified for the heat-up cycle has been completed.

14•4 DATA ACQUISITION SYSTEMS

Data acquisition (DAQ) is the collection, analysis, and storage of information by a computer-based system. Initially these systems were used only with large mainframe computers and installations with thousands of input data channels. Today, however, powerful personal computers and PLCs have opened up the advantages of data acquisition to all manufacturers at a modest cost. A data acquisition or computer interface system allows you to feed data from the real world to your computer. It takes the signals produced by temperature sensors, pressure transducers, flow meters, and so on, and converts them into a form that your processor can understand. With an acquisition system, you can

use your computer to gather, monitor, display, and analyze your data. *Supervisory control and data acquisition (SCADA)* systems have additional *control* output capabilities you can also use to control your processes accurately for maximum efficiency (Fig. 14-22).

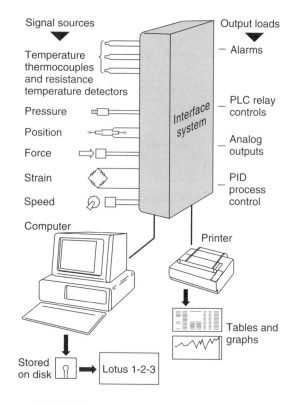

Fig. 14-22

Supervisory control and data acquisition (SCADA) system.

The great advantage of data acquisition is that data are stored automatically in a form that can be retrieved for later analysis without error or additional work. It is easy to examine the effect of factors other than those originally anticipated because the cost of measuring some additional points is low. Measurements are made under processor control and then displayed on screen and stored on disk. Accurate measurements are easy to obtain, and there are no mechanical limitations to measurement speed.

In addition to producing product, PLCs also produce data. The information contained in this data can often be used to improve the efficiency of the process. The new generation of programmable controllers have been designed to accept a wide range of inputs, both analog and digital, and have powerful arithmetic functions and controlling ability, so they are now suitable for some data acquisition applications. Under the control of the user program, the programmable controller can present a word of digital information to a digital-to-analog converter, which will convert the digital signal to its analog equivalent. The resulting analog equivalent can then be used to control an output device such as an actuator, or it can be recorded.

Data acquisition software works with the hardware interface equipment. In addition to controlling the data collection by the interface, the software is designed to display the data in tables like a spreadsheet or in histogram charts, pie charts, or line plots, and also to analyze the data. Most data acquisition systems

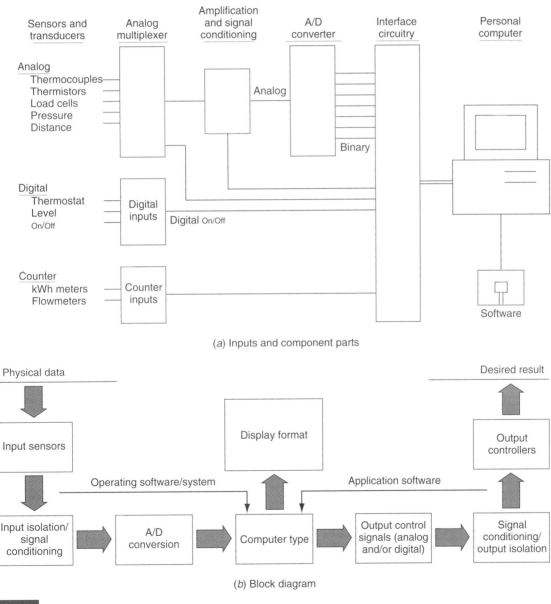

(a) Inputs and component parts

(b) Block diagram

Fig. 14-23

Major components of a data acquisition system.

share a basic overall similarity. Major system components (Fig. 14-23 on p. 391) include:

◆ Sensors and transducers are used to convert a mechanical or electrical signal into a specific signal such as a 4- to 20-mA loop, or a 0- to 5-V signal.

◆ Time-division multiplexer accepts several signals at once and allows the user to pick the signal to be examined (Fig. 14-24), much like a television that has 99 channels but displays only one at a time. It is used to reduce system cost by time-sharing expensive elements. A multiplex system is, in effect, an intelligent connector designed to permit the replacement of large bundles of parallel wiring with a simple twisted pair. Its use simplifies wiring and reduces cost.

◆ Amplifiers are used to increase the magnitude of a signal. Amplifiers can increase the voltage or current a sensor produces.

◆ Signal conditioners change a signal to make it easier to measure or more stable (i.e., current to voltage converter, V/F converter, or a filter).

◆ A/D converters are special types of signal conditioners that convert analog signals into digital signals.

◆ Digital and counter inputs are signals that have a specific number of possible values (1 bit = 2 values; 4 bits = 16 values).

There are two kinds of data acquisition interface systems: *plug-in* and *stand-alone*. These two extremes in providing signal conditioning to a data acquisition system involve either putting the signal conditioning circuitry directly on a data acquisition board, or using a stand-alone data acquisition instrument with built-in signal conditioning.

Figure 14-25 shows a modular plug-in data acquisition system that uses separate modules for signal conditioning and data acquisition. The advantage of a modular system is its ability to use the latest in plug-in data acquisition hardware, software, and PC technology while maintaining the proper signal-conditioning performance of stand-alone data acquisition units.

Most sensors *cannot* be wired directly to a data acquisition board without preconditioning. Special *signal conditioning* is needed. For example, thermocouples require cold junction compensation. RTDs, thermistors, and strain gauges need current or voltage excitation. Strain gauges often need bridge completion resistors. Most sensors need amplification to match the input range of the A/D converter. Isolation and filtering are recommended to protect your equipment and to make accurate measurements in industrial environments. Anti-aliasing filters are useful for filtering out high-frequency noise from a measurement. Input multiplexers are economical for connecting to a large number of signals. External amplification allows low-level signals to be amplified for better transmission to the data acquisition board. Discrete inputs from switches or high-current relays need isolation and conditioning before reaching the computer.

Data acquisition systems can get caught in ground loops (Fig. 14-26). A *ground loop* results when there are two or more connections to ground on the same electrical path, causing different amounts of current to flow through some ground wires. The value received at the data acquisition board represents the voltage from the desired signal, as well as the voltage value from the additional ground currents in the system. If one of the ground references cannot be removed, signal conditioning accessories with isolation circuitry can break the ground loops and eliminate the effects of ground currents on measurement.

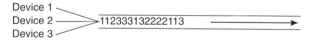

Device 1
Device 2 ⟶ 112333132222113 ⟶
Device 3

Each device shares time on the line.

Fig. 14-24

Time-division multiplexer.

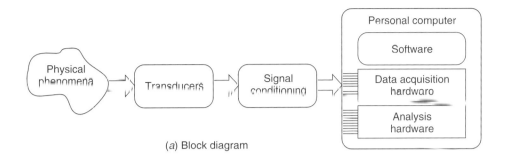

(a) Block diagram

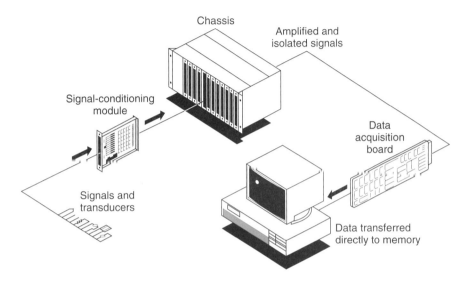

(b) External instrumentation front-end chassis for signal conditioning
boards sends conditioned signals directly to a plug-in data
acquisition board for high-speed input into computer memory

Fig. 14-25

Modular plug-in data acquisition system.

Important terms associated with the selection, operation, and connection of data acquisition systems include:

Absolute signals Signals that do not require a reference quantity.

Accuracy Not equal to resolution (the most common perception is that these two terms are the same). Accuracy is the total of the errors caused by:

Resolution + gain + offset + noise

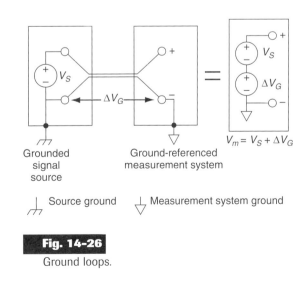

Alias A false lower frequency component that appears in sampled data acquired at a sampling rate that is too low.

Analog A signal that is continuous and has an infinite number of possible values.

Analog ground Provides a reference for signals. No shared circuit currents should flow in an analog ground.

Baud rate Serial communications data transmission rate expressed in bits per second (b/s).

Calibration Ensuring that the output signal from a sensor is proportional (and the proportion is known) to the quantity the sensor is measuring. The manufacturer will usually give the calibration of the transducer on a data sheet or on a nameplate. For example:

> **Distance Sensor**
> Full scale . . . . 2.0 mm
> Output at full scale . . . . 5.0 mV
> Excitation voltage . . . . 10.0 V

◆ In this case, a 5.0 mV output corresponds to a full-scale travel of 2.0 mm.

◆ The offset factor, often called the *zero,* is the output when the sensor is set at 0 mm travel.

Control update interval Refers to how often the computer checks the measured value and adjusts the control output. If it's *too* slow, the control may not work. If it's *much faster* than the rate the system changes, computer time is wasted. The update rate is generally set to five to ten times faster than the time it takes the controlled system to droop by the desired accuracy.

Data storage values Data acquisition systems can usually be programmed to save various values, including:

◆ *Averages:* the average value of all values.

◆ *Maximums:* the maximum value of all measurements made in a certain interval.

◆ *Minimums:* the minimum value of all measurements made in a certain interval.

◆ *Raw samples:* as measured each sample time.

◆ *Totals:* the total of all measurements made in a certain interval.

Floating signals and common mode voltages A *floating signal* is not connected to any other systems or voltages. An example of a truly floating voltage is a standard 12 V car battery with no cables attached:

◆ The voltage between the (−) and (+) terminals is 12 V.

◆ The voltage between either terminal and any other system is 0 (or undefined).

Consider a string of batteries in series with one end connected to ground (as shown in Fig. 14-27):

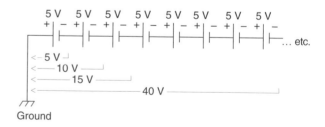

Fig. 14-27

- The voltage at the beginning is 0 with respect to ground.

- The differential voltage across any battery is only 5 V.

- The voltage at the end of the chain is 40 V.

- The *common mode* of a differential measurement across the third battery is 10 V. Common mode signals can be present only on differential signals. Common mode voltage is the average of the voltage on the (+) and (−) differential connections referenced to ground.

- When measuring the voltage across the last battery, the 5 V signal has 40 V of common voltage (Fig. 14-27).

Gain   The factor by which a signal is amplified, sometimes expressed in decibels (dB). A simple way to reduce the effects of system noise is to amplify the signal close to the source and thus boost the analog signal above the noise level before it can be corrupted by noise in the lead wires.

Graphic user interface (GUI)   An intuitive, easy-to-use way of communicating information to and from a computer program by means of graphical screen displays. GUIs can resemble the front panels of instruments or other objects associated with a computer program.

Ground   There are two types of ground in data acquisition. *System* or *power ground* is used to provide a path for current flow. *Analog* or *signal ground* is used to provide a voltage reference for signal conditioners and converters. In general, no power current should flow in analog ground connections. Analog ground should never be connected to each other or to power ground because this would form a loop for current to flow through.

Input/output (I/O)   The transfer of data to or from a computer system via communications channels, operator interface devices, and/or data acquisition and control interfaces.

Isolation   The process of measuring a signal without *direct* electrical connection (Fig. 14-28). Isolation allows a signal to pass from its source to the measurement device without a galvanic or physical connection. Using optical or modulation techniques, high common mode voltages are rejected, ground loops are broken, and the desired differential voltage is passed on for measurement. Isolators also provide an important safety function by protecting against high-voltage surges from sources like power lines, lightning, or high-voltage equipment.

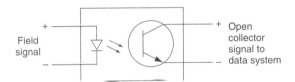

Fig. 14-28

Optical isolator.

Multitasking   A property of an operating system in which several processes can be run simultaneously.

Noise   Undesirable interference with an otherwise normal signal. Common sources of noise include power lines (60 Hz), fluorescent lights, and large motors. Filters can be used to reject unwanted noise within a certain frequency range.

Real time   A property of an event or system in which data are processed as they are acquired instead of being accumulated and processed later.

Resolution   The smallest signal increment that can be detected by a measurement system. Resolution can be thought of as how closely a quantity can be measured. Imagine a 1 ft. ruler. If the only graduations on a 1 ft ruler were inches, the resolution would be 1 in. If the graduations were every ¼ in., the resolution would be ¼ in. The closest we would be able to measure any object would be ¼ in. Resolution can be expressed in bits, proportions, or percentage of full scale. For example, a system can have 12-bit resolution, 1 part in 4096 resolution, and 0.0244% of full scale.

Sample and hold amplifier   This device takes a very fast "snapshot" of the input voltage and freezes it before the A/D converts, which prevents the A/D from measuring a signal that is changing and might be inaccurate. Most fast (kHz) A/D converters use a sample and hold amplifier in the system. With *simultaneous* sample and hold:

◇ More than one channel is captured at the same time.

◇ Each channel needs its own sample and hold amplifier.

◇ A global hold signal freezes all channels.

◇ The A/D converter steps through each channel, one at a time.

◇ The user is thus able to compare data samples taken at the same instant from different sensors.

Save interval or save period   Specifies the time between data saves (Fig. 14-29). Note the following about the save interval:

◇ It can be a constant, always the same.

◇ It can vary with time, or it can vary as a function of certain measurements.

◇ In many data-logging applications, a fixed save rate can result in too much data. A variable save rate can be set up to save data only when something changes.

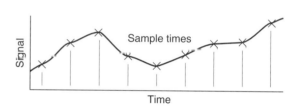

Fig. 14-29

Track-and-hold (T/H)   A circuit that tracks an analog voltage and holds the value on command.

Transfer rate   The rate, measured in bytes, at which data are moved from source to destination after software initialization and set-up operations; the maximum rate at which the hardware can operate.

Triggering   Used to start and/or stop data acquisition. The system waits until it obtains a start signal (i.e., a trigger), and then it measures and saves data and waits for a stop signal.

Various cable types are available for connecting sensors to data acquisition systems. Wires and cables act like antennas, picking up noise from the environment.

The more remote your sensors and signals are, the more attention you will need to pay to your cabling scheme. Unshielded wires or ribbon cables are inexpensive and work well for high-level signals and short to moderate cable lengths. For low-level signals or longer signal paths, you will want to consider shielded or twisted-pair wiring. If the signal has a bandwidth greater than 100 kHz, however, you will want to use coaxial cables. Fiberoptic links are also available. Fiberoptic cables are noise immune because the data are transmitted as light.

Another useful method for reducing noise corruption is to use differential measurement. Because both the (+) and (−) signal lines travel from signal source to the measurement system, they pick up the same noise with differential measurement. A differential input rejects voltages common to both signal lines.

Questions

1. List the three types of processes carried out in industries.

2. State the type of process used for each of the following applications:

    a. Mixing ingredients to manufacture chemically based products

    b. Assembly of television sets

    c. Manufacturing of electronic chassis

3. List three reasons why automatic machines and processes were developed.

4. Compare individual, centralized, and distributive control systems.

5. Define what is meant by the term *process control system.*

6. State the basic function of each of the following as part of a process control system:

    a. Sensors

    b. Operator-machine interface

    c. Signal conditioning

    d. Actuators

    e. Controller

7. Compare open-loop and closed-loop control systems.

8. List four classifications of controllers according to the type of control they provide.

9. Outline the operating sequence of an on/off controller.

10. Explain how a proportional controller eliminates the cycling associated with on/off control.

11. What process error or deviation is produced by a proportional controller?

12. How does integral control work to eliminate offset?

13. What does the derivative action of a controller respond to?

14. List three factors of a system error signal that influence the output response of a proportional-integral-derivative (PID) controller.

15. Why must a PID controller be custom-tuned to each process controlled?

16. Compare manual, autotune, and intelligent tuning of a PID controller.

17. Define *data acquisition.*

18. Outline how a typical data acquisition and control system operates.

19. Explain the function of each of the following devices with reference to a data acquisition system:

   a. Transducer
   b. Analog multiplexer
   c. Amplifier
   d. Signal conditioner
   e. A/D converter

20. Explain each of the following terms as it applies to the operation of a data acquisition system:

   a. Sensor calibration
   b. Floating signal
   c. Gain
   d. Isolation
   e. Real time
   f. Resolution
   g. Save interval
   h. Transfer rate

Problems

1. Give an example of an open-loop control system. Explain why the system is open-loop and identify the controller and final control element.

2. Give an example of a closed-loop control system. Explain why the system is closed-loop and identify the sensor, controller, and final control element.

3. How would an on/off controller respond if the deadband were too narrow?

4. In a home heating system with on/off control, what will be the effect of widening the deadband?

5. A. Calculate the proportional band of a temperature controller with a 5% bandwidth and a set point of 500°F.

   B. Calculate the upper and lower limits beyond which the controller functions as an on/off unit.

   C. Calculate the proportion gain:
   $$\text{Gain} = \frac{100}{\text{percentage bandwidth}}$$

6. What effect, if any, would a higher process gain have on offset errors?

7. Describe the advantages of a 4- to 20-mA current loop as an input signal compared to a 0- to 5-V input signal.

8. How might a data acquisition system be applied to a petrochemical metering station whose purpose is measurement and flow control?

9. If noise were a problem in an application, what types of changes might help alleviate the problem?

15

Computer-Controlled Machines and Processes

After completing this chapter, you will be able to:

◆ Discuss how a computer's operating system is designed to function

◆ Explain how a work cell functions

◆ Compare the methods by which computers communicate with each other

◆ Discuss the principle of operation of computer numerical control

◆ Discuss the principle of operation of robotic computer control

Since personal computers (PCs) are now common in industry, this chapter contains a brief discussion about important PC operations. The use of machine interfacing circuits and with protocol standards are presented. This chapter also introduces computer numerical control (CNC) and robotic computer control.

15·1 COMPUTER FUNDAMENTALS

Manufacturing systems are becoming increasingly computer-based, and understanding of computer fundamentals is essential. Computers, particularly personal computers (PCs), are now accepted as useful devices for machine and process control. I/O cards that plug directly into the expansion slots of most PCs are now available. These cards perform data-acquisition functions as well as data conversion and power control. A PC equipped with special software packages and I/O cards can emulate many of the popular PLCs in use today.

All computers consist of two basic components: hardware and software. The computer *hardware* (Fig. 15-1) includes the physical components that make up the computer system:

◆**Power Supply**
Converts the 120 V ac electricity from the line cord to dc voltages needed by the computer system.

◆**Floppy Drives**
Allows information to be stored and read from removable floppy disks.

◆**Hard Drive**
Allows information to be stored and read from nonremovable hard disks.

◆**Motherboard**
Holds and electrically interconnects the major sections of the entire computer system.

◆**Microprocessor Chip**
Interprets the instructions for the computer and performs the required process for each of these instructions. The main functions include:
- Controlling of data around the system
- Performing all math operations
- Performing all logical operations and decisions
- Performing system control and arbitrating when and how information is transferred

◆**Clock Chip**
Acts as a system stepper (or "metronome") for the computer processing states. In general, an increase in clock rate results in increased performance.

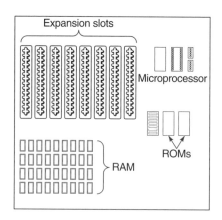

(a) Motherboard holds and electrically interconnects major hardware sections

(b) Plug-in components and cards are easily serviceable; expansion slots allow upgrading of computer capabilities

Fig. 15-1
Computer hardware.

◆ Read-Only Memory (ROM) Chips

Memory programmed at the factory that cannot be changed or altered by the user. It can be read but not written to, and it retains its content when the power in turned off. The ROM contains the basic input/output systems. (BIOS) program that allows you to access I/O devices.

◆ RAM Chips

Read/write memory. Memory used to store computer programs and interact with them. It loses its content when the power is turned off. The more RAM memory held by the computer the larger the application program you can run.

◆ Bus

A group of binary data lines or connecting cables that provides a means of sending and receiving information between different parts of the computer. Usually expressed in multiples of 8. One of the major differences between computers is the number of data lines available.

◆ Peripheral Cards

Allows accessory features and connects the computer to input and output devices such as drives, printers, monitors, and other external devices.

◆ Expansion Slots

Connectors used for the purpose of connecting other circuits to the motherboard.

The computer *software* is what gives the computer "life." The software resides inside the hardware. You can think of software as a computer *program*. A computer program is nothing more than a list of instructions telling the computer what to do. The programs are usually stored on some form of mass storage system, such as a floppy disk, and loaded into the computer's random access memory (RAM) as required.

Disk drives act as both input and output devices, so they are used for placing information into the computer as well as storing information from the computer. There are actually three different kinds of disk drives popularly used in personal computers: the hard (fixed) disk drive, the 5¼-in. floppy disk drive and the 3½-in. floppy disk drive.

The *floppy disk* (diskette) is a magnetic storage device on which information is stored for later retrieval (Fig. 15-2). The information is stored on tracks in sectors of the disk; the location of the information is configured to allow the computer access to a sector to

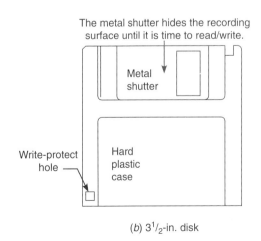

Recording surface is always exposed

Read-write slot

Index hole

Write-protect notch

Label

(*a*) 5¹⁄₄-in. disk

The metal shutter hides the recording surface until it is time to read/write.

Metal shutter

Write-protect hole

Hard plastic case

(*b*) 3¹⁄₂-in. disk

Fig. 15-2

Floppy disk.

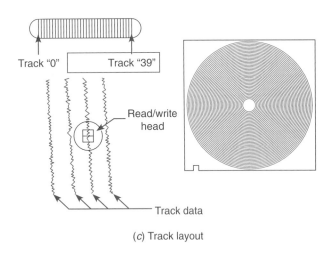

(c) Track layout

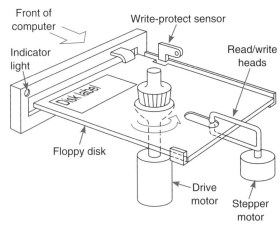

(d) Floppy disk drive

Fig.15-2 (continued)

Floppy disk.

read the information. The disk must be formatted into tracks and sectors using the disk operating system format command before any information can be stored on it.

All disks have a *write-protect*. For a 5¼-in. disk, this involves merely putting an adhesive label over the notch. For a 3½-in. disk, there is a small physical sliding tab on the lower lift side that should be opened if you want to protect the disk from being written on. The smaller 3½-in. micro floppy can generally store more information than its larger 5¼-in. counterpart. This is because of the higher quality of the magnetic surface in the 3½-in. disk.

All disks must be *formatted* before they can be used. Formatting makes your disk compatible with the computer you are using. Once the disk is inserted into the computer, the drive motor spins the disk while the stepper motor moves the read/write head to different positions on the disk. The indicator light indicates whether the disk drive is active.

The disk is inserted in the disk drive with the slot and labels up. The disk drive door must then be shut to allow the disk reading heads to read the information. Care must be taken when handling the disk, or the information stored on the disk may be corrupted.

◆ Never touch the exposed disk at the head slot, or surface.

◆ Never bend the disk.

◆ Never expose the disk to heat or excessive sunlight.

◆ Never expose the disk to electromagnetic or electrostatic fields that are likely to occur in the industrial environment.

◆ Use only labels designed specifically for use with disks. These labels will not get stuck in a disk drive when you remove a disk.

◆ Use only soft-tip pens when labeling the disk.

◆ Never force the disk into the disk drive.

◆ Always make backup disks (copies) and store the original disks in a safe place.

Hard disk drives (Fig. 15-3) store and retrieve data more quickly and can hold much more information than floppy disks. The hard disk mass storage system is located

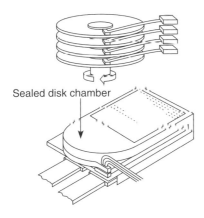

Sealed disk chamber

Fig. 15-3

Hard disk drive unit.

permanently within the computer and can store numerous programs simultaneously. The contents of any floppy disk can be relocated onto the hard disk, and the programs can then be used as required, keeping the floppy disks as backup copies. The hard disk drive is normally referred to as the C: drive (it is conventional to place a colon after the drive letter). If one floppy disk drive is used, it is called the A: drive, and if a second floppy disk drive is used, it is called the B: drive.

There are two major categories of software: system software and application software. *System software* provides the programs that allow you to interact with your computer—to operate the disk drives, the printer, and other devices used by the computer. Operating systems software contains commands and programs that allow the user to interact with the computer. This software is thought of as a translator between hardware devices and application programs or users. *Application software* involves programs written to give the computer a specific application, such as a word processor. Application software makes up the majority of the software available in the market. For application software to work in your computer, the system software must first be loaded into the computer's memory.

The most popular operating system for personal computers is *MS-DOS. MS* refers to the manufacturer, Microsoft, and DOS to its disk operating system. DOS creates a common platform for all the software you use and allows you to:

◆ Control input and output operations

◆ Interpret and execute commands entered from the keyboard

◆ Read, write, and edit files

◆ Back up files

◆ Organize a disk

◆ Manage files

◆ Display the contents of a disk

◆ View error messages

Different versions of DOS have evolved because people wanted the program to do more and because technology gave us faster equipment. VER is a DOS command that will show you the version of DOS your computer is using.

A *file* is a collection of data stored under a single name. When information is stored on the disk, the file is saved or written to the disk. When the information is used, the disk is read or data are loaded into the computer's memory. A computer can move, copy, rename, or delete the file at your command. Some important types of files include:

◆**Executable File**
A list of instructions to the CPU

◆**Data File**
A list of information

◆**Text File**
A series of characters like a letter, number, punctuation mark, word space, etc.

◆**Graphics File**
A picture converted to digital code

After a disk is formatted, writing or reading a file is a process that involves your software, DOS, the PC's basic input/output system (BIOS), and the mechanism of the disk drive itself. The BIOS serves as a link between the computer hardware and the software programs. It is stored permanently in the computer's memory, which serves as a translator between the CPU and all the other entities it interacts with.

When the computer is first turned on, the BIOS causes the computer to look into your floppy disk drives to see whether there is a disk there containing DOS. If there is, the computer will load the DOS into itself. The process of loading DOS into your computer is called *booting*. If there are no floppy disks in your drives that contain DOS, the hard drive (C: drive) will be examined for DOS. If there is no DOS (or no hard drive), then an error message will be displayed (Fig. 15-4).

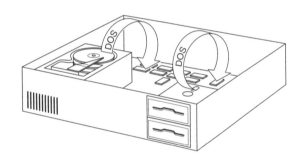

Fig. 15-4

Loading DOS into memory.

Directories are simply a way of organizing all the programs and files of floppies and hard disks. Because disks store large amounts of data, it is useful to divide the storage into uniquely named areas. Each area is used to store a group of related files (e.g., WP, LOTUS, DOS). This setup provides easy organization and maintenance of the disk. The directory structure performs the same basic function as a filing cabinet. Directories are organized according to a tree structure.

The first and main directory of the computer is called the *root directory*. Ideally, the second level of the directory structure contains the general listings of the contents of the disk. *Tree* is a DOS command that will show you the directory structure of your computer.

The *DOS prompt* (A>) is an indicator to you that DOS is loaded and that the computer is ready for you to enter a command. The prompt will stay on the screen until you either enter an instruction or turn the computer off. The letter *A* indicates that the computer is currently ready to access drive A. If you wish to switch to another disk drive, you must tell DOS to do so by entering the new drive letter followed by a colon and then pressing Enter (e.g., A>b:). The screen will then display the new prompt B>.

DOS is simply a long list of *commands* that the computer can use and understand to control, manage, and run other programs. Some of the commands used most often are listed below:

◆ DIR command (directory command)
 - Used to list the contents of any disk
 - Will list all files and subdirectories in the current working directory
 - DIR/P (display one page)
 - DIR/W (display wide removes the data and times; the "/W" means "wide" and gives you a name-only list)
 - DIR/S (list all files in subdirectories)

◆ CD command (change directory command)
 - Allows you to move around in the directory structure
 - Used to change to another subdirectory or the root directory

◆ MD command (make directory command)
 - Used to create a subdirectory

◆ RD command (remove directory command)
 - Used to delete a directory

◆ TREE command
  - Displays the current directory tree structure

◆ CLS command (clear screen command)
  - Used to clear the video display screen and place the prompt and cursor at the top left corner of the screen

◆ COPY command
  - Used to copy files from one location to another, for example, from drive A to drive B

◆ REN command (rename command)
  - Used to rename a file with a new name without changing its contents

◆ DEL command (delete command)
  - Used to delete any unwanted files on a disk

◆ TYPE command
  - Used to display a file on the screen, allowing you to read its contents

◆ FORMAT command
  - Used to prepare a new disk for use with DOS

◆ DISKCOPY command
  - Used to make an exact duplicate of a floppy disk

◆ CHKDSK command (check disk command)
  - Used to check the status of a disk, how many files are on it, and how much of the disk is used

Microsoft Windows® is a graphics-based interface for MS-DOS. It uses graphical images to replace the verbal commands to run DOS. Instead of typing commands, you select graphic pictures (icons) using a mouse. Windows allows for *multitasking*. No computer can actually do two things at once, but it appears to do so by rapidly shuffling back and forth from one program to another. Programs in Windows share the CPU resources, passing control back and forth at times when the computer is not controlled by the user. Windows' multitasking operation is one of the principal factors drawing people away from DOS application development. Instead of developing an application that does everything, with Windows you can use multiple specialized applications that run concurrently and work together.

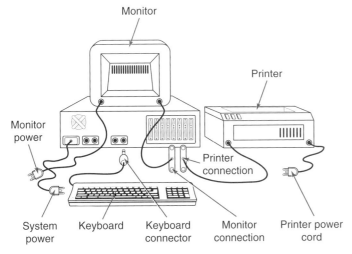

(a) Typical cable connections for computer

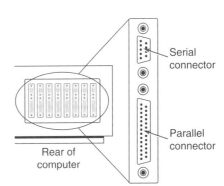

(b) Serial and parallel port connections

Fig. 15-5

Computer peripheral connections.

Ports are the connecting devices that stick out the back end of the computer case (Fig. 15-5 on p. 407). They are actually the back edge of an expansion card, providing the electrical door or gateway connecting the computer and a peripheral. Ports are usually a preset number of pins to allow the proper cable connection and are referred to as male or female (with pins or holes to fit the pins). Ports are also referred to as *serial* (for a mouse connection) or *parallel* (for a printer connection). A serial port sends data one bit at a time over a single, one-way wire; a parallel port can send several bits of data across eight parallel wires simultaneously. Because of its simplicity, the serial port has been used to make a PC communicate with just about any device imaginable. The serial port is often referred to as an *RS-232 port* (Electronics Industries Associations designation). The parallel port is often called a *Centronics port.*

Selecting a computer can be a complex task, because there are numerous vendors and scores of different configurations possible. Key specifications include:

◆**Access Time**
Refers to the length of time it takes for an information storage device to return a piece of information once it is requested. For a hard drive, the typical time is 9 to 28 milliseconds (thousandth of a second). For RAM, typical times are 70 to 120 nanoseconds (billionths of a second).

◆**Baud**
A measure of information transfer speed over modems. One baud represents one signal change per second. The faster the signal, the faster the data can be transferred. Newer modems actually transfer data faster than one bit with every signal change; the transfer of information is measured in bps (bits per second). For example, a modem running at 1200 baud and passing 2 bits of information with

every signal change is transferring data at a rate of 2400 bps, or about 240 characters per second.

◆**Bus**
A wire or group of wires that carries a flow of digital information. When a computer moves information from one component to another, it uses a bus to do it. To increase the speed of data movement, engineers have increased the size of the bundle of wires. The earliest PCs used eight data channels, which were called 8-bit buses because they could carry eight pieces of information at one time, similar to an eight-lane highway. Newer machines use 16-bit and 32-bit buses.

◆**Cache**
A memory cache is a superfast set of memory chips that tries to anticipate what data the CPU will need next and save it so it can be delivered to the CPU quickly (unlike RAM, which provides the information much more slowly). The result is that the CPU spends less time looking for instructions and thus speeds up processing time and overall system performance.

◆**Central Processing Unit (CPU)**
PC/XT/AT/386;486/Pentium/RISC ... With all these terms and different processors, which is the right one for your application? The different terms—PC, XT, and so on—are only marketing names. What is important is the type of processor and the speed at which it operates. In the IBM world, there are multiple processor types: the 8088 8-bit processor, the 80286 16-bit processor, the 80386 32-bit processor, the 80486 32-bit processor family, and several new generation 64-bit processors. Although there are many differences, the key factors are the number of instructions per clock cycle and the number of clock cycles per second of each CPU.

◆ Clock Speed

Refers to the rate at which the CPU clock moves. Every computer has a clock; the CPU uses pulses from the clock to pace its activities and coordinate the execution of instructions with other system components. If the clock beats faster, the system components move faster. Clock speed rates are measured in megahertz (MHz), or millions of oscillations (beats) per second. Typical clock speeds range from 4 MHz through 150 MHz.

◆ Hard Disk Drive

The hard disk drive is the primary mass information storage device of the computer. Although 40 MB of memory used to be a lot, standard drives now seem to be 80 MB, with 600 MB and even 1.2 GB (gigabyte, or billion bytes) available. Cost and the requirements of the programs are the primary concerns in this area.

◆ Monitors

Every screen displays thousands of tiny dots (called *pixels*) just below the screen surface. The greater the number of dots and the closer together the dots are, the sharper the image and the more information displayed. Video graphics array (VGA) and super-VGA are the current industry standards. A VGA monitor has at least 480 rows of dots, with each row consisting of 640 dots.

◆ RAM

Random access memory is the working memory of the computer. How much RAM the computer has is closely related to performance. RAM is referred to in megabytes (MB), and average capacities are 640 K to 16 MB. Without enough RAM, the memory becomes cramped and programs can mysteriously fail to function because the computer has nowhere to write the new instructions.

15·2 COMPUTER-INTEGRATED MANUFACTURING

Today automation is moving rapidly toward a true point of central control that resides in the system operator's office. It is increasingly necessary for system operators to have fingertip control of the process by way of their personal computers. One application in which the computer is used to monitor and control a networked PLC system is shown in Fig. 15-6.

Computer-integrated manufacturing (CIM) systems provide individual machines used in manufacturing with data communication functions and compatibility, allowing the individual devices to be integrated into a *single* system. In general, four levels of computer integration are required for CIM to work: the *cell level, area level, plant level,* and *device level* (Fig. 15-7 on p. 410). Each level has certain tasks within its range of responsibility, which include:

◆ Plant Controller

Used for tasks such as purchasing, accounting, materials management, resource planning, and report generation.

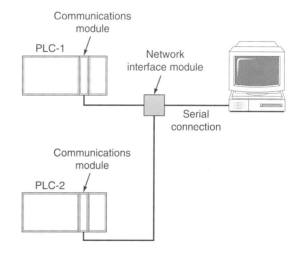

Fig. 15-6

Computer used to monitor and control a networked PLC system.

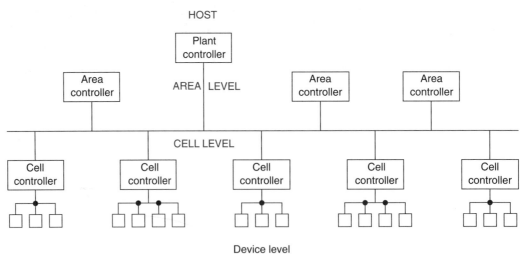

HOST

Plant controller

AREA LEVEL

Area controller   Area controller   Area controller   Area controller

CELL LEVEL

Cell controller   Cell controller   Cell controller   Cell controller   Cell controller

Device level

Devices are task-specific equipment

Fig. 15-7

Levels of control in a computer-integrated manufacturing system.

◆**Area Controller**

Used for tasks such as machine and tool management, maintenance tracking, materials handling and tracking, and computer-assisted simulations.

◆**Cell Controller**

Used for device machine control and data acquisition.

◆**Device Controller**

Used for direct control of equipment that produces or handles the product.

A *work cell* can be defined as a group of machine tools or equipment integrated to perform a unit of the manufacturing process. A typical work cell with associated machines is shown in Fig. 15-8. The computer, or cell controller, is basically a communicator between components. The main difference between the programmable controller and the cell controller is the language used to program the cell controller. The PLC programming language is simple and requires limited programming knowledge; the cell controller requires the programmer to have more programming knowledge. With this stipulation in mind, PLC manufacturers are developing PLCs with greater software

capability. A recent outgrowth of this effort is a computer called a *cell controller*. Cell controllers combine the software sophistication of a PC with the I/O handling capability of a PLC.

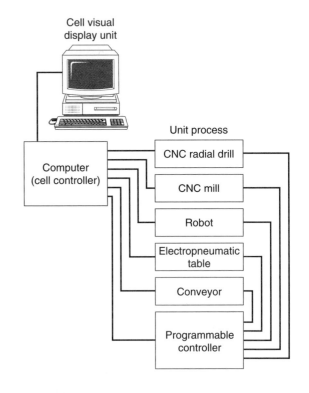

Cell visual display unit

Unit process

Computer (cell controller)

CNC radial drill

CNC mill

Robot

Electropneumatic table

Conveyor

Programmable controller

Fig. 15-8

Work cell with associated machines.

The entire industry of cell control was impossible until the advent of the industrial personal computer. More and more cell control programs are running on the personal computer level of hardware. Cell control involves a group of automated programmable machine controls (programmable controllers, robots, etc.) designed to work together to perform a complete manufacturing or process related task. The function of a cell controller is to coordinate and oversee the operation of the machine controls within the cell through its communication and information-processing capabilities. Even if the devices are unable to communicate with each other, they must be able to communicate with the cell controller. The cell controller must be able to upload/download programs, exchange variable information, start/stop the device, and monitor the performance of each device.

Cell control today involves influencing what happens up or down a production line. Picture an assembly line, for example. Assume you produce a part and that part has to be tested. A robot can take the part, position it for the test, and, depending on the outcome, put it back on the line or reroute the part. If the specs are off, you will want to convey this information up- or downstream, wherever the effects will be. You also want to reroute that part with a bill that describes the problem so that the part doesn't undergo diagnostics testing again but, rather, can be repaired or adjusted and sent back on-line. This process requires communications links to avoid the problems that the defective part will cause if information about it is not relayed.

15·3 DATA COMMUNICATIONS

Data communications refers to the different ways that microprocessor-based systems talk to each other and to other devices. Initially, data communication systems were provided between programmable controllers, but it was apparent that the advantages of integrated manufacture required computer-controlled and numerically controlled machines and robot controllers to be connected to the programmable controller, which in turn would interconnect with a host computer and other equipment via the factory local area network.

Communications between programmable controllers, or between programmable controllers and computers, has become a common application. Local area networks are the backbone of communications networks. The fundamental job of a *local area network* (LAN) is to provide communication between PLCs or between PLCs and computers. In industry, the transmission medium used most often is coaxial cable or optical fiber because of the high noise immunity. The rate at which characters can be transmitted along a communications line depends on the number of bits of binary information that can be sent at a given time. This transfer of information is measured in bits per second. *Data highway,* a term associated with PLC systems, refers to local area networks. LANs come in three basic *topologies* (that is, the physical layout or configuration of the communications network): star, ring, or bus (Fig. 15-9 on p. 412). The points where the devices connect to the transmission medium are known as *nodes,* or stations.

In a *star network,* a central control device or hub is connected to several nodes. This configuration allows for bidirectional communication between the central control device and each node. All transmission must be between the central control device and the nodes since the central control device controls all communication. All transmissions must be sent to the hub, which then sends them to the correct node. One problem with the star topology is that if the hub goes down, the entire LAN is down. This type of system works best when information is transmitted primarily between the main controller and remote PLCs. However, if most communication is to occur between PLCs,

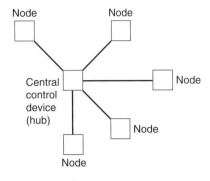

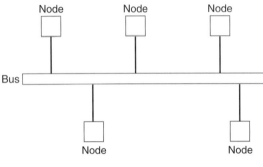

(a) Star network

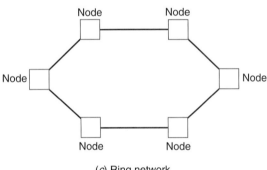

(b) Bus network

(c) Ring network

Fig. 15-9

Network topologies.

centralized or distributed among the nodes. Communication can take place between any two nodes without information having to be routed through a central node. This setup is a sort of parallel communications system where the bus is simply the communications cable that connects the PLCs.

In a *ring network,* each node is connected in series with each other to form a loop. Transmitted messages travel from node to node in the loop. Messages are aimed at a particular node or station number, with each node listening for its own identification number. Signals are passed around the ring and are regenerated at each node. Control can be centralized or distributed. This type of system is popular for *token-passing* protocol. The token, or bit pattern, is used by a node to gain control of the communications channel. The key to successful implementation of computer-integrated manufacturing is communication compatibility among all the controllers involved in the process. Although communication standards do exist, each computer system may use a different standard. In addition, as the level of sophistication of communication increases, the need for more sophisticated standards also increases.

Protocol refers to a predetermined set of conventions that specify the format and timing

the operation speed is affected. Also, the star system can use substantial amounts of communication conductors to connect all remote PLCs to one central location.

Each node is connected to a central bus in a *bus network.* When a node sends a message on the network, the message is aimed at a particular station or node number. As the message moves along the total bus, each node is listening for its own node identification number and accepts only information sent to that number. Control can be either

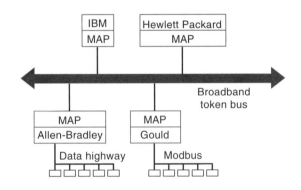

MAP gateway used to connect equipment with different protocols

Fig. 15-10

Network interconnections.

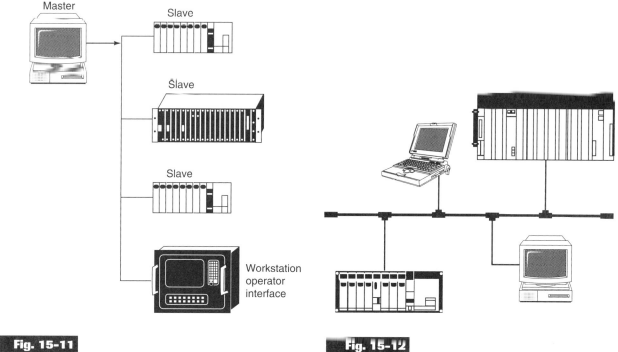

Fig. 15-11

Master-slave access method.

Fig. 15-12

Peer-to-peer or token-passing access method.

of message transmission between two or more communicating devices. It is nothing more than an agreed-upon set of rules by which communication takes place. There are several standards to define the signal protocol. For computer-integrated manufacturing, you need to tie all the different devices together using a *common* protocol. When the protocol is different, additional hardware and programming are required (Fig. 15-10). General Motors identified the need for a common protocol and developed the *manufacturing automation protocol (MAP)* specification, which is available to anyone. It has become an industry standard, and most new devices indicate that they are MAP-compatible or feature MAP interfaces. MAP is a token-passing bus LAN system.

Protocol is the method used by a LAN to establish criteria for receiving and transmitting data through communications channels. It is, in effect, the way a LAN directs traffic on its data highway. The two basic access methods for network communications are master-slave and peer-to-peer. A *master-slave*

system (Fig. 15-11) uses a host computer to manage all network communications between network devices. The programming of the master network device incorporates routines to address each slave device individually. Direct communication among slave devices is impossible. Information to be transferred between slaves must be sent first to the network master unit, which will, in turn, retransmit the message to the designated slave device.

With a peer-to-peer system (Fig. 15-12), each network device has the ability to request use of, and then take control of, the network for the purpose of transmitting information to or requesting information from other network devices. This type of network communication scheme is often described as a *token-passing system* since control of the network can be thought of as a token passed from unit to unit. In the token-passing method, only one station can talk at a time. The node or station must have the token to be able to use the line. The token circulates among the stations until one of them wants to use the line. The node

Computer-Controlled Machines and Processes

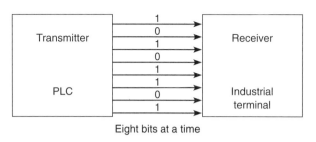

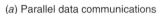

(a) Parallel data communications

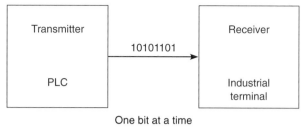

(b) Series data communications

Fig. 15-13

Parallel and series transmission.

then grabs the token and uses the line. The node has use of the token only for the duration of its transmission. Once the node is finished transmitting, the token is recirculated automatically among the nodes in the LAN.

There are two basic ways of transferring information (Fig. 15-13). One is the *parallel* method, where several bits are transferred at the same time; the other is the *series* method, where one bit is transferred at a time. Parallel data communication is satisfactory over short distances of approximately 30 ft. Serial data communication requires only two wires and can be transmitted over greater distances. Each data word in the serial transmission must be denoted with a known start followed by the data bits that contain the intelligence of the data transmission and a stop bit. Often an extra bit, termed a *parity bit,* is used to provide some error-detecting ability.

Modems permit any two devices to communicate over a single pair of dedicated wires or phone lines (Fig. 15-14). A pair of modems can be used to send and receive digital signals by using different tones for logic 1's and logic 0's. If data transmission is required between two distant points, or the public telephone network is to be used, the data bits must be converted to audio tones because data bits represented as voltage levels may be so attenuated that they make the data signal unreadable. A modem is used to convert the transmitted data bits to audio

tones. It modulates the signal at the transmitter and demodulates the signal at the receiver to convert the audio tones back to data bits. The term *modem* is made up of the first two letters of *mo*dulate and the first three letters of *dem*odulate.

There are several different standards for modem transmission. The *duplex* mode refers to two-way communication between two devices (Fig. 15-15). Full-duplex transmission allows the transmission of data in both directions simultaneously. Half-duplex transmission uses a switching arrangement whereby information can be sent in only one direction at a time.

Data communication can be either synchronous or asynchronous. In *synchronous* data transmissions, once the transmission starts, the time from one data byte to the next is known. *Asynchronous* data transmission

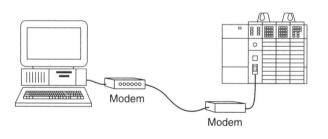

Fig. 15-14

Communications using a pair of modems.

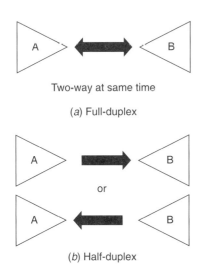

Two-way at same time

(a) Full-duplex

or

(b) Half-duplex

Fig. 15-15

Duplex modes.

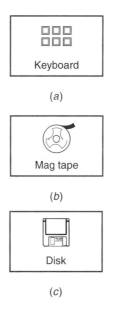

Keyboard

(a)

Mag tape

(b)

Disk

(c)

Fig. 15-16

Methods used to enter an NC program.

means that the time between data bytes is random and cannot be predicted by the sending or receiving system. Special software in the microprocessor must be used to synchronize these random (asynchronous) data words with the microprocessor operation. Most systems use asynchronous data communications.

15·4 COMPUTER NUMERICAL CONTROL (CNC)

In general terms, *numerical control (NC)* is a flexible method of controlling machines automatically through the use of numerical values. Numerical control devices are driven over a continuous path or to various points to manufacture a component or device using a program that coordinates the machine movement and operation numerically. NC machines are program-dependent and bear a close relation to programmable controllers.

Numerical control enables an operator to communicate with machine tools through a series of numbers and symbols. A set of commands makes up the NC program and directs the machine to orient a cutting tool with respect to a workpiece, select different tools, control cutting speed, and direct spindle rotation, as well as perform a range of auxiliary functions such as turning coolant flow on or off. Many languages can be used for writing an NC program, but the one used most is called *automatically programmed tools (APTs)*. Originally, NC programs were stored on punched paper tape, but recently they have been stored using some form of mass storage such as magnetic tape, floppy disk, or solid-state memory (e.g., RAM or ROM; see Fig. 15-16).

Data that are input manually (Fig. 15-16a) can be used to program the control system by setting the dials, switches, pushbuttons, etc. Most NC machine tools can be programmed manually, especially for setup purposes. Magnetic tape (Fig. 15-16b) in a tape cassette is used for some applications. The major advantage of floppy disks (Fig. 15-16c) is that it is much faster to retrieve information from this medium than from any other.

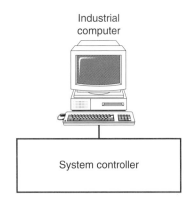

Industrial computer

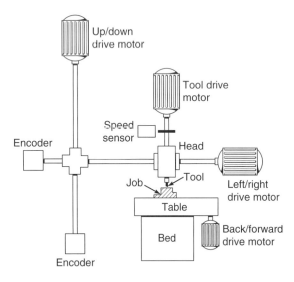

System controller

Up/down drive motor

Tool drive motor

Speed sensor

Encoder

Head

Job

Tool

Left/right drive motor

Table

Bed

Back/forward drive motor

Encoder

Fig. 15-17

Typical numerical control system.

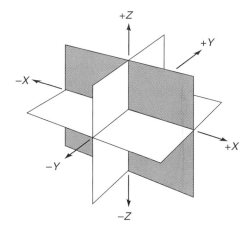

(a) Three-dimensional coordinate planes (axes) used in numerical control

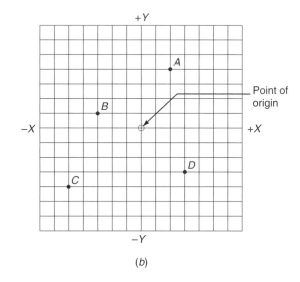

Point of origin

(b)

Fig. 15-18

Locating points using Cartesian coordinates.

Figure 15-17 illustrates a typical numerical control system. The controller reads and interprets the instructions, and directs the machine to perform the operations desired. The machine operator is alerted when material must be inserted, tools must be changed, and so on. The system controller is usually an industrial computer that stores and reads the program electronically and converts the program information into signals that drive motors to control the machine tool. The controller provides signals to either electrical or hydraulic motors that cause the machine head to be driven left or right, as well as up or down. The table can be motor driven to move the workpiece back and forth. Feedback to the controller is provided by the position encoders and speed sensor so that the exact location and speed of the machine head is known. As the accuracy of the electronics exceeds the mechanical accuracy of the system, the repeatability, and therefore consistency, of the manufactured components are within tight tolerances.

Numerical control is ideally suited for operations involving the production of parts made from similar feedstock (raw material) with variations in size and shape. Even if production quantities are in small lots, NC can be economically feasible, but it is necessary that the sequence of operations can be performed on the same NC machine. However, for complete manufacturing of parts involving several dissimilar sequences of operation, several NC machines may be used in production.

If the controlled machine is a three-axis type, the location address of the tool is prefixed with the letter x, y, or z. Using x, y, and z coordinates, the machine can be directed to the correct location. The workpiece is located by using Cartesian coordinates, whereby the position of a point can be defined with reference to a set of axes at right angles to each other, as illustrated in Fig. 15-18. The vertical axis is the y axis and the horizontal axis is the x axis. The point where the two axes cross is the zero, or origin, point. To the left of the point of origin on the x axis and below the point of origin on the y axis, locations are written preceded by a minus (−) sign. Above the point of origin on the x axis, locations are written preceded by a plus (+) symbol. The z axis of motion is parallel to the machine spindle and defines the distance between the workpiece and the machine.

In Fig. 15-18b, x, y, and z words refer to coordinate movement of the machine tool for positioning or machining purposes. Using Cartesian coordinates, each point is located as follows:

A is x = +2 and y = +4
B is x = −3 and y = +1
C is x = −5 and y = −4
D is x = +3 and y = −3

The programming of NC machines can be categorized into two main areas: point-to-point

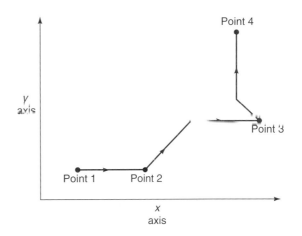

Fig. 15-19

Point-to-point programming.

programming and contour programming. *Point-to-point programming* involves straight-line movements. Figure 15-19 shows an example of point-to-point programming in which four holes are to be drilled. The point where each hole is to be located is identified using x and y coordinates. After each hole is drilled, the machine is instructed to move to the next point where a hole is to be drilled, and so on. The holes are drilled sequentially until the program is completed. The path the machine takes between holes is not important because the tool is in the air between hole locations. The depth of each operation is controlled by the z axis.

Contour (also known as *continuous path*) *programming* involves work like that produced on milling machines, where the cutting tool is in contact with the workpiece as it travels from one programmed point to the next. Continuous path positioning requires the ability to control motions on two or more machine axes simultaneously to keep a constant cutter-workpiece relationship. The method by which contouring machine tools move from one programmed point to the next is called *interpolation*. All contouring controls provide linear interpolation, and most are capable of linear and circular interpolation (Fig. 15-20 on p. 418).

Computer-Controlled Machines and Processes

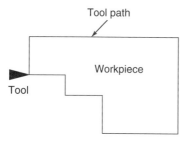

(a) Linear interpolation

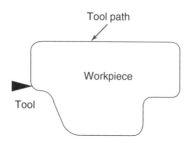

(b) Linear and circular interpolation

Fig. 15-20

Contour programming.

Computerized numerical control (CNC) was introduced to replace the punched tape and hard-wired machine control of older NC units (Fig. 15-21). The CRT screen shows the exact position of the machine table and/or the cutting tool at every position while a part is machined. CNC introduced a flexibility into the manufacturing industry because the microprocessor-based control equipment brought new features to NC machines such as:

◆ Improved program mass storage; disk instead of punched tape

◆ Ease of editing programs

◆ Possibility of more complex contouring because of the computer's capability for mathematical manipulation

◆ Reusable machine pattern that could be stored and retrieved as required

◆ Possibility of plantwide communication with many peripheral devices

Fig. 15-21

Computer numerical control (CNC).
(Courtesy Allen-Bradley Company, Inc.)

Simple CNC programming consists of taking information from a part drawing and converting this information into a computer program. The program design can be carried out on a personal computer. A numerical control programming language must be used to design the program, which simplifies the program writing because the computer will calculate the machine coordinates. Using a computer to assist in program writing allows the program to be:

◆ Stored on a convenient mass storage system

◆ Retrieved and edited as required

◆ Tested off-line prior to using the program to control a machine

◆ Plotted using a plotter connected to the computer to assist in program debugging

•5 ROBOTICS

Robots are computer or PLC controlled devices that perform tasks usually done by humans. The basic industrial robot widely used today is an *arm* or *manipulator* that moves to perform industrial operations. Tasks are specialized and vary tremendously. They include:

◆Handling
Loading and unloading components onto machines

◆Processing
Machining, drilling, painting, and coating

◆Assembling
Placing and locating a part in another compartment

◆Dismantling
Breaking down an object into its component parts

◆Welding
Assembling objects permanently by arc welding or spot welding

◆Transporting
Moving materials and parts

◆Painting
Spray painting parts

◆Hazardous Tasks
Operating under high levels of heat, dust, radioactivity, noise, and noxious odors

A robot is simply a series of mechanical links driven by servomotors. The area at each junction between the links is called a *joint* or *axis*. The axis may be a straight line (linear), circular (rotational), or spherical. Figure 15-22 illustrates a six-axis robot arm. The *wrist* is the name usually given to the last three joints on the robot's arm. Going out along the arm, these wrist joints are known as the *pitch joint, yaw joint,* and *roll joint.* High-technology robots have from 6 to 9 axes. As the technology increases, the number of axes may increase to 16 or more. These robots' movements are meant to resemble human movements as closely as possible.

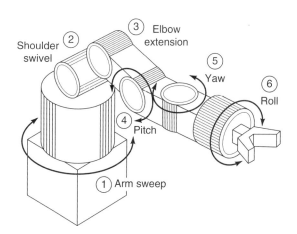

Fig. 15-22

Six-axis robot arm.

The reach of the robot is defined as the *work envelope*. All programmed points within the reach of the robot are part of the work envelope. The shape of a work envelope is determined by the major (nonwrist) types of axes a robot has (Fig. 15-23). A robot that has two linear major axes and one rotational major axis has a cylindrical work envelope. A robot that has three rotational major axes has a work envelope very much like the motion range of a human body from waist to shoulder to elbow. Understanding the work envelope of the robot with which you work will help you avoid personal injury or potential damage to equipment.

Most applications require that additional tooling be attached to the robot. End-of-arm tooling, commonly called the *end effector,* varies depending on the type of work the robot does. Grippers or hands are used in materials handling and assembly. Spot welding and arc welding require their own tooling, as do painting and dispensing (Fig. 15-24).

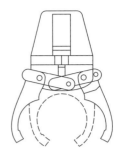

(*a*) Gripper

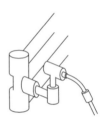

(*b*) Grinder

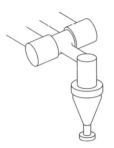

(*c*) Gas welding torch

Fig. 15-24

End-of-arm tooling devices.

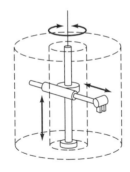

(*a*) Cylindrical

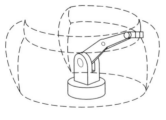

(*b*) Articulated

Fig. 15-23

Robot work envelope.

Robots usually have one of three possible sources of manipulator or muscle power: electric motors, hydraulic actuators (pistons driven by oil under pressure), or pneumatic actuators (pistons driven by compressed air). Robots powered by compressed air are lightweight, inexpensive, and fast-moving but generally not strong. Robots powered by hydraulic fluid are stronger and more expensive, but they may lose accuracy if their hydraulic fluid changes temperature.

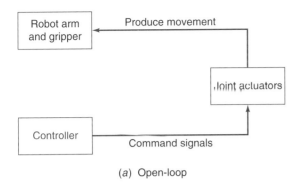

(a) Open-loop

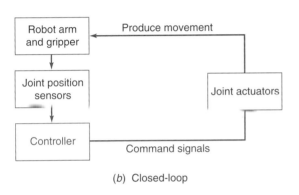

(b) Closed-loop

Fig. 15-25

Open-loop and closed-loop robotic control systems.

Originally all robots used hydraulic servodrives. Driven mostly by the level of service required to maintain hydraulic servosystems in these early industrial robots, engineers developed the articulated robot with dc electric servodrive motors. The industrial robot has since evolved from dc electric to ac electric. The benefits of ac servomotors over dc motors were significant. The ac servomotor incorporated brushless, maintenance-free designs, and incremental encoders are servoposition feedback.

There are two types of robot control systems: closed loop and open loop (Fig. 15-25). In the *open-loop system,* there are no sensors or feedback signals that measure how the manipulator actually moved in response to the command signals. The *closed-loop system* uses feedback signals from joint position sensors. The controller compares the actual positions of the arm joints with the programmed positions. It then issues command signals that minimize or eliminate any discrepancies or errors.

The term *servo robot* is often used to refer to a closed-loop (feedback) system, and the term *nonservo robot* refers to an open-loop (no feedback) system. Nonservo robots are generally small and designed for light payloads. They have only two positions (open or closed) for each joint and they operate at high speed. They are often called bang-bang robots because of the way they bang from position to position. They are programmed for a task by setting adjustable mechanical limit stops. Nonservo robots are excellent for pick-and-place operations such as loading and unloading parts from machines. Unlike nonservo robots, which have no control of velocity and operate with jerky motions, servo robots have smooth motions. Servo robots can control velocity, acceleration, and deceleration of each link as the manipulator moves from point to point. They can use programs that may branch to different sequences of motions depending on some condition measured at the time the robot is working.

Each axis of a servo robot is fundamentally a closed-loop servo control system. An example of a simple servo operation is illustrated in Fig. 15-26 on p. 422. In this example, the servo amplifier is responsible for amplifying the difference between the command voltage and the feedback voltage. The error signal produced is used to operate the servomotor, which is mechanically connected to the end effector and feedback potentiometer.

The *controller* contains the power supply, operator controls, control circuitry, and memory that direct the operation and

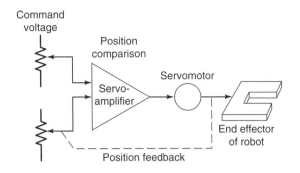

Fig. 15-26

Simple servo operation.

motion of the robot and communication with external devices. The controller has three major functional tasks to perform:

◆ Provide motion control signals for the manipulator unit (also known as signal processing)

◆ Provide storage for programmed events

◆ Interpret input/output signals, including operator instructions

Different controller configurations are used (Fig. 15-27). In general, the controller includes the following devices to operate the system:

◆**Operator Panel**
Equipped with lights, buttons, and keyswitches. Performs tasks such as powering up and powering down the system, calibrating the robot, resetting the controller after an error occurs, holding robot motion, starting or resuming automatic operation, and stopping the robot in an emergency.

◆**Teach Pendant**
Equipped with a keypad and liquid crystal display (LCD) screen and connected to the controller by a cable that plugs into the computer RAM board inside the controller. Performs tasks such as jogging the robot; teaching positional data; testing program execution;

recovering from errors; displaying user messages, error messages, prompts, and menus; displaying and modifying positional data, program variables, and inputs and outputs; and performing some operations on computer files.

◆**CRT Screen and Keyboard**
Resembles a standard computer terminal. Performs tasks such as performing computer file operations; displaying status and diagnostic information; and entering, translating, and debugging programs.

The robotic controller is a microprocessor-based system that operates in conjunction with input and output cards or modules. An increasing number of robots are controlled from programmable controllers. The programming format used can be either the coil and contact or the drum/sequencer type Before programming the PLC to control the

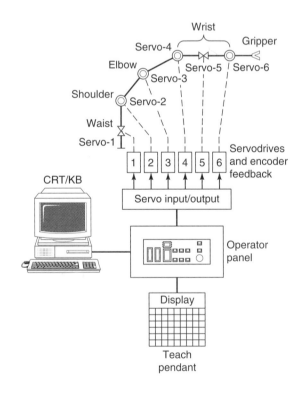

Fig. 15-27

Robot control system configuration.

robot, you must develop a scheme to connect and interface the PLC with the robot.

With the growing use of computers and PLCs in industry, the robot controller has become more important than the manipulator it controls. The robot controller is now required to communicate with devices outside itself, such as PLCs and plant computer systems. All major robot manufacturers have developed their own specialized program language for use with their robots. High-level, language-based robot controllers are usually ASCII-based systems that are compatible with the DOS environment. Compatibility here means that robot application engineers can develop their robot programs off-line in an office environment and later download their program to the robot using an RS-232-C serial interface from their personal computer.

Alternate programming methods called *walk-through* and *teach-through* can also be used. In walk-through teaching, the programmer actually takes hold of the manipulator and physically moves it through the maneuvers it is intended to learn. The controller records the moves for playback later, perhaps at a much higher speed. Teach-through programming involves using a joystick or teach pendant to guide the robot along the planned path. The controller calculates a smooth path for the robot using computer-based teach software.

Questions

1. Explain the function of each of the following computer hardware components:

a. Power supply	**d.** Motherboard	**g.** RAM chips
b. Floppy drives	**e.** Microprocessor chip	**h.** Peripheral cards
c. Hard drive	**f.** ROM chips	**i.** Expansion slots

2. What are the kinds of disk drives popularly used in personal computers?

3. List five precautions to be observed when handling floppy disks.

4. Compare the storage capacity and speed of hard disk drives with those of floppy disks.

5. How are the different disk drives of a computer identified?

6. Compare system software and application software.

7. List five tasks DOS allows you to do.

8. Describe the contents of each of the following types of files:

a. Executable	**c.** Text
b. Data	**d.** Graphics

9. Explain the function of the computer BIOS.

10. What does the term *booting* refer to?

11. Explain the function of directories.

12. What does the DOS prompt indicate?

13. Explain the function of each of the following DOS commands:

a. DIR	**e.** DEL	**i.** CD
b. CLS	**f.** TYPE	**j.** CHKDSK
c. COPY	**g.** FORMAT	
d. REN	**h.** DISKCOPY	

14. Explain how Windows allows for multitasking.

15. Name the two types of ports found on the back end of the computer case, and describe how each sends and retrieves data.

16. Explain the importance of each of the following in the selection of a computer:

a. Access time	**d.** Cache	**g.** Hard disk drive
b. Baud	**e.** Clock speed	**h.** Monitor
c. Bus	**f.** CPU	**i.** RAM

17. Describe the makeup of an industrial work cell.

18. List three types of LAN configurations or topologies.

19. Explain the importance of protocol in establishing communications between devices.

20. Compare master-slave and peer-to-peer communication formats.

21. Give a brief description of how numerical control devices operate.

22. What is the most popular NC programming language?

23. Numerical control is best suited for what types of industrial operations?

24. Explain the function of Cartesian coordinates in NC programming.

25. Compare the motions involved in NC point-to-point and contour programming.

26. List five improvements brought about by the introduction of computerized numerical control.

27. What does simple CNC programming consist of?

28. List five specialized tasks commonly performed by an industrial robot.

29. Give a brief description of the makeup of a robot.

30. Define the robot work envelope.

31. List three types of power that can be used to operate the robot manipulator.

32. What are the major tasks performed by a robotic system controller?

33. Describe three ways in which robots are programmed.

Problem

1. Determine as many of the following specifications as possible about the computer available to you for lab experiments:

   a. Number of floppy disk drives
   b. Type of disk drives
   c. Label designation for each drive
   d. Storage capability of each drive
   e. Access time of each drive
   f. Baud rate
   g. Bus structure
   h. Memory cache storage
   i. Clock speed
   j. Type of processor
   k. Type of monitor
   l. RAM memory capacity
   m. Version of DOS used

Glossary

Access To locate data stored in a programmable logic controller system or in computer related equipment.

Accumulated value The number of elapsed timed intervals or counted events.

Actuator An output device normally connected to an output module. Examples are an air valve and cylinder.

Address A code that indicates the location of data to be used by a program, or the location of additional program instructions.

Algorithm Mathematical procedure for problem solving.

Alphanumeric Term describing character strings consisting of any combination of letters, numerals, and/or special characters (e.g., A15$) used for representing text, commands, numbers, and/or code groups.

Alternating current (ac) input module An input module that converts various alternating current signals originating at user devices to the appropriate logic level signal for use within the processor.

Alternating current (ac) output module An output module that converts the logic level signal of the processor to a usable output signal to control a user alternating current device.

Ambient temperature The temperature of the air surrounding a module or system.

American National Standard Code for Information Interchange (ASCII) An 8-bit (7 bits plus parity) code that represents all characters of a standard typewriter keyboard, both uppercase and lowercase, as well as a group of special characters used for control purposes.

American National Standards Institute (ANSI) A clearinghouse and coordinating agency for voluntary standards in the United States.

American wire gauge (AWG) A standard system used for designating the size of electrical conductors. Gauge numbers have an inverse relationship to size; larger numbers have a smaller diameter.

Analog device Apparatus that measures continuous information (e.g., voltage-current). The measured analog signal has an infinite number of possible values. The only limitation on resolution is the accuracy of the measuring device.

Analog input module An input circuit that employs an analog-to-digital converter to convert an analog value, measured by an analog measuring device, to a digital value that can be used by the processor.

Analog output module An output circuit that employs a digital-to-analog converter to convert a digital value, sent from the processor, to an analog value that will control a connected analog device.

Analog signal Signal having the characteristic of being continuous and changing smoothly over a given range, rather than switching suddenly between certain levels, as with discrete signals.

Analog-to-digital (A/D) converter A circuit for converting a varying analog signal to a corresponding representative binary number.

AND (logic) A boolean operation that yields a logic 1 output if all inputs are 1, and a logic 0 if any input is 0.

Arithmetic capability The ability to do addition, subtraction, multiplication, division, and other advanced math functions with the processor.

Array A combination of panels, such as LEDs, coordinated in structure and function.

ASCII-input module Converts ASCII-code input information from an external peripheral into alphanumeric information a PLC can understand.

ASCII-output module Converts alphanumeric information from the PLC into ASCII code to be sent to an external peripheral.

Automatic control A process in which the output is kept at a desired level by using feedback from the output to control the input.

Auxiliary power supply A power supply not associated with the processor. Auxiliary power supplies are usually required to supply logic power to input/output racks and to other processor support hardware, and are often referred to as *remote power supplies.*

Backplane A printed circuit board, located in the back of a chassis, that contains a data bus, power bus, and mating connectors for modules to be inserted in the chassis.

BASIC A computer language using brief English-language statements to instruct a computer or microprocessor.

Battery indicator A diagnostic aid that provides a visual indication to the user, and/or an internal processor software indication, that the memory power-fail support battery is in need of replacement.

Baud A unit of signaling speed equal to the number of discrete conditions or signal events per second. Often defined as the number of binary digits transmitted per second.

BCD-input module Allows the processor to accept 4-bit BCD digital codes.

BCD-output module Enables a PLC to operate devices that require BCD-coded signals to operate.

Binary A number system using 2 as a base. The binary number system requires only two digits, zero (0) and one (1), to express any alphanumeric quantity desired by the user.

Binary-coded decimal (BCD) A system of numbering that expresses each individual decimal digit (0 through 9) of a number as a series of 4-bit binary notations. The binary-coded decimal system is often referred to as 8421 code.

Binary word A related group of 1's and 0's that has meaning assigned by position, or by numerical value in the binary system of numbers.

Bit An abbreviated term for the words *binary digit.* The bit is the smallest unit of information in the binary numbering system. It represents a decision between one of two possible and equally likely values or states. It is often used to represent an off or on state as well as a true or false condition.

Bit manipulation instructions A family of programmable logic controller instructions that exchange, alter, move, or otherwise modify the individual bits of single or groups of processor data memory words.

Bit storage   A user-defined data table area in which bits can be set or reset without directly affecting or controlling output devices. However, any storage bit can be monitored as necessary in the user program.

Block diagram   A method of representing the major functional subdivisions, conditions, or operations of an overall system, function, or operation.

Block format   A PLC screen format with a vertical rectangle on the right. Output and internal operation data are inserted into the rectangle during programming.

Block transfer   An instruction that copies the contents of one or more contiguous data memory words to a second contiguous data memory location. An instruction that transfers data between an intelligent input/output module or card and specified processor data memory locations.

Boolean algebra   A mathematical shorthand notation that expresses logic functions, such as AND, OR, EXCLUSIVE OR, NAND, NOR, and NOT.

Branch   A parallel logic path within a rung.

Buffer   In software terms, a register or group of registers used for temporary storage of data; a buffer is used to compensate for transmission rate differences between the transmitter and receiving device. In hardware terms, an isolating circuit used to avoid the reaction of one circuit with another.

Bug   A system defect or error that causes a malfunction. Can be caused by either software or hardware.

Burn   The process by which information is entered into programmable read-only memory.

Bus   A group of lines used for data transmission or control. Power distribution conductors.

Byte   A group of adjacent bits usually operated on as a unit, such as when moving data to and from memory. There are 8 bits per byte.

Cascading   A technique used when programming timers and counters to extend the timing or counting range beyond what would normally be available. This technique involves the driving of one timer or counter instruction from the output of another similar instruction.

Cathode-ray tube (CRT) terminal   A portable enclosure containing a cathode-ray tube, a special-purpose keyboard, and a microprocessor used to program a programmable logic controller.

Cell controller   A specialized computer used to control a work cell through multiple paths to the various cell devices.

Central processing unit (CPU)   That part of the programmable logic controller that governs system activities, including the interpretation and execution of programmed instructions. The central processing unit is also referred to as the *processor* or the *CPU*.

Character   A symbol that is one of a larger group of similar symbols and that is used to represent information on a display device. The letters of the alphabet and the decimal numbers are examples of characters used to convey information.

Chassis   A housing or framework used to hold assemblies. When the chassis is filled with one or more assemblies, it is often referred to as a *rack*.

Chip A tiny piece of semiconductor material on which electronic components are formed. Chips are normally made of silicon and are typically less than 1/4 in. square and 1/100 in. thick.

Clear An instruction or a sequence of instructions that removes all current information from a programmable logic controller's memory.

Clock A circuit that generates timed pulses to synchronize the timing of computer operations.

Clock rate The speed at which the microprocessor system operates.

Coaxial cable A transmission line constructed such that an outer conductor forms a cylinder around a central conductor. An insulating dielectric separates the inner and outer conductors, and the complete assembly is enclosed in a protective outer sheath. Coaxial cables are not susceptible to external electric and magnetic fields and generate no electric or magnetic fields of their own.

Code A system of communications that uses arbitrary groups of symbols to represent information or instructions. Codes are usually employed for brevity or security.

Coil Represents the output of a programmable logic controller. In the output devices it is the electrical coil that, when energized, changes the status of its corresponding contacts.

Coil format A PLC screen format with function coils on the right. Output and internal operational data are inserted into and around the coil.

Communication module Allows the user to connect the PLC to high-speed local networks that may differ from the network communication provided with the PLC.

Compare An instruction that compares the contents of two designated data memory locations of a programmable logic controller for equality or inequality.

Compatibility The ability of various specified units to replace one another with little or no reduction in capability. The ability of units to be interconnected and used without modification.

Complementary metal-oxide semiconductor (CMOS) CMOS-based logic offers lower power consumption and high-speed operation.

Computer Any electronic device that can accept information, manipulate it according to a set of preprogrammed instructions, and supply the results of the manipulation.

Computer integrated manufacturing (CIM) A manufacturing system controlled by an easily reprogrammable computer for flexibility and speed of change-over.

Computer interface A device designed for data communication between a programmable logic controller and a computer.

Contact The current-carrying part of an electric relay or switch; the contact engages to permit power flow and disengages to interrupt power flow to a load device.

Contact bounce The uncontrollable making and breaking of a contact during the initial engaging or disengaging of the contact.

Contact histogram An instruction sequence that monitors a designated memory bit or a designated input or output point for a change of state. A listing is generated by the instruction sequence that displays how quickly the monitored point is changing state.

Contactor A special-purpose relay designed to establish and interrupt the power flow of high-current electric circuits.

Contact symbology A set of symbols used to express the control program with conventional relay symbols.

Continuous current per module The maximum current for each module. The sum of the output current for each point should not exceed this value.

Continuous current per point The maximum current each output is designed to supply continuously to a load.

Control logic The control plan for a given system. The program.

Control relay A relay used to control the operation of an event or a sequence of events.

Core memory A type of memory system that employs ferrite cores to store information. Core memory operates by magnetizing the ferrite core in one direction to represent a 1, on, or true state, and in the opposite direction to represent a 0, off, or false state. This form of memory is nonvolatile.

Counter An electromechanical device in relay-based control systems that counts numbers of events for the purpose of controlling other devices based on the current number of counts recorded. A programmable logic controller instruction that performs the functions of its electromechanical counterpart.

Cross reference In ladder diagrams, letters or numbers to the right of coils or functions. The letters or numbers indicate where on other ladder lines contacts of the coil or function are located.

Crosstalk Undesired energy appearing in one signal path as a result of coupling from other signal paths or using a common return line.

Current The rate of electrical electron movement, measured in amperes.

Current-carrying capacity The maximum amount of current a conductor can carry without heating beyond a predetermined safe limit.

Current sinking Refers to an output device (typically an NPN transistor) that allows current flow from the load through the output to ground.

Current sourcing Output device (typically a PNP transistor) that allows current flow from the output through the load and then to ground.

Cursor The intensified or blinking element in the user program or file display. A means for indicating on the screen of a cathode-ray tube the point where data entry or editing occurs.

Cycle A sequence of operations repeated regularly. The time it takes for one such sequence to occur.

Data Information encoded in a digital form, which is stored in an assigned address of data memory for later use by the processor.

Data address A location in memory where data can be stored.

Data file A group of data memory words acted on as a group rather than singly.

Data highway A communications network that allows devices such as PLCs to communicate. They are normally proprietary, which means that only like devices of the same brand can communicate over the highway.

Data link The equipment that makes up a data communications network.

Data manipulation The process of exchanging, altering, or moving data within a programmable logic controller or between programmable logic controllers.

Data manipulation instructions A classification of processor instructions that alter, exchange, move, or otherwise modify data memory words.

Data table The part of processor memory that contains input and output values as well as files where data is monitored, manipulated, and changed for control purposes.

Data transfer The process of moving information from one location to another, i.e., from register to register, from device to device, and so forth.

Data transmission line A medium for transferring signals over a distance.

Debouncing The act of removing intermediate noise states from a mechanical switch.

Debug The process of locating and removing mistakes from a software program or from hardware interconnections.

Decimal number system A number system that uses ten numeral digits (decimal digits), 0, 1, 2, 3, 4, 5, 6, 7, 8, 9. Each digit position has a place value of 1, 10, 100, 1000, and so on, beginning with the least significant (rightmost) digit. Base 10.

Decrement The act of reducing the contents of a storage location or value in varying increments.

Diagnostic program A user program designed to help isolate hardware malfunctions in the programmable logic controller and the application equipment.

Diagnostics The detection and isolation of an error or malfunction.

Digital device One that processes discrete electric signals.

Digital gate A device that analyzes the digital states of its inputs and outputs and an appropriate output state.

Digital signals A system of discrete states: high or low, on or off, 1 or 0.

Digital-to-analog converter An electrical circuit that converts binary bits to a representative, continuous analog signal.

Dip switch A group of small, in-line on-off switches. From *dual in-line package.*

Discrete I/O A group of input and/or output modules that operate with on/off signals, as contrasted to analog modules that operate with continuously variable signals.

Disk drive The device that writes or reads data from a magnetic disk.

Diskette The flat, flexible disk on which a disk drive writes and reads.

Display The image that appears on a cathode-ray tube screen or on other image projection systems.

Display menu The list of displays from which the user selects specific information for viewing.

Distributed control A method of dividing process control into several subsystems. A PLC oversees the entire operation.

Divide A programmable logic controller instruction that performs a numerical division of one number by another.

Documentation An orderly collection of recorded hardware and software data such as tables, listings, and diagrams to provide reference information for programmable logic controller application operation and maintenance.

Down counter A counter that starts from a specified number and increments down to zero.

Download Loading data from a master listing to a readout or another position in a computer system.

Dry-contact-output module Enables a PLC's processor to control output devices by providing a contact isolated electrically from any power source.

Edit The act of modifying a programmable logic controller program to eliminate mistakes and/or simplify or change system operation.

Electrically erasable programmable read-only memory (EEPROM) A type of programmable read-only memory that is programmed and erased by electrical pulses.

Electrical optical isolator A device that couples input to output using a semiconductor light source and detector in the same package.

Electromagnetic interference (EMI) A phenomenon responsible for noise in electric circuits.

Element A single instruction of a relay ladder diagram program.

Emergency stop relay A relay used to inhibit all electric power to a control system in an emergency or other event requiring that the controlled hardware be brought to an immediate halt.

Enable To permit a particular function or operation to occur under natural or preprogrammed conditions.

Enclosure A steel box with a removable cover or hinged door used to house electric equipment.

Encoder A rotary device that transmits position information. A device that transmits a fixed number of pulses for each revolution.

Energize The physical application of power to a circuit or device to activate it. The act of setting the on, true, or 1 state of a programmable logic controller's relay ladder diagram output device or instruction.

Erasable programmable read-only memory (EPROM) A programmable read-only memory that can be erased with ultraviolet light, then reprogrammed with electrical pulses.

Error signal A signal proportional to the difference between the actual output and the desired output.

Even parity When the sum of the number of 1's in a binary word is always even.

Examine if closed (XIC) Refers to a normally open contact instruction in a logic ladder program. An *examine if closed* instruction is true if its addressed bit is on (1). It is false if the bit is off (0).

Examine if open (XIO) Refers to a normally closed contact instruction in a logic ladder program. An *examine if open* instruction is true if its addressed bit is off (0). It is false if the bit is on (1).

Exclusive OR gate A logic device requiring one or the other, but not both, of its inputs to be satisfied before activating its output.

Execution The performance of a specific operation accomplished through processing one instruction, a series of instructions, or a complete program.

Execution time The total time required for the execution of one specific operation.

False As related to programmable logic controller instructions, a disabling logic state.

Fault Any malfunction that interferes with normal operation.

Fault indicator A diagnostic aid that provides a visual indication and/or an internal processor software indication that a fault is present in the system.

Fault-routine file A special subroutine that, if assigned, executes when the processor has a major fault.

Feedback In analog systems, a correcting signal received from the output or an output monitor. The correcting signal is fed to the controller for process correction.

File A formatted block of data treated as a unit.

Floppy disk A recording disk used with a computer disk drive for recording data. The disk is flexible, not rigid.

Flow chart A graphical representation for the definition, analysis, or solution of a problem. Symbols are used to represent a process or sequence of decisions and events.

Force function A mode of operation or instruction that allows an operator to override the processor to control the state of a device.

Force off function A feature that allows the user to reset an input image table file bit or de energize an output, independent of the programmable logic controller program.

Force on function A feature that allows the user to set an image table file bit or energize an output, independent of the programmable logic controller program.

Full duplex A mode of data communications in which data may be transmitted and received simultaneously.

Functional block instruction set A set of instructions that moves, transfers, compares, or sequences blocks of data.

Gate A circuit having two or more input terminals and one output terminal, where an output is present when and only when the prescribed inputs are present.

GET instruction A programmable logic controller instruction that fetches the contents of a specified data memory word. GET instructions are often used to fetch data prior to programming mathematical and data manipulation instructions.

Glitch A voltage or current spike of short duration that adversely affects the operation of a PLC.

Gray code A binary coding scheme that allows only 1 bit in the data word to change state at each increment of the code sequence.

Gray-encoder module Converts the gray-code signal from an input device into straight binary.

Ground A conducting connection between an electric circuit or equipment chassis and the earth ground.

Ground potential Zero voltage potential with respect to the ground.

Half duplex A mode of data transmission that communicates in two directions, but in only one direction at a time.

Handshaking The method by which two digital machines establish communication.

Hard contacts Any type of physical switch contacts.

Hard copy Any form of a printed document such as a ladder diagram program listing, paper tape, or punched cards.

Hardware The mechanical, electric, and electronic devices that make up a programmable logic controller and its application.

Hardwired The physical interconnection of electric and electronic components with wire.

Hexadecimal A number system having a base of 16. This numbering system requires 16 elements for representation, and thus uses the decimal digits zero (0) through nine (9) and the first six letters of the alphabet, A through F.

High-level language A powerful set of user-oriented instructions in which each statement may translate into a series of instructions or subroutines in machine language.

High = true A signal type where the higher of two voltages indicates a logic state of on (1).

Histogram A graphic representation of the frequency at which an event occurs.

Host computer A main computer that controls other computers, PLCs, or computer peripherals.

Image table An area in programmable logic controller memory dedicated to input/output data. Ones and zeros (1's and 0's) represent on and off conditions, respectively. During every input/output scan, each input controls a bit in the input image table file; each output is controlled by a bit in the output image table file.

Immediate input instruction A programmable logic controller instruction that temporarily halts the user program scan so that the processor can update the input image table file with the current status of one or more user-specified input points.

Immediate output instruction A programmable logic controller instruction that temporarily halts the user program scan so that the current status of one or more user-specified output points can be updated to current output image table file status by the processor.

Impedance The total resistive and inductive opposition that an electric circuit or device offers to a varying current at a specified frequency. Impedance is measured in ohms (Ω) and is denoted by the symbol Z.

Increment The act of increasing the contents of a storage location or value in varying amounts.

Inductance   A circuit property that opposes any current change. Inductance is measured in henrys and is represented by the letter H.

Industrial terminal   The device used to enter and monitor the program in a PLC.

Input   Information transmitted from a peripheral device to the input module, and then to the data table.

Input devices   Devices such as limit switches, pressure switches, pushbuttons, and analog and/or digital devices that supply data to a programmable logic controller.

Input/output (I/O) address   A unique number assigned to each input and output. The address number is used when programming, monitoring, or modifying a specific input or output.

Input/output (I/O) module   A plug-in type assembly that contains more than one input or output circuit. A module usually contains two or more identical circuits. Normally it contains 2, 4, 8, or 16 circuits.

Input/output (I/O) scan time   The time required for the processor to monitor inputs and control outputs.

Input/output (I/O) update   The continuous process of revising each and every bit in the input and output tables, based on the latest results from reading the inputs and processing the outputs according to the control program.

Instruction   A command that causes a programmable logic controller to perform one specific operation. The user enters a combination of instructions into the programmable logic controller's memory to form a unique application program.

Instruction set   The set of general-purpose instructions available with a given controller. In general, different machines have different instruction sets.

Integrated circuit (IC)   A circuit in which all components are integrated on a single tiny silicon chip.

Intelligent input/output module   A microprocessor-based module that performs processing or sophisticated closed-loop application functions.

Interface   A circuit that permits communication between the central processing unit and a field input or output device. Different devices require different interfaces.

Interlock   A system for preventing one element or device from turning on while another device is on.

Internal coil instruction   A relay coil instruction used for internal storage or buffering of an on/off logic state. An internal coil instruction differs from an output coil instruction because the on/off status of the internal coil is not passed to the input/output hardware for control of a field device.

Inversion   Conversion of a high level to a low level, or vice versa.

Inverter   The digital circuit that performs inversion.

I/O group   A logically addressed unit consisting of sixteen input points and sixteen output points. Eight I/O groups make up one rack.

Isolated input/output (I/O) circuits   Input and output circuits that are electrically isolated from any and all other circuits of a module. Isolated input/output circuits are designed to allow field devices that are powered from different sources to be connected to one module.

Jumper A short length of conduit used to make a connection between terminals around a break in a circuit.

Jump instruction An instruction that permits the bypassing of selected portions of the user program. *Jump* instructions are conditional whenever their operation is determined by a set of preconditions, and unconditional whenever they are executed to occur every time they are programmed.

K $2^{10} = 1K = 1024$. Used to denote size of memory and can be expressed in bits, bytes, or words. Example: 2K = 2048.

k Kilo. A prefix used with units of measurement to designate quantities 1000 times as great.

Keyboard The alphanumeric keypad on which the user types instructions to the PLC.

Keying Keying bands installed on backplane connectors to ensure that only one type of module can be inserted into a keyed connector.

Label instruction A programmable logic controller instruction that assigns an alphanumeric designation to a particular location in a program. This location is used as the target of a *jump, skip,* or *jump to subroutine* instruction.

Ladder diagram An industry standard for representing relay logic control systems. The diagram resembles a ladder because the vertical supports of the ladder appear as power feed and return buses, and the horizontal rungs of the ladder appear as series and/or parallel circuits connected across the power lines.

Ladder diagram programming A method of writing a user program in a format similar to a relay ladder diagram.

Ladder matrix A rectangular array of programmed contacts that defines the number of contacts that can be programmed across a row and the number of parallel branches allowed in a single ladder rung.

Language A set of symbols and rules for representing and communicating information among people, or between people and machines. The method used to instruct a programmable device to perform various operations. Examples include boolean and ladder contact symbology.

Language module Enables the user to write programs in a high-level language. BASIC is the most popular language module. Other language modules available include C, FORTH, and PASCAL.

Latching relay A relay that maintains a given position by mechanical or electrical means until released mechanically or electrically.

Latch instruction One-half of an instruction pair (the second instruction of the pair being the *unlatch* instruction) that emulates the latching action of a latching relay. The *latch* instruction for a programmable logic controller energizes a specified output point or internal coil until it is de-energized by a corresponding *unlatch* instruction.

Leakage The small amount of current that flows in a semiconductor device when it is in the off state.

Least significant bit (LSB) The bit that represents the smallest value in a nibble, byte, or word.

Least significant digit (LSD) The digit that represents the smallest value in a byte or word.

Light-emitting diode (LED) A semiconductor PN-type junction that emits light when biased in the forward direction.

Light-emitting diode (LED) display A display device incorporating light-emitting diodes to form the segments of the displayed characters and numbers.

Limit switch An electric switch actuated by some part and/or motion of a machine or equipment.

Line A component part of a system used to link various subsystems located remotely from the processor. The source of power for operation. Example: 120V alternating current line.

Line-powered sensor Normally, three-wire sensors, although four-wire sensors also exist. The line-powered sensor is powered from the power supply. A separate wire (the third) is used for the output line.

Liquid-crystal display (LCD) A display device using reflected light from liquid crystals to form the segments of the displayed characters and numbers.

Load The power used by a machine or apparatus. To place data into an internal register under program control. To place a program from an external storage device into central memory under operator control.

Load-powered sensor A load-powered sensor has two wires. A small leakage current flows through the sensor even when the output is off. The current is required to operate the sensor electronics.

Load resistor A resistor connected in parallel with a high-impedance load so that the output circuit driving the load can provide at least the minimum current required for proper operation.

Local area network (LAN) A system of hardware and software designed to allow a group of intelligent devices to communicate within a fairly close proximity.

Local input/output (I/O) A programmable logic controller whose input/output distance is physically limited. The PLC must be located near the process; however, the PLC may still be mounted in a separate enclosure.

Local power supply The power supply used to provide power to the processor and a limited number of local input/output modules.

Location In reference to memory, a storage position or register identified by a unique address.

Logic A process of solving complex problems through the repeated use of simple functions that can be either true or false. The three basic logic functions are AND, OR, and NOT.

Logic diagram A diagram that represents the logic elements and their interconnections.

Logic level The voltage magnitude associated with signal pulses representing 1's and 0's in binary computation.

Loop control A control of a process or machine that uses feedback. An output status indicator modifies the input signal effect on the process control.

Loop resistance The total resistance of two conductors measured at one end (conductor and shield, twisted pair, conductor and armor).

Low A state of being off, 0, or false.

Low = true A signal type where the lower of two voltages indicates a logic state of on (1).

Machine language A programmable language using the binary form.

Magnetic disk A flat, circular plate with a magnetic surface on which data can be stored by selective polarization.

Magnetic tape Tape made of plastic and coated with magnetic material; used to store information.

Malfunction Any incorrect function within electronic, electric, or mechanical hardware.

Manipulation The process of controlling and monitoring data table bits, bytes, or words by means of the user program to vary application functions.

Manufacturing automation protocol (MAP) Standard developed to make industrial devices communicate more easily.

Masking A means of selectively screening out data. Masking allows unused bits in a specific instruction to be used independently.

Mass storage A means of storing large amounts of data on magnetic tape, floppy disks, etc.

Master control relay (MCR) A mandatory hardwired relay that can be de-energized by any series-connected emergency stop switch. Whenever the master control relay is de-energized, its contacts open to de-energize all application input and output devices.

Master control relay (MCR) zones User program areas where all nonretentive outputs can be turned off simultaneously. Each master control relay zone must be delimited and controlled by master control relay fence codes (master control relay instructions).

Matrix A logic network that is an intersection of input and output connection points.

Memory That part of the programmable logic controller where data and instructions are stored either temporarily or semipermanently. The control program is stored in memory.

Memory map A diagram showing a system's memory addresses and what programs and data are assigned to each section of memory.

Memory protect A hardware circuit incorporated into PLC systems to protect user programs. Generally, a key switch mechanism.

Menu A list of programming selections displayed on a programming terminal.

Metal-oxide semiconductor (MOS) A semiconductor device in which an electric field controls the conductance of a channel under a metal electrode called a *gate*.

Metal oxide varistor (MOV) Used for suppressing electrical power surges.

Microprocessor A central processing unit manufactured on a single integrated-circuit chip (or several chips) by utilizing large-scale integration technology.

Microsecond One millionth of a second $= 1 \times 10^{-6}$ second $= 0.000001$ second.

Millisecond One thousandth of a second $= 1 \times 10^{-3}$ second $= 0.001$ second.

Mnemonic A term, usually an abbreviation, that is easy to remember and pronounce.

Mnemonic code A code in which information is represented by symbols or characters.

Mode A term used to refer to the selected operating method such as automatic, manual, TEST, PROGRAM, or diagnostic.

Module An interchangeable, plug-in item containing electronic components.

Module addressing A method of identifying the input/output modules installed in a chassis.

Module group Two or more modules that, as a group, perform a specific function or operation, or are thought of as a single unit.

Monitor Any display device incorporating a cathode-ray tube as the primary display medium. The act of listening to or observing the operation of a system or device.

Most significant bit (MSB) The bit representing the greatest value of a nibble, byte, or word.

Most significant digit (MSD) The digit representing the greatest value of a byte or word.

Motor controller or starter A device or group of devices that serve to govern, in a predetermined manner, the electric power delivered to a motor.

Motor starter A special relay designed to provide power to motors; it has both a contactor relay and an overload relay connected in series and prewired so that, if the overload operates, the contactor is de-energized.

Multiplexing The time-shared scanning of a number of data lines into a single channel. Only one data line is enabled at any time. The incorporation of two or more signals into a single wave from which the individual signals can be recovered.

Multiply instruction A programmable logic controller instruction that provides for the mathematical multiplication of two numbers.

Multiprocessing A method of applying more than one microprocessor to a specific function to speed up operation time and reduce the possibility of system failure.

National Electrical Code (NEC) A set of regulations, developed by the National Fire Protection Association, that govern the construction and installation of electric wiring and electric devices. The National Electrical Code is recognized by many governmental bodies, and compliance is mandatory in much of the United States.

National Electrical Manufacturers Association (NEMA) An organization of electric device and product manufacturers. The National Electrical Manufacturers Association issues standards relating to the design and construction of electric devices and products.

NEMA Type 12 enclosure A category of industrial enclosures intended for indoor use and designed to provide a degree of protection against dust, falling dirt, and dripping noncorrosive liquids. They do not provide protection against conditions such as internal condensation.

Nested branches A branch that begins or ends within another branch.

Network A series of stations or devices connected by some type of communications medium.

Node In hardware, a connection point on the network. In programming, the smallest possible increment in a ladder diagram.

Noise Random, unwanted electric signals, normally caused by radio waves or electric or magnetic fields generated by one conductor and picked up by another.

Noise filter or **noise suppressor** An electronic filter network used to reduce and/or eliminate any noise that may be present on the leads to an electric or electronic device.

Noise immunity A measure of insensitivity of an electronic system to noise.

Noise spike A short burst of electric noise with more magnitude than the background noise level.

Nonretentive output An output controlled continuously by a program rung. Whenever the rung changes state (true or false), the output turns on or off. Contrasted with a retentive output, which remains in its last state (on or off) depending on which of its two rungs, latch or unlatch, was last true.

Nonvolatile memory A memory designed to retain its data while its power supply is turned off.

NOR The logic gate that results in zero unless both inputs are zero.

Normally closed contact (NC) A contact that is conductive when its operating coil is not energized.

Normally open contact (NO) A contact that is nonconductive when its operating coil is not energized.

NOT A logical operation that yields a logic 1 at the output if a logic 0 is entered at the input, and a logic 0 at the output if a logic 1 is entered at the input. The NOT, also called the *inverter,* is normally used in conjunction with the AND and OR functions.

Octal numbering system A numbering system that uses only the digits 0 through 7. Also called base 8.

Odd parity Condition when the sum of the number of 1's in a binary word is always odd.

Off-delay timer An electromechanical relay with contacts that change state a predetermined time period after power is removed from its coil; on re-energization of the coil, the contacts return to their shelf state immediately. A programmable logic controller instruction that emulates the operation of the electromechanical off-delay relay.

Off-line Equipment or devices that are not connected to, or do not directly communicate with, the central processing unit.

Off-line programming and/or **off-line editing** A method of programmable logic controller programming and/or editing where the operation of the processor is stopped and all output devices are switched off. Off-line programming is the safest way to develop or edit a programmable logic controller program since the entry of instructions does not affect operating hardware until the program can be verified for accuracy of entry.

On-delay timer An electromechanical relay with contacts that change state a predetermined time period after the coil is energized; the contacts return to their shelf state immediately on de-energization of the coil. A programmable logic controller instruction that emulates the operation of the electromechanical on-delay timer.

One-shot A programmed technique that sets a storage bit or output for only one program scan.

On-line Equipment or devices that communicate with the device they are connected to.

On-line data change Allows the user to change various data table values using a peripheral device while the application is operating normally.

On-line programming and/or **on-line editing** The ability of a processor and programming terminal to make joint user-directed additions, deletions, or changes to a user program while the processor is actively solving and executing the commands of the existing user program. Extreme care should be exercised when performing on-line programming to ensure that erroneous system operation does not result.

Operand A number used in an arithmetic operation as an input.

Operating system The fundamental software for a system that defines how it will store and transmit information.

Operational amplifier (op-amp) A high-gain dc amplifier used to increase signal strength for devices such as analog input modules.

Optical coupler A device that couples signals from one circuit to another by means of electromagnetic radiation, usually infrared or visible. A typical optical coupler uses a light-emitting diode to convert the electric signal of the primary circuit into light, and a phototransistor in

the secondary to reconvert the light back into an electric signal. Sometimes referred to as optical isolation.

Optical isolation Electrical separation of two circuits with the use of an optical coupler.

OR A logical operation that yields a logic 1 output if one of any number of inputs is 1, and a logic 0 if all inputs are 0.

Output Information sent from the processor to a connected device via some interface. The information could be in the form of control data that will signal some device such as a motor to switch on or off, or vary the speed of a drive.

Output device Any connected equipment that will receive information or instructions from the central processing unit, such as control devices (e.g., motors, solenoids, alarms) or peripheral devices (e.g., line printers, disk drives, displays). Each type of output device has a unique interface to the processor.

Output image table file A portion of a processor's data memory reserved for the storage of output device statuses. A 1, on, or true state in an output image table file storage location is used to switch on the corresponding output point.

Output instruction The term applied to any programmable logic controller instruction capable of controlling the discrete or analog status of an output device connected to the programmable logic controller.

Output register or **output word** A particular word in a processor's output image table file where numerical data are placed for transmission to a field output device.

Overflow Exceeding the numerical capacity of a device such as a timer or counter. The overflow can be either a positive or negative value.

Overload A load greater than the one that a component or system is designed to handle.

Overload relay A special-purpose relay designed so that its contacts transfer whenever its current exceeds a predetermined value. Overload relays are used with electric motors to prevent motor burnout due to mechanical overload.

Parallel circuit A circuit in which two or more of the connected components or contact symbols in a ladder program are connected to the same pair of terminals so that current may flow through all the branches. Contrasted with a series connection, where the parts are connected end to end so that current flow has only one path.

Parallel instruction A programmable logic controller instruction used to begin and/or end a parallel branch of instructions programmed on a programming terminal.

Parallel operation A type of information transfer where all bits, bytes, or words are handled simultaneously.

Parallel transmission A computer operation in which two or more bits of information are transmitted simultaneously.

Parity The use of a self-checking code employing binary digits in which the total number of 1's is always even or odd.

PC Personal computer.

Peripheral equipment Units that communicate with the programmable logic controller but are not part of the programmable logic controller. Example: a programming device or computer.

PID module Proportional integral-derivative closed-loop control lets the user hold a process variable at a desired set point.

Pilot-type device Used in a circuit as a control apparatus to carry electric signals for directing performance. This device does not carry primary current.

Polarity The directional indication of electrical flow in a circuit. The indication of charge as either positive or negative, or the indication of a magnetic pole as either north or south.

Port A connector or terminal strip used to access a system or circuit. Generally ports are used for the connection of peripheral equipment.

Positive logic The use of binary logic in such a way that 1 represents a positive logic level (e.g., 1 = +5 V, 0 = 0 V). This is the conventional use of binary logic.

Power supply A device used to convert an alternating current or direct current voltage of specific value to one or more direct current voltages of a specified value and current capacity. The power supplies designed for use with programmable logic controllers convert 120 or 240 V alternating current to the direct current voltages necessary to operate the processor and input/output hardware.

Preset value (PR) The number of time intervals or events to be counted.

Pressure switch A switch activated at a specified pressure.

Printed circuit board A glass-epoxy card with copper foils for electric conductors and electronic components.

Process A continuous manufacturing operation.

Program A sequence of instructions to be executed by the processor to control a machine or process.

Program files The area of processor memory where the ladder logic programming is stored.

Programmable controller A computer that has been hardened to work in an industrial environment and is equipped with special I/O and a control programming language.

Programmable read-only memory (PROM) A retentive memory used to store data. This type of memory device can be programmed only once and cannot be altered afterward.

Programming terminal Also known as a *programmer*. A combination of keyboard and CRT used to insert, modify, and observe programs stored in a PLC.

Proportional-integral derivative (PID) A mathematical formula that provides a closed-loop control of a process. Inputs and outputs are continuously variable and typically will be analog signals.

Protocol A formal definition of establishing criteria for receiving and transmitting data through communications channels.

Proximity switch An input device that senses the presence or absence of a target without physical contact.

Pulse A short change in the value of a voltage or current level. A pulse has a definite rise and fall time and a finite duration.

PUT instruction A programmable logic controller instruction that places the data retrieved by a get instruction in a data memory location specified as part of the put instruction.

Rack A housing or framework used to hold assemblies. A plastic and/or metal assembly that supports input/output modules and provides a means of supplying power and signals to each input/output module or card.

Rack fault A red diagnostic indicator that lights to signal a loss of communication between the processor and any remote input/output chassis. The condition based on the loss of communication.

Random-access memory (RAM) A memory system that permits the random accessing of any storage location for the purpose of either storing (writing) or retrieving (reading) information. Random-access memory systems allow the data to be retrieved and stored at speeds independent of the storage locations being accessed.

Read The accessing of information from a memory system or data storage device. The gathering of information from an input device or devices or a peripheral device.

Read-only memory (ROM) A permanent memory structure where data are placed in the memory's storage locations at time of fabrication, or by the user at a speed much slower than it will be read. Information entered in a read-only memory is usually not changed once it is entered.

Read/write memory Memory where data can be stored (write mode) or accessed (read mode). The write mode replaces previously stored data with current data; the read mode does not alter stored data.

Real-time clock (RTC) A device that continually measures time in a system without respect to what tasks the system is performing.

Rectifier A solid-state device that converts alternating current to pulsed direct current.

Register A memory word or area for the temporary storage of data used within mathematical, logical, or transfer functions.

Relay An electrically operated device that mechanically switches electric circuits.

Relay contacts The contacts of a relay that are either opened and/or closed according to the condition of the relay coil. Relay contacts are designated as either normally open or normally closed in design.

Relay logic A representation of the program or other logic in a form normally used for relays.

Remote input/output (I/O) system Any input/output system that permits communication between the processor and input/output hardware over a coaxial or twin axial cable. Remote input/output systems permit the placement of input/output hardware at any distance from the processor.

Report An application data display or printout containing information in a user-designed format. Reports can include operator messages, part records, and production lists. Initially entered as messages, reports are stored in a memory area separate from the user program.

Report generation The printing or displaying of user-formatted application data by means of a data terminal. Report generation can be initiated by means of either a user program or a data terminal keyboard.

Resolution The smallest distinguishable increment into which a quantity is divided.

Response time The amount of time required for a device to react to a change in its input signal, or to a request.

Retentive instruction Any programmable logic controller instruction that does not need to be continuously controlled for operation. Loss of power to the instruction does not halt execution or operation of the instruction.

Retentive timer An electromechanical relay that accumulates time whenever the device receives power, and maintains the current time should power be removed from the device. Loss of power to the device after reaching its preset value does not affect the state of the contacts.

Retentive timer instruction A programmable logic controller instruction that emulates the timing operation of the electromechanical retentive timer.

Retentive timer reset instruction A programmable logic controller instruction that emulates the reset operation of the electromechanical retentive timer.

Reverse video A cathode-ray tube display characterized by black alphanumeric characters on a white background, as contrasted to the standard display of white alphanumeric characters on a black background.

Routine A series of instruction that performs a specific function or task.

RS-232 An Electronic Industries Association (EIA) standard for data transfer and communication for serial binary communication circuits.

Run The single, continuous execution of a program by a programmable logic controller.

Rung A group of programmable logic controller instructions that controls an output or storage bit, or performs other control functions such as file moves, arithmetic, and/or sequencer instructions. This is represented as one section of a ladder logic diagram.

Scan time   The time required to read all inputs, execute the control program, and update local and remote input and output statuses. This is effectively the time required to activate an output controlled by programmed logic.

Schematic   A diagram of graphic symbols representing the electrical scheme of a circuit.

Screen   The viewing surface of a cathode-ray tube, where data is displayed.

Search function   Allows the user to display quickly any instruction in the programmable logic controller program.

Self-diagnostic   The hardware and firmware within a controller that monitors its own operation and indicates any fault it can detect.

Sensor   A device used to gather information by the conversion of a physical occurrence to an electric signal.

Sequencer   A mechanical, electric, or electronic device that can be programmed so that a predetermined set of events occurs repeatedly.

Sequence table   A table or chart indicating the sequence of operation of output devices.

Sequential control   A process that dictates the correct order of events and allows one event to occur only after the completion of another.

Serial communication   A type of information transfer where the bits are handled sequentially. Contrasted with parallel communication.

Series circuit   A circuit in which the components or contact symbols are connected end to end, and all must be closed to permit current flow.

Servo module   The device whose feedback is used to accomplish closed-loop control. Though programmed through a PLC, once programmed it can control a device independently without interfering with the PLC's normal operation.

Shield   A barrier, usually conductive, that substantially reduces the effect of electric and/or magnetic fields.

Shift   To move binary data within a shift register or other storage device.

Shift register   A PLC function capable of storing and shifting binary data.

Short circuit   An undesirable path of very low resistance in a circuit between two points.

Short-circuit protection   Any fuse, circuit breaker, or electronic hardware used to protect a circuit or device from severe overcurrent conditions or short circuits.

Signal   The event or electrical quantity that conveys information from one point to another.

Significant digit   A digit that contributes to the precision of a number. The number of significant digits is counted beginning with the digit contributing the most value, called the *most significant digit* (leftmost), and ending with the digit contributing the least value, called the *least significant digit* (rightmost).

Silicon-controlled rectifier (SCR) A semiconductor device that functions as an electronic switch.

Single-scan function A supervisory instruction that causes the control program to be executed for one scan, including input/output update. This troubleshooting function allows step-by-step inspection of occurrences while the machine is stopped.

Sink mode output A mode of operation of solid-state devices in which the device controls the current from the load. For example, when the output is energized, it connects the load to the negative polarity of the supply.

Snubber A circuit generally used to suppress inductive loads; it consists of a resistor in series with a capacitor (RC snubber) and/or a MOV placed across the alternating current load.

Software Programs that control the processing of data in a system, as contrasted to the physical equipment itself (hardware).

Solid state Any circuit or component that uses semiconductors or semiconductor technology for operation.

Solid-state switch Any electronic device incorporating a transistor, silicon-controlled rectifier, or triode alternating current semiconductor switch to control the on/off flow of electric power is often referred to as containing a solid-state contact or switch.

Source mode output A mode of operation of solid-state output devices in which the device controls the current to the load. For example, when the output is energized, it connects the load to the positive polarity of the supply.

State The logic 0 or 1 condition in programmable logic controller memory or at a circuit input or output.

Station Any programmable logic controller, computer, or data terminal connected to, and communicating by means of, a data highway.

Status The operating condition of a device, usually on or off.

Stepper-motor module Provides pulse trains to a stepper-motor translator that enables control of a stepper motor.

Storage bit A bit in a data table word that can be set or reset, but is not associated with a physical input or output terminal point.

Subroutine A portion of a larger user program that may be accessed and executed any number of times during a single scan of the programmable logic controller.

Subtract A programmable logic controller instruction that performs the mathematical subtraction of one number from another.

Suppression device A unit that attenuates the magnitude of electrical noise.

Surge A transient wave of current or power.

Synchronous shift register A shift register where only one change of state occurs per control pulse.

Synchronous transmission A type of serial transmission that maintains a constant time interval between successive events.

Syntax Rules governing the structure of a language.

System A set of one or more programmable logic controllers, input/output devices and modules, and computers, with associated software, peripherals. terminals, and communication networks, that together provide a means of performing information processing for controlling machines or processes.

Tasks A set of instructions, data, and control information capable of execution by a central processing unit to accomplish a specific purpose.

Terminal address The alphanumeric address assigned to a particular input or output point. It is also related directly to a specific image table bit address.

Thermocouple A temperature-measuring device that utilizes two dissimilar metals for temperature measurement. As the junction of the two dissimilar metals is heated, a proportional voltage difference, which can be measured, is generated.

Thumbwheel switch A rotating switch used to input numeric information into a controller.

Time base A unit of time generated by a microprocessor's clock circuit and used by PLC timer instructions. Typical time bases are 0.01, 0.1, and 1.0 seconds.

Timed contact A normally open and/or normally closed contact that is actuated at the end of a timer's time-delay period.

Timer In relay-panel hardware, an electromechanical device that can be wired and preset to control the operating interval of other devices. In a programmable logic controller, a timer is internal to the *processor;* that is, it is controlled by a user-programmed instruction.

Toggle switch A panel-mounted switch with an extended lever; normally used for on/off switching.

Token The logical right to initiate communications in a communication network.

Token passing A technique where tokens are circulated among nodes in a communication network.

Topology The structure of a communications network, i.e., ring, bus, etc.

Transducer A device used to convert physical parameters such as temperature, pressure, and weight into electric signals.

Transformer An electric device that converts a circuit's electrical energy into a circuit or circuits with different voltages and current ratings.

Transistor A three-terminal active semiconductor device composed of silicon or germanium that is capable of switching or amplifying an electric current.

Transistor-transistor logic (TTL) A semiconductor logic family in which the basic logic element is a multiple-emitter transistor. This family of devices is characterized by high speed and medium power dissipation.

Transitional contact A contact that, depending on how it is programmed, will be on for one program scan every 0 to 1 transition, or every 1 to 0 transition of the referenced coil.

Transmission line A system of one or more electric conductors used to transmit electric signals or power from one place to another.

TRIAC A solid-state component capable of switching alternating current.

True As related to programmable logic controller instructions, an on, enabled, or 1 state.

Truth table A table listing that shows the state of a given output as a function of all possible input combinations.

TTL-input module Enables devices that produce TTL-level signals to communicate with a PLC's processor.

TTL-output module Enables a PLC to operate devices requiring TTL-level signals to operate.

Two's complement A numbering system used to express positive and negative binary numbers.

Unlatch instruction One-half of a programmable logic controller instruction pair that emulates the unlatching action of a latching relay. The unlatch instruction de-energizes a specified output point or internal coil until re-energized by a latch instruction. The output point or internal coil remains de-energized regardless of whether or not the unlatch instruction is energized.

Ultraviolet-erasable programmable read-only memory An erasable programmable read-only memory that can be cleared (set to 0) by exposure to intense ultraviolet light. After being cleared, it may be reprogrammed.

Up counter An event that starts from 0 and increments up to the preset value.

Variable A factor that can be altered, measured, or controlled.

Variable data Numerical information that can be changed during application

operation. It includes timer and counter accumulated values, thumbwheel settings, and arithmetic results.

Volatile memory A memory structure that loses its information whenever power is removed. Volatile memories require a battery backup to ensure memory retention during power outages.

Watchdog timer Monitors logic circuits controlling the processor. If the watchdog timer, which is reset every scan, ever times out, the processor is assumed to be faulty, and is disconnected from the process.

Word A grouping or a number of bits in a sequence treated as a unit.

Word length The total number of bits that comprise a word. Most programmable logic controllers use either 8 or 16 bits to form a word.

Work cell A group of machines that work together to manufacture a product. Normally includes one or more robots. The machines are programmed to work together in appropriate sequences. Often controlled by one or more PLCs.

Write Refers to the process of loading information into memory. Can also refer to block transfer, i.e., a transfer of data from the processor data table to an intelligent input/output module.

Zone control last (ZCL) state instructions A user-programmed fence for zone control last state zones.

Zone control last (ZCL) state zones
Assigned program areas that may control the same outputs through separate rungs, at different times. Each such zone is bound and controlled by zone control last state instructions. If all zone control last state zones are disabled, the outputs will remain in their most recent controlled state.

Index

S

Safety
- and equipment grounding, 356
- ESTOP (emergency stop) relay, 354
- forcing external I/O addresses, 243–44
- hardwired master control relay, 245–46
- in PLC programs (motor starter example), 246
- preventive maintenance checklist, 361
- safety wiring diagram, 244–45
- STOP buttons, 246–47

Sample and hold amplifier, 396

Save interval/save period, 396

Scale data (SCL) instruction, 311

Seal-in circuits, 145–46

Search function, 359

Selectable timed enable/disable, 247–48

Selector switch, 127

Sensors, proximity
- introduction, 131
- actuation of, 134–35
- applications, typical, 131, 132
- capacitive sensors, 133–34
- current-sourcing/current-sinking, 132, 133
- flow measurement, 140, 142
- hysteresis in, 133, 134
- inductive sensors, 131–32, 133
- leakage current in, 133
- photoelectric (photovoltaic) sensors, 135–37
- RPM/velocity sensors, 142
- strain/weight sensors, 140
- temperature sensors, 140, 141
- typical connections, 134
- ultrasonic sensors, 139
- *See also* Bar codes

Sequencer instructions
- introduction, 325–26
- sequencer input (SQI) instruction
  - introduction, 335
  - SQI instruction program, 335
  - SQI/SQO instructions in one rung, 335–36
- sequencer load (SQL) instruction
  - introduction, 336
  - SQL instruction program, 336–37
- sequencer output (SQO) instruction
  - binary step information, 326–27, 328
  - event-driven output program, 334
  - event-driven traffic lights, 332
  - instruction block parameters, 335–36
  - mask word, use of, 327, 329
  - output formats (block, coil), 329
  - PLC-2 output function blocks, 329
  - sequencer steps, 326, 328
  - SQO/SQI instructions in one rung, 335–36
  - time-driven output program, 334–35
  - time-driven traffic lights, 331

Sequencer programs
- event-driven traffic lights, 332
- sequencer chart (matrix), 330
- sequencer data worksheets, 331, 332
- SQO instruction block parameters, 335–36

Sequencer programs (continued)
- time-driven output program, 334–35
- time-driven traffic lights (SLC-150), 330–31

Sequencers, mechanical
- introduction, 324
- cam-operated sequencer, 324, 325
- dishwasher application, 324, 325–27
- drum-operated sequencer, 324, 325
- sequencer data table, 324, 325

Sequential control process, 149–50

Series data transmission, 414

Series instructions, sequencing of, 152

Servo drives/servomotors, 421

Servo I/O modules, 32

Set point control
- introduction, 287
- closed-loop control, 288, 290
- proportional control, 290–91
- proportional-integral control, 291
- temperature control (PLC-2), 287–88, 289
- temperature control (PLC-5, SLC-500), 288, 289

Shielded cables, use of, 357

Shift registers
- introduction, 337
- basic concept, 337
- bit shift registers, synchronous
  - bit shift left (BSL) program, 339
  - bit shift left (BSL) register, 337–38
  - bit shift right (BSR) program, 340
  - bit shift right (BSR) register, 337–38
  - circulating shift register, 337
  - shift instruction program data, 338–39
  - spray-painting application, 341–42
  - 16-station machine flow application, 341, 343
- word shift registers, asynchronous
  - data indexing (in/out), 344–45
  - FIFO load/unload (FFL/FFU) instruction pair, 343
  - FIFO stack program data, 343–44
  - LIFO load/unload (LFL/LFU) instruction pair, 345
  - LOAD/UNLOAD pulses in, 342–43

Sign bit
- in positive/negative numbers, 52, 53
- *See also* Positive/negative numbers

Signal conditioning, 392

Single-ended PLCs, 14–15

Sinking/sourcing
- in discrete output modules, 28, 29
- in proximity sensors, 132, 133

Size categories, PLC, 14

SKIP instruction. *See* Instructions [JUMP (JMP) instruction]

Snubbing circuits
- varistors in, 359
- *See also* Surge suppression

Software categories, PLC, 14

Solar cells, 135, 136

Index

461

XIC (examine if closed), 105, 112–13

XIO (examine if open), 106, 112–13

XOR (exclusive-OR) gate, 76

Zener diodes
    in optical isolators, 25, 26
    varistors as, 358, 359

Zone control last (ZCL) instruction. *See under* Instructions

UHI
Millennium
Institute

Please return/renew this item by the last date shown

Tillibh/ath-chlaraidh seo ron cheann-latha mu dheireadh